AF889126

Advanced and Intelligent Manufacturing in China

Series Editor

Jie Chen, Tongji University, Shanghai, China

This is a set of high-level and original academic monographs. This series focuses on the two fields of intelligent manufacturing and equipment, control and information technology, covering a range of core technologies such as Internet of Things, 3D printing, robotics, intelligent equipment, and epitomizing the achievements of technological development in China's manufacturing sector. With Prof. Jie Chen, a member of the Chinese Academy of Engineering and a control engineering expert in China, as the Editorial in Chief, this series is organized and written by more than 30 young experts and scholars from more than 10 universities and institutes. It typically embodies the technological development achievements of China's manufacturing industry. It will promote the research and development and innovation of advanced intelligent manufacturing technologies, and promote the technological transformation and upgrading of the equipment manufacturing industry.

Bingqi Chen · Siyao Chen

Machine Vision Technology

Chemical Industry Press Co., Ltd.

Springer

Bingqi Chen
College of Engineering
China Agricultural University
Beijing, China

Siyao Chen
The University of Tokyo
Tokyo, Japan

ISSN 2731-5983 ISSN 2731-5991 (electronic)
Advanced and Intelligent Manufacturing in China
ISBN 978-981-95-1183-9 ISBN 978-981-95-1184-6 (eBook)
https://doi.org/10.1007/978-981-95-1184-6

Jointly published with Chemical Industry Press
The print edition is not for sale in China (Mainland). Customers from China (Mainland) please order the print book from: Chemical Industry Press.

© Chemical Industry Press 2026

This work is subject to copyright. All rights are solely and exclusively licensed by the Publisher, whether the whole or part of the material is concerned, specifically the rights of reprinting, reuse of illustrations, recitation, broadcasting, reproduction on microfilms or in any other physical way, and transmission or information storage and retrieval, electronic adaptation, computer software, or by similar or dissimilar methodology now known or hereafter developed.
The use of general descriptive names, registered names, trademarks, service marks, etc. in this publication does not imply, even in the absence of a specific statement, that such names are exempt from the relevant protective laws and regulations and therefore free for general use.
The publishers, the authors, and the editors are safe to assume that the advice and information in this book are believed to be true and accurate at the date of publication. Neither the publishers nor the authors or the editors give a warranty, express or implied, with respect to the material contained herein or for any errors or omissions that may have been made. The publishers remain neutral with regard to jurisdictional claims in published maps and institutional affiliations.

This Springer imprint is published by the registered company Springer Nature Singapore Pte Ltd.
The registered company address is: 152 Beach Road, #21-01/04 Gateway East, Singapore 189721, Singapore

If disposing of this product, please recycle the paper.

Foreword

Manufacturing is the mainstay of the national economy, the foundation of a nation, the tool for national prosperity, and the basis for national strength. Over the past decade, China's manufacturing industry has developed rapidly, with its comprehensive strength continuously increasing and its international status significantly enhanced, making it the largest manufacturing country in the world. However, China is still in the process of industrialization, with prominent issues of being "large but not strong", and there is still a considerable gap compared to advanced countries. To address the issues of being "large but not strong", weak independent innovation capabilities, and high dependence on foreign key core technologies and high-end equipment that restrict China's development, the State Council released the "China Manufacturing 2025" national plan on May 8, 2015. Subsequently, the Ministry of Industry and Information Technology released the "China Manufacturing 2025" plan, which proposed a three-step strategy for the development of a strong manufacturing nation, along with the goals, guiding principles, and strategic routes for 2025. This plan included nine strategic tasks and ten key development areas. On August 19, 2016, the Ministry of Industry and Information Technology, the National Development and Reform Commission, the Ministry of Science and Technology, and the Ministry of Finance jointly released the implementation guidelines for five major projects under the "China Manufacturing 2025": manufacturing innovation center, industrial foundation strengthening, green manufacturing, intelligent manufacturing, and high-end equipment innovation.

In response to the major strategic deployment made by the Central Committee of the Communist Party of China and the State Council to build a strong manufacturing nation, local governments, enterprises, and research departments are actively exploring and implementing strategies. Accelerating the integration and development of new-generation information technology and manufacturing technology, promoting the transformation of China's manufacturing model from "Made in China" to "Intelligent Manufacturing in China," and accelerating the transformation of China's manufacturing industry from large to strong has become our new historical mission. Currently, the information revolution is rapidly evolving, with technologies such as the Internet of Things, cloud computing, big data, and artificial intelligence widely permeating all fields of the economy and society, and the prosperity of the information economy has become an important indicator of national strength. Technologies in fields such as additive manufacturing (3D printing), robotics and intelligent manufacturing, control and information technology, and artificial intelligence are continuously making significant breakthroughs, driving the fragmentation and transformation of the traditional industrial system and reshaping the international division of labor in manufacturing. The integration and development of manufacturing technology and information technology such as the Internet have become a major trend and main feature of the new round of technological revolution and industrial transformation. Against this backdrop of great development and transformation in China's manufacturing industry, Chemical Industry Press has proactively aligned with the trends of technology and industrial development, organized the publication of the "China Manufacturing 2025" Publishing Project series, which can be said to be both pioneering and timely.

The book series *China Manufacturing 2025 Publishing Project* is a set of high-level, original academic monographs closely centered around the action program for the first decade of implementing the strategy of building a strong manufacturing nation, as issued by the State Council—"China Manufacturing 2025". The series focuses on the two major fields of intelligent manufacturing and equipment, and control and information technology, covering a range of core technologies such as the Internet of Things, big data, 3D printing, robotics, intelligent equipment, industrial network security, knowledge automation, and artificial intelligence. The selection and planning of the series are closely aligned with the "China Manufacturing 2025" plan and its 11 supporting implementation guides, action plans, or special plans. Each volume is organized around core technologies in various fields, concentrating on the technological development achievements in the domestic manufacturing sector, aiming to strengthen the research, promotion, and application of advanced technologies, providing targeted direction guidance and systematic technical reference for the implementation of the "China Manufacturing 2025" action program.

This set of books embodies the following major characteristics:

Firstly, the content of the series strives for originality, focusing on networked and intelligent technologies, gathering many cutting-edge technologies, reflecting some of the latest technological achievements both domestically and internationally, especially highlighting the relevant original technological achievements in China. These books include many new technologies that have received numerous national and provincial-level scientific and technological awards, and the publication of these books greatly facilitates the promotion and application of new technologies! These contents not only help technical personnel solve practical problems but also provide new directions and expand new ideas for research.

Secondly, each volume of the series not only introduces new technologies, theories, and methods across various professional fields but also prioritizes those with strong application potential, providing concrete examples of their implementation and adoption to accelerate the transformation of outstanding scientific research achievements into industrial applications.

The series is led by Academician Sun Youxian, an expert in control engineering in China, who serves as the chairman of the editorial board, with the participation of several academicians such as Wu Cheng, Wang Tiannian, and Zheng Nanning in the planning and organization work. Many young and middle-aged scholars, including Changjiang Scholars, Outstanding Youth, and Excellent Youth, participate in the specific writing work, ensuring a high academic level and writing quality.

It is believed that the publication of this series of books has positive significance for promoting the implementation of the important national strategic plan "China Manufacturing 2025", effectively promoting the research and innovation of intelligent manufacturing technology in China, driving the technological transformation and upgrading of the equipment manufacturing industry, improving the design capability and technical level of products, thereby enhancing the core competitiveness of China's manufacturing industry from multiple perspectives.

中国工程院院士 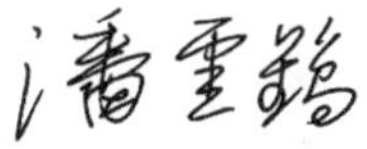

Hangzhou, China
Yunhe Pan

Editorial Committee

Director

Sun Youxian (Academician)

Deputy Directors

Wang Tiannan (Academician)
Yang Huayong (Academician)
Wu Cheng (Academician)
Chen Chun (Academician)
Chen Jie (Academician)
Zheng Nanning (Academician)
Gui Weihua (Academician)
Qian Feng (Academician)
Guan Xiaohong (Academician)

Members

Ma Zhengxian
Wang Dayi
Wang Tiannan
Wang Rongming
Wang Yaonan
Tian Yantao
Gong Shuili
Qiao Fei

Ren Chunnian
Yi Tingfeng
Liu Min
Liu Yanjun
Liu Huicong
Liu Lijun
Sun Changyin
Sun Youxian
Du Yulei
Li Li
Li Hui
Li Shaoyuan
Li Yajiang
Li Jianing
Yang Weimin
Yang Huayong
Wu Fei
Wu Cheng
Wu Weiguo
Song Hao
Zhang Ping
Zhang Jing
Zhang Congpeng
Zhang Yuru
Zhang Yongde
Zhang Jinsheng
Chen Wei
Chen Gang
Chen Chun
Chen Jie
Chen Wanmi
Chen Changjun
Chen Huajun
Chen Bingqi
Chen Moai
Chen Jiwen
Chen Zengqiang
Luo Ying
Luo Xueke
Zheng Nanning
Fang Lijin
Zhao Chunhui
Hu Changhua
Hu Fuwen
Jiang Jingang

Fei Yanqiong
Gui Weihua
Chai Yi
Qian Feng
Xu Jining
Guo Tongying
Cao Jujiang
Kang Rui
Jiao Zhiwei
Zeng Xianwu
Xie Ying
Xie Shengli
Cai Deng
Guan Xiaohong
Wei Qingsong
"China Manufacturing 2025" Publishing Project
Machine Vision Technology
Compiled by Bingqi Chen

化学工业出版社
·北京·

Introduction

The core content of "China Manufacturing 2025" is the informatization and intelligence of equipment production and application, and machine vision is the key technology to achieve this goal. When it comes to "machine vision" or its software counterpart, "image processing," many people are somewhat familiar with the terms. However, those without specialized knowledge often mistakenly equate "image processing" with software like Photoshop, which is primarily used for image editing. In reality, there is a fundamental difference between the two. Image processing in machine vision involves a computer analyzing and interpreting existing images, then performing actions or processing based on the analysis. In contrast, Photoshop relies on human judgment to manually alter an image's color, shape, or composition through direct editing. In other words, one involves the machine analyzing and interpreting images to automatically execute actions, while the other relies on humans to analyze and interpret images and manually make modifications. This distinction is the fundamental difference between the two. This book explores the components of machine vision, the theoretical algorithms behind image processing, and their application systems.

Currently, there are quite a few books on image processing on the market, generally focusing on explaining image processing algorithm theories or programming methods. The author has also compiled two books on image processing VC++ programming and one on machine vision theory and application examples, mainly aimed at image processing programmers. However, only a minority of people are currently involved in image processing programming. In the future, an increasing number of individuals will focus on operations and technical services related to robots and intelligent equipment. Currently, there are relatively few machine vision educational books in China tailored to this group. In recent years, local polytechnic institutes have frequently sought advice on the construction of image processing laboratories, focusing on teaching image processing theory rather than programming. This focus makes it challenging to recommend suitable textbooks or configure image processing laboratories for their needs. To meet the needs of this large group, this book aims at popularizing education, explaining complex theoretical algorithms in a simple and easy-to-understand, illustrated manner, providing practical processing cases for each algorithm, to help general learners realize that machine vision is not as complex as it seems and to boost the confidence of those who may engage in machine vision project development in the future.

This book compiles many of the most popular algorithms in image processing, offering significant reference value for both professional image processing researchers and programmers.

During the writing of this book, various degrees of assistance were received from Tian Hao, Ouyang Di, Zeng Baoming, Wang Qiao, Yang Ming, Qiao Yan, Zhu Deli, Liang Xihuizi, Chen Hongmi, Dai He, and technical support from Beijing Hyundai Fubo Technology Co., Ltd., to whom heartfelt thanks are extended!

Due to the limitations of the author's abilities, shortcomings in the book are inevitable. We sincerely invite readers and experts to provide criticism and corrections.

Beijing, China

编著者：陈兵旗 中国北京

Bingqi Chen

Contents

Part I
Machine Vision Theory and Algorithms

Chapter 1
Machine Vision

1.1 The Role of Machine Vision

When mentioning "vision", it naturally brings to mind "human eyes". In simple terms, machine vision is "machine eyes". Similarly, from the role of human eyes in the human body, one can also infer the role of machine vision in machines. However, although the functions are largely similar, there are some fundamental differences. The similarity lies in both being the "organs" for obtaining external information for the subject (human or machine), while the difference lies in the ability to acquire and process information.

For the human eye, a single glance allows a person to instantly recognize what they are seeing, including its type, quantity, colour, shape, and distance, with remarkable accuracy. Due to long-term accumulation from childhood, people can effortlessly state the general information of what they see, which is an advantage of the human eye. However, it is important to note that the information perceived by the human eye is only "general" and not precise. For example, you can quickly discern how many people are in your field of vision, even identifying how many are men or women, their body shapes, and attire, as well as the surrounding environment, but you cannot accurately determine their height, waist circumference, or distance from you. At most, you can say "approximately XX," which is a disadvantage of the human eye. If a person were to select defective parts on a factory production line, even on a slow production line that a person can react to, they would soon complain, "This is not a job for humans," and it is even less feasible for fast production lines. Indeed, these tasks are not suited for the human eye but are tasks for machine vision.

For machine vision, the aforementioned factory online inspection is its strong suit. It can not only detect product defects but also accurately measure product dimensions. As long as the camera resolution is sufficient, achieving precision of 0.001 mm or even higher is not an issue. However, unlike humans, who can instantly judge all items in their field of vision, machine vision generally lacks this capability. Machine vision does not automatically store previously "seen" items like the human

© Chemical Industry Press 2026

B. Chen and S. Chen, *Machine Vision Technology*, Advanced and Intelligent Manufacturing in China, https://doi.org/10.1007/978-981-95-1184-6_1

eye does; without inputting relevant analysis and judgment programs, it is "blind" and knows nothing. Of course, it can be made to learn continuously by inputting learning programs, but it cannot understand everything like humans do, at least not at the current level.

In summary, machine vision is the eye of the machine, capable of analysing and judging target objects through programming, detecting defects, measuring dimensions and colours, and providing specific precise information for specific machine actions.

Machine vision has broad application prospects and can be used in various aspects of social production and people's lives. In terms of replacing human labour, all tasks that require observation and judgment by the human eye can be accomplished with machine vision, making it most suitable for tasks involving repetitive actions (such as workpiece quality inspection) and judgments that easily tire the eyes (such as circuit board inspection). Machine vision can also achieve accurate measurements, fine judgments, and microscopic recognition that the human eye cannot. Table 1.1 provides examples of machine vision applications in different fields.

1.2 Hardware Composition of Machine Vision

The hardware composition of the human eye can be broadly described as the eyeball and the brain, while the hardware composition of machine vision can be roughly described as a camera and a computer, as shown in Fig. 1.1. In addition to the camera, other image acquisition equipment includes image capture cards and light sources. The following provides a more detailed explanation from the perspectives of computers and image input acquisition equipment.

1.2.1 Computer

There are many types of computers, including desktop computers, laptops, tablets, industrial computers, and microprocessors, but their core components are all central processing units, memory, hard drives, and displays, differing only in the shape, size, and performance of these core components.

(1) Central Processing Unit

The central processing unit, also known as the CPU (central processing unit), as shown in Fig. 1.2, is the core part of the computer, equivalent to the human brain, with the main function of executing computer instructions and processing data in computer software. Its development has been very rapid, with the computing speed of personal computers now surpassing that of supercomputers from ten years ago.

(2) Hard Drive

Table 1.1 Machine vision application fields and examples

Field of application	Use cases
Medicine	Diagnosis and treatment based on X-ray images, ultrasound images, microscope images, magnetic resonance imaging (MRI) images, CT images, infrared images, three-dimensional images of human organs, etc. Patient monitoring and care
Remote sensing	Satellite images are used for earth resource survey, topographic survey, mapping, weather forecast, as well as agriculture, fishery, environmental pollution investigation, urban planning, etc.
Space exploration	Compression, transmission, recovery and processing of massive cosmic images
Military	Moving target tracking, precise positioning and guidance, warning system, automatic fire control, anti-camouflage, UAV reconnaissance
Public security and transportation	Surveillance, face recognition, fingerprint recognition, traffic flow monitoring, vehicle violation judgment and license plate recognition, vehicle size detection, auto navigation
Industry	Circuit board inspection, Computer Aided Design (CAD), Computer Aided Manufacturing (CAM), product quality online inspection, assembly robot visual inspection, handling robot visual navigation, production process control
Agriculture, forestry, biology	Fruit and vegetable picking, fruit and vegetable grading, farmland navigation, crop growth monitoring and 3D modeling, pest detection, forest fire detection, microbial detection, animal behavior analysis
Post and telecommunications, communications, network	Mail automatic sorting, image data compression, transmission and recovery, TV phone, video chat, mobile phone image wireless network transmission and analysis
Sports	Human motion measurement, ball trajectory tracking measurement
Film and television, entertainment	3D film, virtual reality, advertisement design, film special effects design, online game
Office	Character recognition, text scanning input, handwritten input, fingerprint password
Services	Nurse robot, cleaning robot

The main storage medium of a computer, used to store files, programs, data, etc. It consists of one or more disks made of aluminium or glass coated with ferromagnetic material, as shown in Fig. 1.3.

Types of hard drives include solid-state drives (SSD), hard disk drives (HDD), and hybrid hard drives (HHD). SSDs use flash memory chips for storage, HDDs use

Fig. 1.1 Simple machine vision system

Fig. 1.2 Central processing unit (CPU)

Fig. 1.3 Hard disk

magnetic disks, and HHDs integrate magnetic hard drives and flash memory into one. The vast majority of hard drives are fixed drives, permanently sealed and fixed in the hard drive enclosure.

Digital image data, like computer program data, is stored on a computer's hard drive. After processing by the computer, the image is displayed on the monitor or saved back onto the hard drive for future use. In addition to the computer's built-in hard drive, there are also external hard drives connected via USB, the most common being what is typically referred to as a USB flash drive. As computer performance continues to improve, hard drive capacity is also expanding, with most modern computer hard drives having a capacity in the terabyte range, where 1 TB equals 1024 GB.

(3) Memory

Memory, also known as RAM (Random Access Memory), as shown in Fig. 1.4, is used to temporarily store data for CPU operations and data exchanged with external storage like hard drives. As long as the computer is running, the CPU will transfer the data needed for operations into the memory for processing. Once the operations are complete, the CPU will output the results, such as copying image data from memory to the display's storage area for viewing. Therefore, the performance of memory significantly impacts the computer's performance.

Nowadays, digital images are generally quite large. For example, a 9-megapixel camera captures images with a maximum size of $3456 \times 2592 = 8{,}957{,}952$ pixels. Each pixel consists of three bytes for red, green, and blue (RGB), totalling 8,957,952

Fig. 1.4 Memory

Fig. 1.5 Display

$\times$ 3 = 26,873,856 bytes, which is approximately 26,873,856 $\div$ 1024 $\div$ 1024 $\approx$ 25.63 MB of memory. However, the actual JPEG image file viewed is around 2 MB, much smaller due to data compression when stored as a JPG file. During image processing, the data must first be decompressed and then read into the computer's memory. Thus, image data heavily consumes computer memory resources, and more memory is beneficial for the computer's operation. Currently, 32-bit computers typically have a minimum of 1 GB and a maximum of 4 GB (2^{32} bytes) of memory, while 64-bit computers generally have a minimum of 8 GB and can reach up to 128 GB (2^{64} bytes).

(4) Display

A display, commonly referred to as a monitor, as shown in Fig. 1.5, is a computer's I/O device, meaning input/output device, available in various sizes and types. Based on the manufacturing materials, displays can be categorised into cathode ray tube (CRT) displays, plasma display panels (PDP), and liquid crystal displays (LCD), among others. Displays can be set to various pixel and colour modes, ranging from 640 $\times$ 480 pixels with 256 colours to 1600 $\times$ 1200 pixels and higher resolutions with 32-bit true colour.

1.2.2 Image Acquisition Devices

Image acquisition devices include cameras, image capture cards, and light sources. Currently, most are digital imaging devices, with a wide variety including PC cameras, industrial cameras, surveillance cameras, scanners, camcorders, and mobile phones, as shown in Fig. 1.6. Additionally, microscopes for microscopic viewing and telescopes for astronomical observation are also image input devices.

The key component of a camera is the lens, as shown in Fig. 1.7. The smaller the focal length, the clearer the close-up view; the larger the focal length, the clearer the

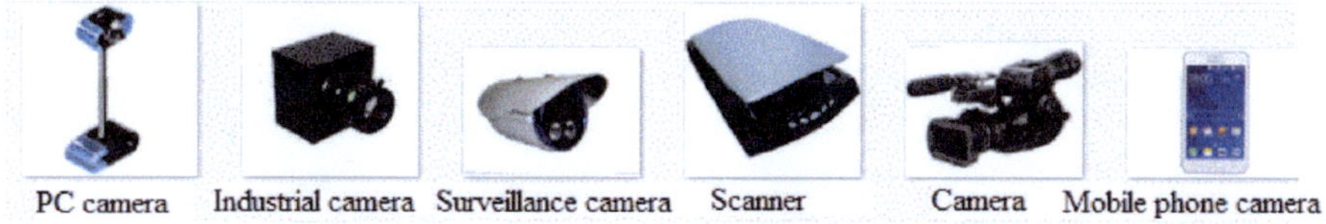

Fig. 1.6 Camera device

distant view, similar to the human eye's cornea. For general imaging equipment, the lens focal length is fixed; common PC cameras and surveillance cameras typically have a focal length of 4 ~ 12 mm. Industrial and scientific instrument lenses may have fixed or adjustable focus lenses.

Cameras are generally connected to computers via dedicated image capture cards, IEEE1394 interfaces, and USB interfaces, as shown in Fig. 1.8. Computer motherboards are equipped with USB interfaces, and some portable computers also have IEEE1394 interfaces in addition to USB. When using IEEE1394 interface digital imaging devices for image input on desktop computers, if the motherboard lacks an IEEE1394 interface, an additional IEEE1394 image capture card is required. As IEEE1394 image capture cards are international standard cards, they are very affordable, with market prices ranging from tens to several hundred Chinese yuan. The image capture frame rate of IEEE1394 interfaces is relatively stable and generally unaffected by computer configuration, whereas the frame rate of USB interfaces is significantly influenced by computer performance. Nowadays, with the continuous improvement of computer and USB interface performance, most digital devices tend to use USB interfaces, while IEEE1394 interfaces are more commonly used for high-performance imaging equipment. For specialised high-performance industrial cameras, such as those with a frame rate of over a thousand frames per second, they typically come with a dedicated image capture card.

For image inspection on indoor production lines, a set of light sources is generally required. The appropriate light source can be selected based on the state of the object being inspected, which not only reduces software development difficulty but also enhances image processing speed. Image processing light sources generally require DC power sources, especially for high-speed image acquisition, as AC power sources can cause flickering, resulting in images that intermittently appear bright

Fig. 1.7 Lens

Fig. 1.8 Image input interface

Fig. 1.9 Light source

and dark. DC light sources typically use light-emitting diodes (LEDs), depending on the specific usage, can be made into different shapes such as circular, rectangular, square, or strip-shaped, as shown in Fig. 1.9. There are companies that specialize in developing and selling light sources specifically for image processing, and such specialized light sources are generally very expensive, with prices ranging from several thousand to tens of thousands of Chinese yuan.

1.3 Software and Programming Tools for Machine Vision

Connecting the hardware of machine vision together, even if powered on, will not function without software. To use a human analogy, the hardware of machine vision is equivalent to the physical structure of the human eye. For the human eye to function, it must first belong to a living person, meaning the heart must be beating to supply blood, which is akin to supplying power to a computer. However, merely being alive is not enough; if brain-dead, the human eye cannot function. The software functionality of machine vision is equivalent to the function of the human brain. The human brain's functions can be divided into basic and special functions. Basic functions generally refer to innate human abilities that do not require learning, while special functions require learning to be realized. Image processing software constitutes the special function of machine vision, which must be developed by developers or users, whereas the computer's operating system (such as Windows) and software development tools are provided by professional companies and can be regarded as the computer's basic functions. The software mentioned here refers to machine vision software development tools and the developed image processing application software.

Computer software development tools include C, C++, Visual C++, C#, Python, Java, BASIC, FORTRAN, etc. Due to the large amount of data processing involved in image processing and analysis, and the need to write complex computational programs, C and C++ are considered the best programming languages for image processing and analysis in terms of computational speed and programming flexibility. Most current image processing and analysis algorithm programs are implemented using these two computer languages. C++ is an upgrade of C, transforming C from a purely procedural language into a complex object-oriented language. C++ fully encompasses C, meaning C language programs can run normally in a C++ environment. Visual C++ is an upgrade of C++, turning the non-visual C++ into a visual type. Programs written in C and C++ can be fully executed in the Visual C++ environment. The most popular version currently is Visual C++ 10, officially known as Microsoft Visual Studio 2010 (also referred to as VC++2010, VS2010, etc.). There are some software providing general image processing algorithms, such as the foreign OpenCV and MATLAB, and the domestic general image processing system ImageSys development platform, all of which can be used on the Visual C++ platform.

1.4 Machine Vision, Robots, and Intelligent Equipment

When mentioning robots, most people think of humanoid robots, and some may believe that only machines that resemble humans in appearance and function can be called robots. However, this is not the case; humanoid robots are just one type of robot and are not yet widespread. Many robots come in various shapes and have different specialized functions, also known as intelligent equipment, as shown in Fig. 1.10. The distinction between machines that can be called robots or intelligent equipment and those that cannot is whether they possess analytical and judgment functions similar to the human brain. The extent or sophistication of these functions is not particularly critical.

The human eye (vision) is the main channel through which the human brain acquires information from the outside world, accounting for over 70% of the total

Fig. 1.10 Different robots and intelligent equipment

information, in addition to the skin (touch), ears (hearing), nose (smell), and mouth (taste). Correspondingly, machine vision is the main channel through which a machine's computer acquires information from the outside world, alongside contact sensors, photoelectric sensors, ultrasonic sensors, electromagnetic sensors, etc. This illustrates the importance of machine vision for robots or intelligent equipment.

1.5 Functions and Precision of Machine Vision

The functions of machine vision, similar to the human eye, can be simply described as judgment and measurement. Each function encompasses a wealth of content. Judgment functions can be divided into determining presence, correctness, defects, etc., and generally do not require tools. Measurement functions include the measurement of geometric parameters such as size, shape, and angle, as well as motion parameters such as speed and acceleration. Like the human eye, measurement functions generally require tools. For example, to achieve a size error of 0.1 mm, human eye measurement typically requires a caliper with a precision of 0.1 mm or more. In machine vision measurement, in addition to needing a 0.1 mm caliper (calibration object), the camera must also have sufficient resolution, meaning the actual size represented by one pixel must be less than or equal to 0.1 mm. For different functions, although the concept of precision varies, the common characteristics during measurement are the need for a fixed lens focal length and pre-calibration. The precision of each function is explained as follows.

(1) Judgment Function

The judgement function also has precision issues. As shown in Fig. 1.11, only when the size of the defect is visible to the human eye on the image can automatic judgement be made. For static images, a defect can be judged as long as its area exceeds the texture structure of the object itself. However, for dynamic judgement on a production line, in addition to the static size of the defect, the relationship between the production line speed and the camera's frame rate must also be considered. For example, if the production line speed is 100 mm per second (100 mm/s) and the camera's image capture frame rate is 100 frames per second (100 fps), then the displacement between each frame image is 1 mm, making it impossible to judge defects smaller than 1 mm.

(2) Precision Measurement

As shown in Fig. 1.12, precision measurement is generally used for measuring the dimensions of static targets. The camera is positioned perpendicular to the target being measured to capture images, and a ruler is placed on the measurement platform for camera calibration.

Figure 1.13 is an example of camera calibration. The white line from "2" to "3" on the image represents an actual distance of 1 cm, with a total of 146 pixels, thus determining that one pixel represents 1/146 (0.00685) cm.

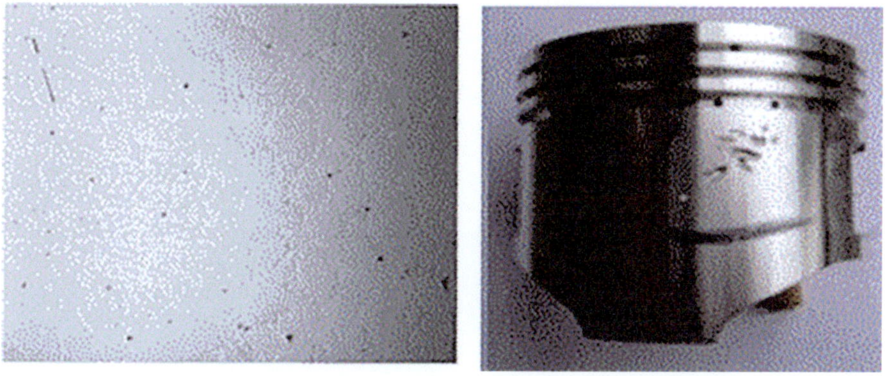

Fig. 1.11 Defective image

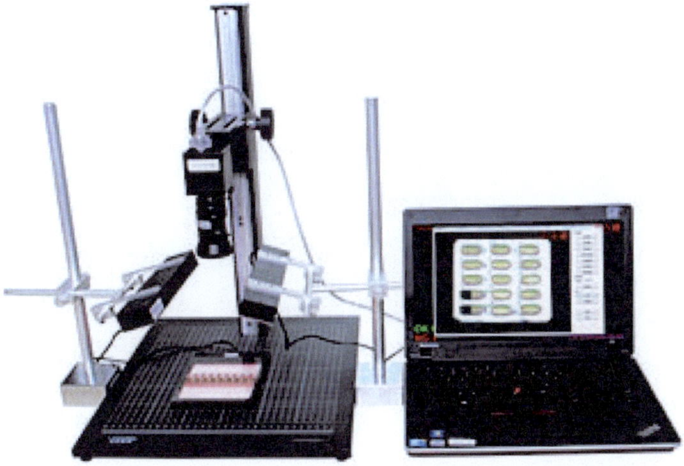

Fig. 1.12 Precision measurement

(3) Photogrammetry

Photogrammetry (also known as camera measurement) is categorized into monocular and binocular measurement, typically encompassing measurements of position, distance, angle, and so forth. Monocular measurement involves using a single camera to capture an image and calculating measurement data based on calibration data, as shown in Fig. 1.14a, where a calibration object is laid flat on the ground near the centre of the camera's field of view. Binocular photogrammetry uses two cameras to simultaneously capture two images, calculating the three-dimensional data of the measured object based on calibration data and the image data, as shown in Fig. 1.14b, where several vertical rods serve as calibration objects.

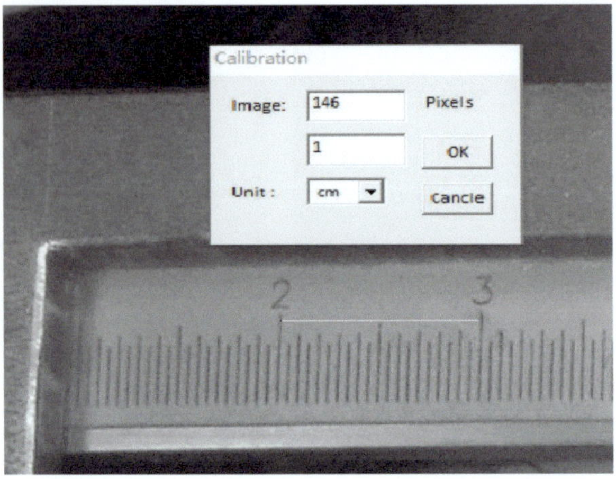

Fig. 1.13 Calibration diagram

(a) Monocular (b) Binocular

Fig. 1.14 Camera measurement

The biggest difference between photogrammetry and the aforementioned precision measurement is that the camera in photogrammetry is generally angled towards the measured object. Due to the camera's tilt angle and generally larger field of view, the actual size represented by a pixel at a certain location cannot be simply used as a calibration value. Geometric perspective transformation is required to calculate the calibration matrix, which also determines that photogrammetry generally does not have very high precision. The precision of photogrammetry is customarily expressed as relative precision in percentage terms, such as a 1% error, rather than using absolute precision in millimeters or centimeters. Based on experience, measurement errors

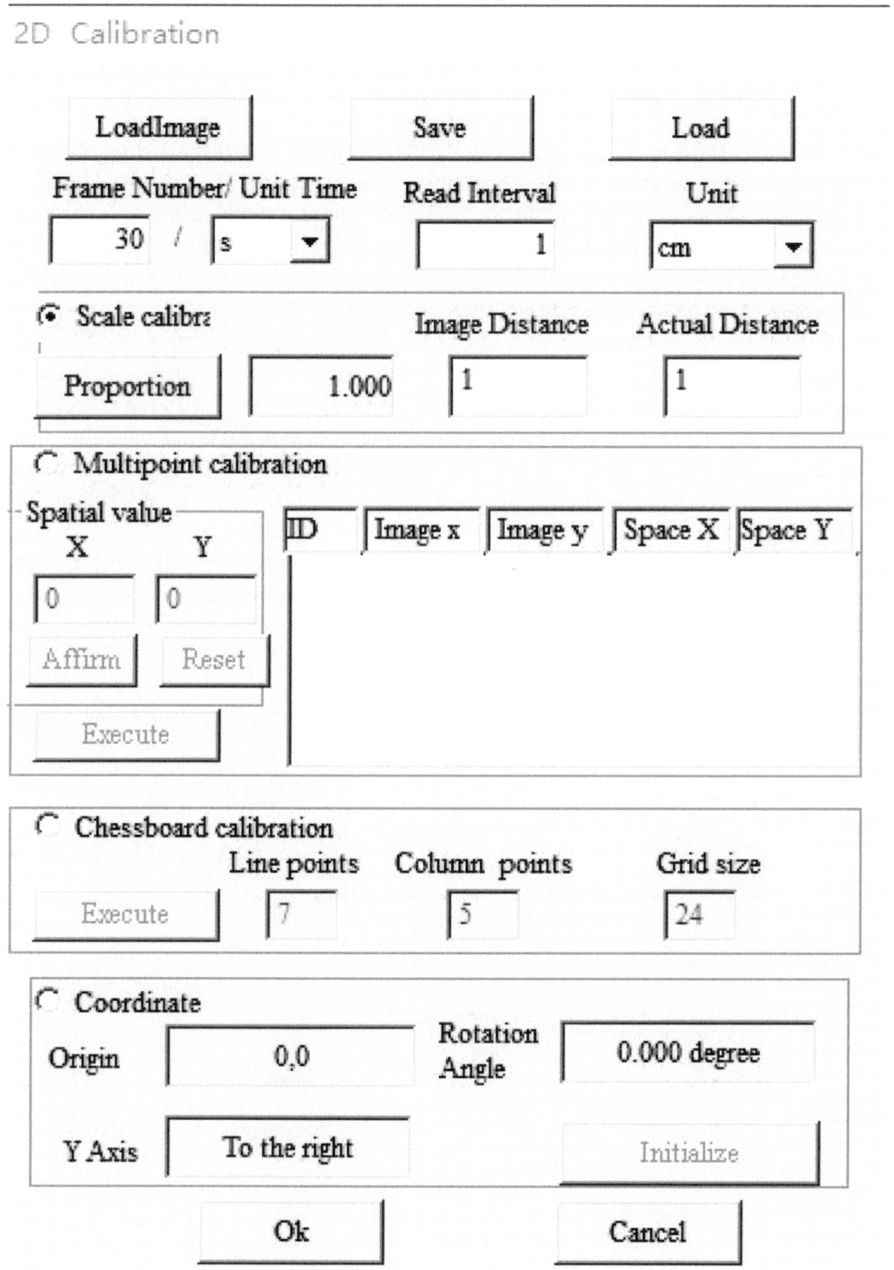

Fig. 1.15 Two-dimensional motion calibration interface

within 10 m are typically within 5%, with errors escalating as the distance increases and as the measured object deviates further from the calibration object.

(4) Motion Measurement

The content of motion measurement generally includes position, distance, speed, acceleration, angle, angular velocity, and angular acceleration. The position, distance, and angle are the contents of the aforementioned camera measurement. Therefore, it can also be said that motion measurement is continuous camera measurement of a moving target. Motion parameters such as speed, acceleration, angular velocity, and angular acceleration are obtained by combining the position, distance, and angle data of the target in each frame with the time difference between frames. The time difference between frames is the frame rate, for example, the time difference between frames at a 30 fps frame rate is 1/30 s (0.03333 s). Figure 1.15 is a calibration interface for two-dimensional motion measurement, which includes distance scale calibration, time (frame rate) calibration, and coordinate transformation for origin selection. Three-dimensional calibration is more complex and will be explained in later chapters.

The precision of motion measurement is similar to that of photogrammetry, generally not high, and is also described using relative precision.

Chapter 2
Image Processing

Image processing is the core software function of machine vision. This chapter introduces the development process and basic knowledge of image processing.

2.1 The Evolution of Image Processing [1, 2]

Image processing is an ancient field. Image processing for the purpose of recording and promotion can be traced back to the Palaeolithic era of the Altamira cave paintings in Spain. Many examples can be found in ancient civilizations, including Egypt and Mesopotamia. The history of Chinese painting can also be traced back to the Neolithic era of primitive society, with a history of at least seven thousand years. Craftsmen made unique contributions to the development of engraving technology and image processing technology through manual painting and printmaking. From the perspective of image information processing technology, the invention of movable type printing (around 1445) and the photocopier (around 1938) can be considered the starting points of image processing. These technologies laid the technical foundation for today's electronic typesetting, scanners, cameras, and other electronic devices. What is now commonly referred to as image processing generally refers to image processing conducted through electronic devices, with the form of processed images having evolved from analogue to digital images.

In 1925, mechanical scanning television emerged, followed by electronic scanning cathode ray tube receivers in 1928, and electronic scanning camera tubes in 1933. These developments, leading up to today's electronic scanning technology, collectively laid the foundation for television technology. Initially, images from electronic devices were analogue, including analogue televisions, analogue cameras, analogue video cameras, and X-ray cameras. These were based on electronic scanning tube technology, using film as the primary recording material and displayed by electronic cathode ray tubes. The main aspects of analogue image processing included:

© Chemical Industry Press 2026

B. Chen and S. Chen, *Machine Vision Technology*, Advanced and Intelligent Manufacturing in China, https://doi.org/10.1007/978-981-95-1184-6_2

① processing related to the input, output, recording, and representation of images; ② utilizing the properties of film and lenses to enhance contrast, define edges more clearly, and modify density characteristics during the development and fixing of photographs; ③ using analogue circuits to emphasise the edges of television images and suppress ghosting. Many of these processes are also used in today's digital image processing.

In the 1940s, digital computers appeared, with the third-generation computer IBM360 in 1964 and the mini-computer DEC/PDP-8 in 1965. With the rapid development of computer technology, the computing environment necessary for digital image processing greatly improved.

The application of digital image processing began with the processing of satellite images. In 1965, the United States National Aeronautics and Space Administration (NASA) released images of Mars taken by the Mariner 4 satellite, and in 1969, the Apollo 11 mission transmitted images of the lunar surface after landing. These were unprecedented applications of digital image processing. In this field, due to harsh environments, the quality of transmitted images was very low, requiring extensive digital image processing before use.

Simultaneously, digital image processing was attempted in the medical field. For example, research was conducted on the measurement and diagnosis of microscope images, blood cell classification, chromosome classification, and cell diagnosis. Around 1965, initial attempts were made to process chest X-ray images, including improving image quality, identifying objects, feature extraction, classification measurement, and pattern recognition. However, unlike satellite images, these were analogue images that first needed digitisation, and many difficulties existed as it was still in the foundational research stage. During this period, in the field of physics, photographs of particle tracks inside accelerators were automatically analysed.

In the latter half of the 1960s, digital image processing began to be applied to general scenes and three-dimensional objects. Research during this period was centred at the Massachusetts Institute of Technology's Artificial Intelligence Laboratory. The "block world" problem, which involved interpreting simple block images captured by television cameras, became a representative research topic in the early field of artificial intelligence. Subsequently, research topics such as image analysis, computer vision, object recognition, scene analysis, and robotic processing emerged in this field. The two-dimensional analogue recognition research of this era focused on character recognition, a vast research project. In Japan, the domestic character recognition device developed in conjunction with the adoption of the postal code system in 1968 became a major factor in accelerating the progress of character recognition research. Many algorithms developed during this time, such as thinning, threshold processing, and shape feature extraction, became important components of basic image processing algorithms and were widely used. In 1968, the earliest international symposium proceedings on image processing was established.

In the early 1970s, digital image processing advanced rapidly with the advent of computed tomography (CT) in the medical field and Earth observation satellites. These involved complex digital image processing from the imaging stage, with vast amounts of data. CT is an instrument that reconstructs cross-sectional images from multiple projection images, based on the mathematical foundation of the Radon transform proposed by Radon in 1917. Fifty years later, advancements in computers and related technologies led to the development of practical applications. CT not only had a revolutionary impact on medicine but also greatly promoted the entire image processing technology, paving the way for obtaining three-dimensional digital images. Approximately 20 years later, technology emerged for virtual three-dimensional reconstruction of the human body using multiple CT images within a computer. This innovation allowed the viewing angle of the three-dimensional image to be adjusted freely, enabling observation of the human body from any orientation to assist in diagnosis and treatment.

Earth observation satellites operate in orbit above the Earth at regular intervals, using sensors to detect reflected energy from the Earth's surface across various spectral bands. The collected data is continuously transmitted back to the ground, where it is processed to reconstruct detailed images of the Earth's surface, which are then made publicly accessible worldwide. Various algorithms for extracting information from these images were developed. Subsequently, remote sensing image processing was formed by synthesising images from ocean observation satellites, meteorological observation satellites, and others, and was widely applied in fields such as geology, plant protection, meteorology, agriculture, forestry, fisheries, marine science, and urban planning. CT images and remote sensing images hold extremely significant importance at the application level. To process them, numerous algorithms have been developed. For instance, for CT images, image reconstruction algorithms were initially developed to improve image quality through spatial frequency processing and grey level processing. Various image measurement algorithms were also developed. On this basis, algorithms for three-dimensional image processing representing the human body's three-dimensional structure were further developed. Regarding remote sensing images, processing algorithms such as image geometric transformation, tilt correction, colour synthesis, classification, structural processing, and domain segmentation have been developed. With technological advancements, the precision of CT images and remote sensing images has continuously improved. Currently, CT resolution can reach below 0.5 mm, and satellite observation of the Earth's surface has achieved a resolution of less than 1 m.

In other fields, to achieve automation in detection, save labour, and improve product quality, large-scale production applications have begun to enter the practical stage. For example, image processing technology has been widely applied in the design and inspection of integrated circuits. With continuous research investment, its practical process has been advanced. However, from the overall perspective of industrial applications, successful examples of practical application are relatively limited. Meanwhile, applications focused on object recognition and scene analysis have spurred research in the field of artificial intelligence, aiming to recognize and interpret general three-dimensional scenes. However, the problems of object recognition and scene analysis are more challenging than anticipated, and even now, there are very few practical application examples.

Office automation processing of drawings, maps, textbooks, etc., closely related to the aforementioned text recognition, has also become an important field of image processing. For instance, fax communication and photocopiers use numerous algorithms such as binary image compression, encoding, geometric transformation, and correction. Japan began developing map databases in 1974, and these accumulated technologies are now widely applied in geographical information systems (GIS) and automotive navigation fields.

In the medical field, besides the aforementioned CT, the commercialisation of blood cell classification devices was first realised, and trial production of cell diagnostic devices began. These early practical devices for simulating image recognition attracted widespread attention. Additionally, computer diagnostic research was conducted to diagnose silicosis, heart disease, tuberculosis, and cancer based on chest X-ray photographs. Simultaneously, auxiliary diagnosis of ultrasound images, X-ray images, angiographic images, and radioactive isotope (RI) images also became research subjects. In these studies, related algorithms for grey-scale image processing, such as differential filtering, distance transformation, thinning, contour detection, and region generation, were developed, forming the foundation for subsequent image processing algorithms.

In terms of hardware, several important developments occurred in the 1970s. For example, the emergence and popularisation of frame storage brought convenience to image processing. On the other hand, the development of digital signal processors (DSP) opened new avenues for advanced processing, including fast Fourier transform (FFT). With the development and advancement of charge-coupled device (CCD) image input devices, rangefinders using laser distance measurement emerged. In computer technology, in the early 1970s, Intel Corporation's microprocessors i4004 and i8008 were introduced, which laid the foundation for subsequent microcomputers such as Altair 1975, Apple II and PET 1977, PC8001 1979. In 1973, Xerox Corporation developed Alto, known as the first workstation. The emergence of the large supercomputer Cray-1 in 1976 expanded the range of processor scale and capability choices, contributing to the development of various scales of image processing systems.

In terms of software, foundational algorithms such as parallel processing and binary image processing were gradually proposed. Among these foundational theories, image transformations (such as discrete Fourier transform, discrete orthogonal transform, etc.), digital graphic geometry, and many methods based on these formed a system, and some general-purpose image processing packages were developed.

In summary, the value and development prospects of image processing during this period were widely recognised, and various application fields realised its utility, initiating foundational research. By the latter half of the period, it entered an era of comprehensive expansion. Especially in the research of foundational methods, processing program frameworks, algorithms, and methodologies, rapid development occurred. In fact, many problems in fields that are now practical or continue to be researched were basically solved during this era. The foundational methods supporting its development mostly began in the 1960s and 1970s.

The image processing that was widely developed in the 1970s further rapidly popularised in the 1980s, with several application fields of image processing introduced earlier entering the practical and popularisation stage. The combination of workstations, memory, and CCD input devices formed specialised systems with better cost-performance ratios at the time, enabling the commercialisation of diverse image processing systems. Many general software tools were developed, and many users' technical personnel were able to develop processing algorithms for various problems. In the 1980s, the core of image processing hardware was workstations equipped with dedicated image processing equipment.

Entering the 1990s, the rapidly globalising internet had a significant impact on image processing. Moreover, during the 1990s, significant advancements in personal computer performance and their widespread adoption brought unprecedented information processing power and diverse image acquisition methods. This progress enabled access to an image processing environment comparable to that of earlier supercomputers, from virtually any location. The need for high-speed transmission of large volumes of images over the network spurred active research in image encoding and compression, leading to the establishment of global standards for image compression methods such as JPEG (Joint Photographic Experts Group) and MPEG (Motion Picture Experts Group). Nowadays, one can freely access various web addresses and download desired images via the internet from home. For example, NASA's web homepage publicly shares various planetary images captured by satellites, which anyone can freely access via the internet, and animations can be watched in real-time during rocket launches.

In the latter half of the 1990s, with the widespread availability of high-performance, affordable digital cameras and image scanners, digital image processing became even more prevalent. Today, the widely available computer environment allows sound, text, images, and video to be freely converted into digital data, ushering in the era of multimedia processing.

Another significant event of the 1990s was the emergence of virtual reality (VR), whose design concept and essence gained worldwide recognition from the early 1990s. The purpose of virtual reality is not merely to "let the world see and understand things recorded there," but to "record, represent things, and experience the world," conceptually changing the way image information is utilised.

In some fields, with the establishment of foundational theories, systems gradually formed and were confirmed. For instance, in digital geometry, which includes three-dimensional digital image forms, monocular and binocular image generation, and stereophotometry, theoretical explanations were provided regarding the relationship between objects (or scenes) in three-dimensional space and their two-dimensional image recordings, focusing on shape and grayscale distribution. Feasible image analysis methods based on these theories were subsequently proposed. Simultaneously, the nature of recording time series of moving images of three-dimensional objects (video images) and the basic methods of video image processing were clarified. As objects became more complex, the importance of "utilising knowledge related to objects" was emphasised, advocating for knowledge-based computer vision and conducting research and experiments on methods of utilising and managing object-related knowledge. Meanwhile, attempts were made to create a knowledge base for image processing methods themselves, leading to the development of various image processing expert systems. In addressing numerous issues in artificial intelligence, such as exploratory space analysis, optimisation, modelling, and learning functions, new concepts and methods like fractal dimensions, chaos, neural networks, and genetic algorithms emerged as technical tools. Additionally, image processing began to explore the study of perceptual information processing from a new perspective of perceptual information.

In the field of applications, medical image processing in the early 1980s transitioned from relying on X-rays to the practical use of magnetic resonance imaging (MRI) facilitated by CT technology. From the late 1980s to the 1990s, advancements such as ultra-fast X-ray CT and spiral CT were introduced successively. Various advancements, represented by the practical use of digital radiography, propelled the overall digitisation of medical imaging, promoting research and popularisation of unified management and telemedicine of medical images. These advancements are systematic technologies that integrate the broad aspects of image transmission, recording, compression, and restoration. Particularly, the application of reconstructing a patient's three-dimensional image within a computer based on spiral CT, known as the "virtual human body," made surgical demonstrations and virtual endoscopy possible. Between 1995 and 1998, Japan and the United States respectively achieved visualisation of human engineering based on full-body X-ray CT and MRI images. In the 1990s, significant efforts were devoted to computer diagnosis of X-ray images, particularly in chest, stomach, and mammography, with some reaching practical levels by the late 1990s. In 1998, the United States announced the first commercial device for computer diagnosis of medical X-ray images.

In the industrial sector, the scope of practical applications has been broadly expanded, and it has started to deliver results. It was not only used for inspecting product appearance dimensions, scratches, and surface shapes but also applied in non-destructive testing of X-ray images, robotic vision judgement, assembly automation, agricultural and aquatic product processing, grading automation, and operations in harsh environments like nuclear reactors.

In the field of remote sensing, the 1980s saw multiple countries launching various Earth observation satellites, leading to a leap in the types and quantities of satellite images available to users. Additionally, due to advancements in computer technology, even low-cost systems could perform data analysis, rapidly expanding users' horizons. In the first half of the 1990s, synthetic aperture radar equipped with active microwave sensors was launched. Satellites equipped with radar and SAR have been launched one after another, and many people have engaged in research on SAR data processing and analysis technologies. Among these, significant progress has been made in the study of using microwave phase information observed by two sets of antennas for ground elevation measurement and earth deformation measurement. In 1999, the high-resolution commercial satellite IKONOS-1 was launched, marking the era of 1 m resolution in satellite remote sensing.

The development of file and textbook processing, the widespread use of fax communication, graphic processing for computer handwriting input, automatic reading of design drawings, and automatic input of documents have gradually evolved in response to continuous demand.

In the areas of monitoring and communication, intelligent encoding with high image compression ratios, environmental monitoring, facial recognition, behavior recognition, and human–computer interaction have found widespread applications across various fields.

In video image processing, as an application of machine vision, visual systems have been mounted on cars and tractors to achieve unmanned driving. In intelligent transportation systems (ITS), processing of highway surveillance videos automatically alerts traffic congestion conditions. Automatic video image editing technology has emerged, reaching a level where general users can operate it. The main technologies in video processing include image compression encoding, decoding, feature extraction, and generation. The concept of intelligent encoding has been proposed, and the analysis, recognition, and communication of video images have begun to develop rapidly.

In the latter half of the 1990s, attention began to focus on constructing a composite reality that freely combines the real world, real images, and computer graphics (CG) with virtual images. CG and image recognition play important roles as key technologies in this area, and real-time interaction with three-dimensional virtual spaces has now been realised. Additionally, in these developments, the recognition that "the computer is a medium" has been established, and the positioning, utilisation methods, and role of image media in multimedia processing will become keywords in future image processing.

The emergence of various software modules in three-dimensional CAD (computer-aided design) has made the application of CAD commonplace in manufacturing, construction, and urban planning. Furthermore, in the process of popularising digital images through various media, to prevent illegal copying and improper use of images, the 1990s saw the emergence of important issues regarding image copyright and its protection, leading to extensive research in electronic watermarking technology and other areas.

The foundation of image processing technology is the environment of computers and communication. As the network environment continues to develop, advancements in high-speed signal processing, large-capacity data recording, data transmission, mobile computing, wearable computing, and other technologies, including ubiquitous computing, will bring greater changes to the image processing environment.

In imaging technology, the practical application of CT, new developments in MRI and ultrasound imaging, indicate the development prospects of imaging technology related to the human body. Scanners, digital cameras, digital camcorders, digital televisions, and mobile phones with digital cameras all allow easy acquisition of image data, meaning that the methods of obtaining image data have become widespread.

In terms of software, the level of intelligence in processing systems is increasing. Various integrations are gradually forming between image recognition and cognition, generation, transmission and storage, or between virtual environments and the real world and their recorded images. For example, machine pets and humanoid robots have appeared, and computer-aided diagnosis (CAD) and computer-aided surgery

(CAS) in medical applications have become practical. As an important means of intelligent recognition, positioning, tracking, and monitoring of objects, image processing is also a crucial component of Internet of Things technology.

In the 1980s and 1990s, with the popularisation of personal computers and the internet, people's production and lifestyle underwent significant changes. In the twenty-first century, the events that will influence human survival will be the promotion and popularisation of various types of robots, with machine vision, as the "eyes" of robots, playing a pivotal role in the new era.

2.2 Sampling and Quantisation of Digital Images [3]

Within a computer, all information is represented as a series of 0s and 1s (binary strings). Each binary digit (bit) has two states, 0 and 1, and eight binary digits can combine to form 256 ($2^8 = 256$) states, which is known as a byte. This means a byte can represent 256 states from 00000000 to 11111111, with each state corresponding to a symbol, including English characters, Arabic numerals, and punctuation marks. Chinese characters encoded with the national standard GB 2312—1980 are 2 bytes, capable of representing $256 \times 256 \div 2 = 32{,}768$ Chinese characters. Standard digital image data is also represented using a byte's 256 states.

Images in computers and digital devices like digital cameras are digital images. When taking photos or scanning documents, the input is a continuous analogue signal, which needs to be converted into a final digital signal through the two steps of sampling and quantisation.

(1) Sampling

Sampling is the process of dividing a continuous image into a grid of discrete pixels. As shown in Fig. 2.1, the finer the sampling, the smaller the pixels, and the more accurately the image can be represented. There are various settings for sampling precision, such as images with 256 horizontal pixels × 256 vertical pixels, 512 horizontal pixels × 512 vertical pixels, and 640 horizontal pixels × 480 vertical pixels. Currently, 12-megapixel smartphone cameras (4000 horizontal pixels × 3000 vertical pixels) are very common. We can observe a pattern where the number of pixels in the length and width of an image is a multiple of 8, which means that a byte is the smallest unit, and this is the standard operation method within computers.

(2) Quantization

Quantization is the process of converting the brightness (gray level) of pixels into discrete integer values. The simplest method is to use two values, black (0) and white (1), i.e., 1 bit (2 levels), to quantize, which is called a binary image. Figure 2.2 illustrates the relationship between the number of quantization bits and image quality. The finer the quantization (the greater the number of bits), the richer the representation of grey levels. For images with more than 6 bits (64 levels), the human eye can hardly discern any difference. In computers, the brightness value of images is generally

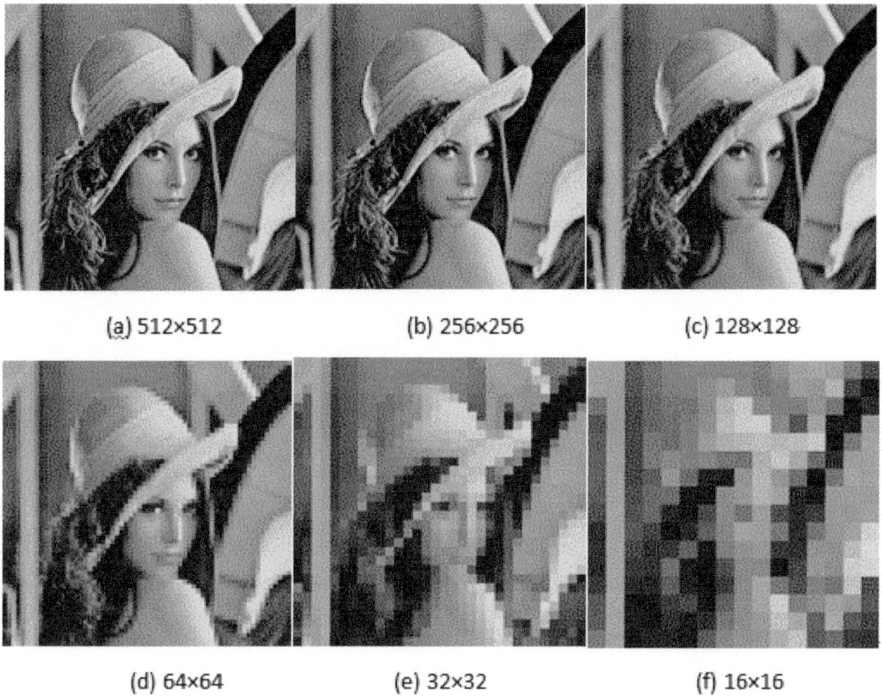

Fig. 2.1 Image effects of different spatial resolutions

represented using 8 bits ($2^8 = 256$ levels), which is one byte, meaning the brightness of a pixel is a value between 0 and 255, where 0 represents the darkest and 255 represents the brightest.

2.3 Colour Images and Grey Scale Images [3]

(1) Colour Images

All colours are composed of three primary colours: R (red), G (green), and B (blue). Each monochrome is artificially divided into 256 levels from 0 to 255. Therefore, different combinations of R, G, and B can represent $256 \times 256 \times 256 = 16{,}777{,}216$ colours, known as full-colour images or true-colour images. If a full-colour image is uncompressed, the file will be very large. For example, a 640×480 pixel full-colour image, with each pixel represented by 3 bytes for the R, G, and B components, requires $640 \times 480 \times 3 = 921{,}600$ (approximately 1 MB) bytes to store.

In addition to full-colour images, there are also 256-colour, 128-colour, 32-colour, 16-colour, 8-colour, and 2-colour images. To reduce the number of bytes stored, these non-full-colour images generally use a palette or a colour table (look-up table, LUT)

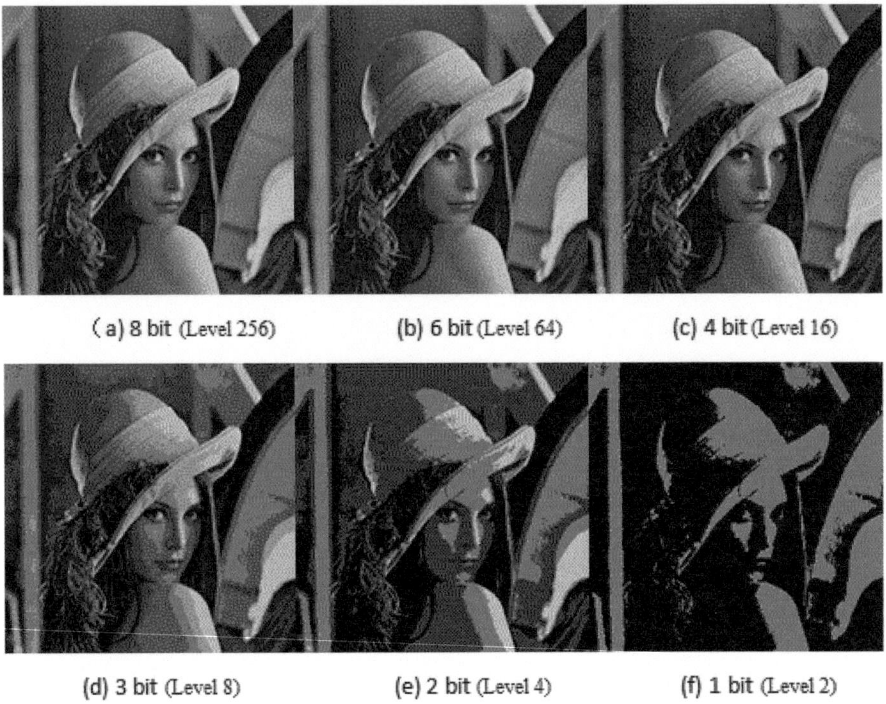

Fig. 2.2 The effect of gray resolution

for storage. Each row in the colour table records the R, G, B values of a colour, i.e., (R, G, B). For example, the first row represents red (255, 0, 0), so when a pixel is red, it only needs to indicate index 0, thus reducing the number of bytes used to represent the image through colour indexing. For example, for a 16-colour image, using colour indexing to represent 16 states can be done with 4 bits (2^4), which is half a byte, requiring $640 \times 480 \times 0.5 = 153{,}600$ bytes for the entire image data, plus the bytes for a colour table. The colour table in Windows has a fixed structure format with 4 parameters, each occupying one byte, where the first 3 parameters represent R, G, B, and the fourth is reserved. Thus, a colour table for 16 colours requires $4 \times 16 = 64$ bytes. Using a colour table to represent a 16-colour image requires a total of $153{,}600 + 64 = 153{,}664$ bytes, which is about 1/6 of the storage space required by the aforementioned method, saving a lot of storage space. Historically, due to limited memory in computers and digital devices, non-full-colour images were more commonly used to save storage space. Nowadays, all digital colour cameras use full-colour images.

Images represented by the three primary colours R, G, and B are known as bitmaps, which have compressed and uncompressed formats with the suffix BMP. Besides bitmaps, there are many other image formats. For example, TIFF images are generally used for satellite image compression formats, where data is not distorted during compression; JPEG images are widely used by digital cameras and other devices, where some signal distortion occurs during compression.

(2) Grey Scale Images

Grey scale images refer to images that contain only brightness information and no colour information. In the BMP format, there is no concept of grey scale images, but if each pixel's R, G, and B are exactly the same, i.e., R = G = B, the image is a grey scale image (or monochrome image).

A colour image can be converted into a grey scale image using Eq. (2.1), where Y is the grey value, and the coefficients for each colour are determined by the International Telecommunication Union (ITU) based on human eye adaptability.

$$Y = 0.299R + 0.587G + 0.114B \tag{2.1}$$

The components of a colour image, R, G, and B, can also be regarded as three grayscale images. Depending on the actual situation, processing just one of the components is sufficient, and it is unnecessary to use formula (2.1) for conversion. This is especially true for real-time image processing, where such an approach can significantly improve processing speed. Figure 2.3 shows the grayscale image converted from a colour image using formula (2.1) and the images of each component R, G, and B. It can be seen that the grayscale image is quite similar to the component images of R, G, and B.

In addition to the components of a colour image and the grayscale image obtained through transformation, there are digital cameras specifically designed for capturing grayscale images. These grayscale cameras are generally used for online image inspection in factories. Historically, black-and-white televisions and cameras also displayed and captured grayscale images. The grayscale images from such equipment were analogue and have now been phased out.

2.4 Image File and Video File Formats

(1) Image File Formats

There are many image file formats, including: BMP, ICO, JPG, JPEG, JNG, KOALA, LBM, MNG, PBM, PBMRAW, PCD, PCX, PGM, PGMRAW, PNG, PPM, PPMRAW, RAS, TARGA, TIFF, WBMP, PSD, CUT, XBM, XPM, DDS, GIF, HDR, IF, etc. The main formats are: BMP, TIFF, GIF, JPEG, etc. The main formats are explained below.

① BMP format. BMP (bitmap) is the standard Windows image format for DOS and Windows-compatible computer systems. BMP format supports RGB, indexed colour, grayscale, and bitmap colour modes. BMP format supports 1, 4, 24, and 32-bit RGB bitmaps. There are uncompressed and compressed formats, with most being uncompressed. File extension: bmp.

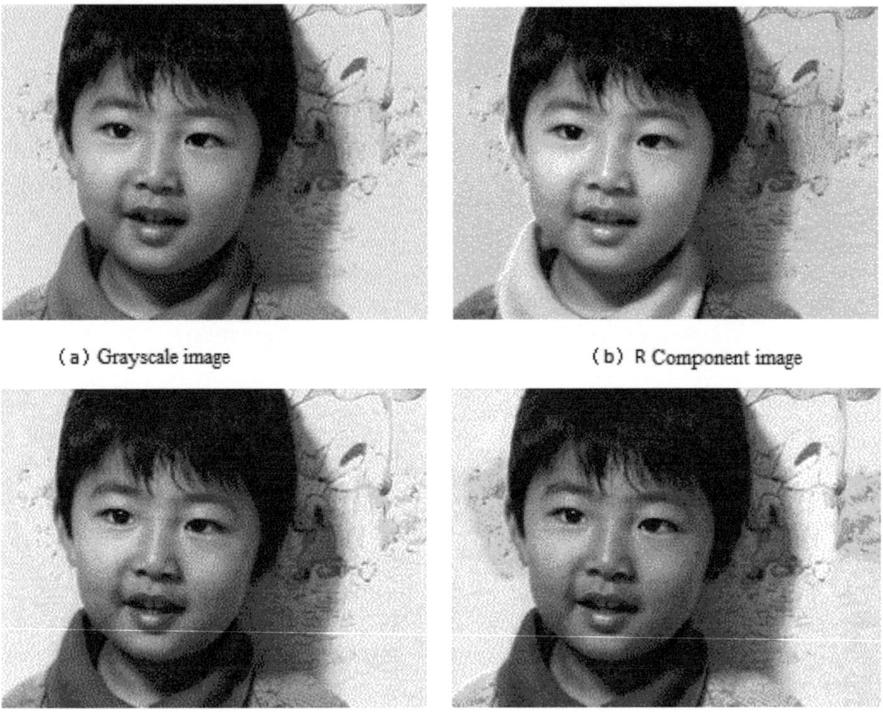

(a) Grayscale image

(b) R Component image

(c) G Component image

(d) B Component image

Fig. 2.3 Gray image and each component image

② TIFF format. TIFF (tag image file format) is used for exchanging files between applications and computer platforms. TIFF is a flexible image format supported by all painting, image editing, and page layout applications. Almost all desktop scanners can generate TIFF images. It is a lossless compression file format. File extensions: tiff, tif.

③ GIF format. GIF (graphic interchange format) is a compressed format used to minimise file size and reduce electronic transmission time. In network HTML (hypertext markup language) documents, GIF file format is commonly used for displaying indexed colour and images, and it also supports grayscale mode. File extension: gif.

④ JPEG format. JPEG (joint photographic experts group) is currently the format with the highest compression rate among all formats. Most colour and grayscale images now use JPEG format for image compression, with a high compression ratio (about 95%), and it supports multiple levels of compression. In network HTML documents, JPEG is used to display pictures and other continuous tone image documents. JPEG format retains all colour information in RGB images and compresses files by selectively discarding data. JPEG is a widely adopted image compression format for digital devices. File extensions: jpeg, jpg.

(2) Video File Formats

Common video formats include: AVI, WMV, MPEG, etc. The following are explanations of each.

① AVI format. AVI (audio video interleaved) is a video format developed by Microsoft, with a long history. AVI format is convenient to use, has good image quality, and allows for flexible compression standards, making it the most widely used format. File extension: avi.

② WMV format. WMV (windows media video) is a general term for a set of digital video codec formats developed by Microsoft, and it is an upgrade of the ASF (advanced systems format) format. File extensions: wmv, asf, wmvhd.

③ MPEG format. MPEG (moving picture experts group) format is a media encapsulation format recognised by the International Organization for Standardization (ISO), including various video formats such as MPEG1, MPEG2, and MPEG4, supported by most machines.

MPEG1 and MPEG2 use first-generation data compression coding technologies such as predictive coding, transform coding, entropy coding, and motion compensation based on Shannon's information theory. MPEG1 has a resolution of 352 $\times$ 240 pixels and a frame rate of 25 frames per second (PAL), widely used in VCD production, games, and online video. Using MPEG1's compression algorithm, a 120-min movie can be compressed to about 1.2 GB. MPEG2 is mainly used in DVD production and also has considerable applications in HDTV (high-definition television broadcasting) and high-demand video editing and processing. Using the MPEG2 compression algorithm, a 120-min movie can be compressed to a size of 5–8 GB, with image quality that far surpasses that of MPEG1.

MPEG4 (ISO/IEC 14496) is an international standard based on second-generation compression coding technology. It uses audiovisual media objects as basic units and employs content-based compression coding to achieve the integration of digital audio and video, graphic synthesis applications, and interactive multimedia.

The MPEG series of standards have had a significant and profound impact on the development of the information industry, including audiovisual consumer electronics like VCDs and DVDs, as well as digital television and high-definition television (DTV and HDTV), and multimedia communications. MPEG's control functions are rich, allowing for multiple videos (i.e., angles), audio tracks, subtitles (bitmap subtitles), etc. A simplified version of MPEG, 3GP, is also widely used on 3G mobile phones. File extensions include: dat (for DVDs), vob, mpg/mpeg, 3gp/3g2/mp4 (for mobile phones), etc.

2.5 Computer Representation of Digital Images

Figure 2.4 shows the R, G, B colour values of pixels centred around the "+" symbol in a 7×7 range. In computers, images are composed of such pixels, and the colour values of each pixel are integerized (or digitized). Figure 2.5 shows the image after local magnification, where the individual small squares in the image, which are pixels, can be seen.

Fig. 2.4 Pixel value

Fig. 2.5 Locally enlarged image

2.6 Common Image Processing Algorithms and Their Generality Issues

The basic algorithms for image processing include: image enhancement, noise reduction processing, image segmentation, edge detection, feature extraction, geometric transformation, etc. Classic algorithms include: Hough transform, Fourier transform (FFT), wavelet transform, pattern recognition, neural networks, genetic algorithms, etc. These algorithms also contain many processing details. The biggest challenge in image processing is that no single algorithm can independently complete the vastly different image processing tasks. For different processing objects, a combination and modification of multiple image processing algorithms is required. Different processing objects and environments present different challenges in image processing. For example, the challenge in online image detection in industrial production is to meet the rapid flow detection of the production line; for agricultural field robots, the challenge in image processing is to adapt to the complex and changing natural environment and lighting conditions. An excellent image processing algorithm developer can design ingenious algorithm combinations and processing methods to make image processing both accurate and fast. It is precisely due to the complexity and variability of image processing algorithms that the challenge becomes more enjoyable.

References

1. Sun Weidong. Image Processing Technology Manual [M]. Beijing: Science Press, 2007.
2. Chen Bingqi. Machine Vision Technology and Application Examples Detailed Explanation [M]. Beijing: Chemical Industry Press, 2014.
3. Chen Bingqi. Practical Digital Image Processing and Analysis [M]. 2nd Edition. Beijing: China Agricultural University Press, 2014.

Chapter 3
Object Extraction

3.1 How to Extract Target Objects

The first step in determining what the target is or measuring its size involves extracting the target from a complex image. For example:

- Extracting only people from street scenes;
- Recognising vehicle number plates and traffic signs in intelligent traffic systems;
- Searching for postal codes in mail for classification;
- Using surveillance cameras to send alerts when someone enters unexpectedly;
- Extracting components on a moving production line;
- Determining the size of crop fruits and classifying them based on their size, etc.

When the human eye searches for target objects in cluttered images, it primarily relies on differences in colour and shape, a process completed unconsciously by people, actually utilising the common knowledge accumulated over years of living. Similarly, machine vision also relies on differences in colour and shape when extracting objects, but since computers do not have this accumulated knowledge, people need to use computer language (programs) to input or calculate the knowledge of target objects through some method, forming the basis for judgment.

The following introduces methods for extracting target objects using shape and colour.

© Chemical Industry Press 2026

B. Chen and S. Chen, *Machine Vision Technology*, Advanced and Intelligent Manufacturing in China, https://doi.org/10.1007/978-981-95-1184-6_3

3.2 Target Extraction Based on Threshold [1]

3.2.1 Binarization Processing

Binarization processing is a method for extracting the target object from an image. There are many methods of binarization processing, the simplest of which is called thresholding. This involves assigning each pixel of the input image to white (255) or black (0) in the output image, depending on whether its grey value is above or below a certain set value (known as the threshold). This can be expressed using Eq. (3.1) or Eq. (3.2).

$$g(x, y) = \left\{ \begin{array}{ll} 255 & f(x, y) \geq t \\ 0 & f(x, y) < t \end{array} \right\} \tag{3.1}$$

$$g(x, y) = \left\{ \begin{array}{ll} 255 & f(x, y) \leq t \\ 0 & f(x, y) > t \end{array} \right\} \tag{3.2}$$

where $f(x, y)$ and $g(x, y)$ are the grey values of the pixels at (x, y) before and after processing, respectively, and t is the threshold.

Depending on the image, sometimes it is necessary to extract the part between two thresholds, as shown in Eq. (3.3). This method is called double threshold binarization processing.

$$g(x, y) = \left\{ \begin{array}{ll} 255 & t_1 \leq f(x, y) \leq t_2 \\ 0 & other \end{array} \right\} \tag{3.3}$$

3.2.2 Determination of the Threshold

We know that the maximum value of a greyscale image pixel is 255 (white), and the minimum value is 0 (black). From 0 to 255, there are 256 levels. Counting the number of pixels at each level in an image (which a computer program can do instantly) and plotting it results in a histogram. As shown in Fig. 3.1, the horizontal axis of the histogram represents pixel levels ranging from 0 to 255, while the vertical axis represents the number of pixels or the proportion of total pixels. Calculating the histogram is one of the important steps in greyscale image target extraction.

For images with a uniform background, there are typically two peaks in the histogram: one for the background and one for the target object. For example, Fig. 3.2a is a *G* component greyscale image of a rice seed, and Fig. 3.2c is its histogram. The high peak on the left side of the histogram (dark area) is the background peak, with more pixels, while the small peak on the right (bright area) is the grain, with fewer pixels. For such images with obvious bimodal histograms, setting

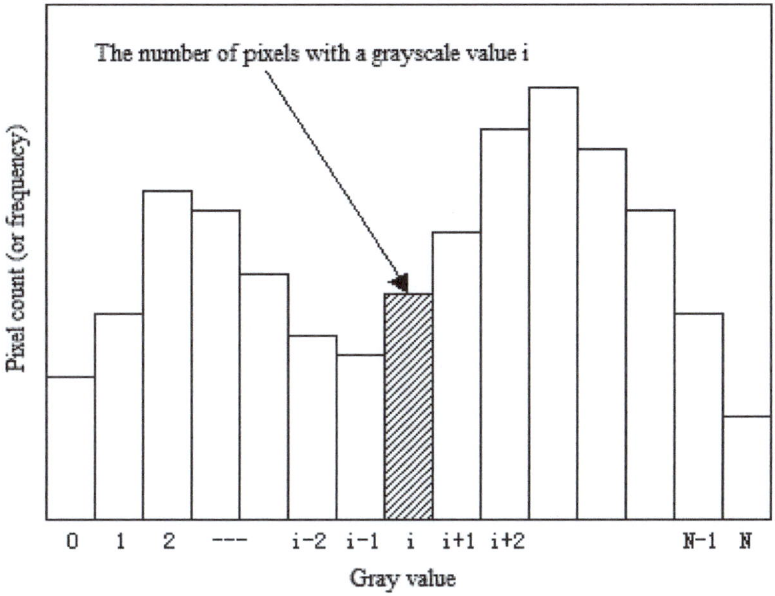

Fig. 3.1 Histogram

the threshold at the trough between the two peaks can effectively extract the target object. Figure 3.2b is a binary image with the threshold set at the trough of 50 between the two peaks, showing a good extraction effect.

If the original image's histogram is highly irregular, it is difficult for a computer program to determine the position of the trough. To more easily find the trough, it is

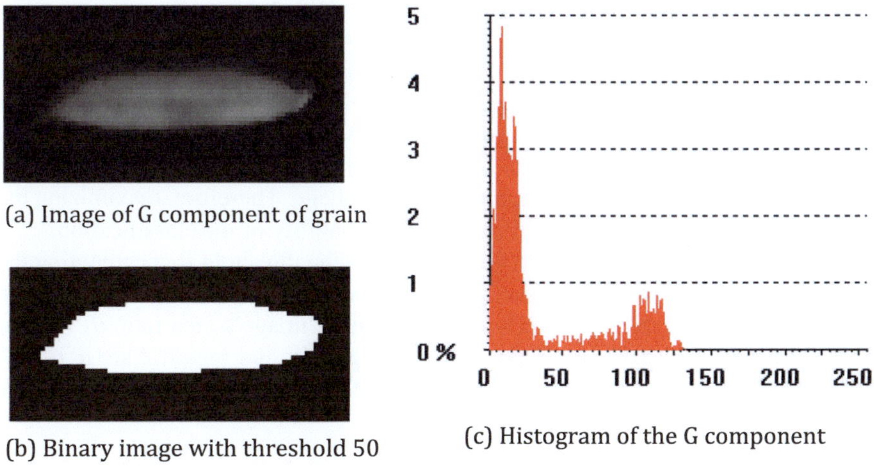

Fig. 3.2 Seed image and its histogram

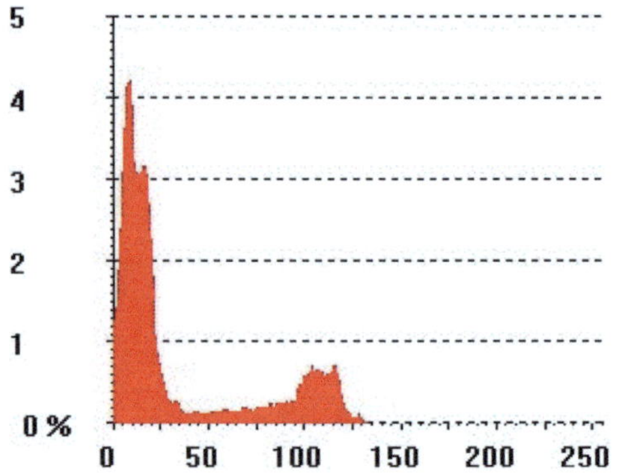

Fig. 3.3 Histogram smoothing

common to average the neighbourhood points on the histogram to reduce its irregularity. Figure 3.3 is the histogram of Fig. 3.2c after averaging five neighbourhood points, making it easier to find the trough position through algorithm writing. This method of taking the trough of the histogram as the threshold is called the "mode method".

In addition to the mode method, there are other methods for determining the threshold, such as the *p*-tile method, discriminant analysis method, variable thresholding, and OTSU method. The p-tile method is used when the proportion of the object in the entire image is known (e.g., *p*%), and the threshold is set at the point where the cumulative number of pixels from the dark grey (or bright grey) side accounts for *p*% of the total pixels on the histogram. The discriminant analysis method determines the threshold by analysing the statistics of the object and background parts when the histogram is divided into two parts. The variable thresholding method is used when the background grey level varies greatly, setting different thresholds for different parts of the image.

Among them, the Otsu method has been widely applied in various image processing tasks. Below is a detailed introduction to the Otsu method.

The Otsu method, also known as the maximum between-class variance method, was introduced by the Japanese scholar Otsu in 1979. It divides an image into background and target parts based on the grayscale characteristics of the image. The larger the between-class variance between the background and the target, the greater the difference between the two parts that make up the image. Therefore, maximizing the between-class variance means minimizing the probability of misclassification.

Assume there are two types of regions, with t as the threshold for segmenting the two regions. From the histogram statistics, it can be obtained that the areas of region 1 and region 2, separated by t, account for the entire image's area ratio θ_1 and θ_2, and the average grayscale of the entire image, region 1, and region 2 are μ, μ_1, μ_2. The relationship between the average grayscale of the entire image and the average grayscale values of region 1 and region 2 is:

$$\mu = \mu_1\theta_1 + \mu_2\theta_2 \tag{3.4}$$

The same region often has similar grayscale characteristics, while different regions exhibit significant grayscale differences. When the grayscale difference between the two regions separated by the threshold t is large, the difference between the average grayscale μ_1, μ_2 of the two regions and the average grayscale μ of the entire image is also large. The variance between regions is an effective parameter to describe this difference, and its expression is:

$$\sigma_B^2(t) = \theta_1(\mu_1 - \mu)^2 + \theta_2(\mu_2 - \mu)^2 \tag{3.5}$$

In the formula, $\sigma_{\mathrm{B}}^2(t)$ represents the variance between the two regions after the image is segmented by the threshold t. Clearly, different t values will result in different between-region variances, meaning that the between-region variance, the mean of region 1, the mean of region 2, the area ratio of region 1, and the area ratio of region 2 are all functions of the threshold t. Therefore, the above formula can be rewritten as:

$$\sigma_B^2(t) = \theta_1(t)[\mu_1(t) - \mu]^2 + \theta_2(t)[\mu_2(t) - \mu]^2 \tag{3.6}$$

Through mathematical derivation, the between-region variance can be expressed as:

$$\sigma_B^2(t) = \theta_1(t)\theta_2(t)[\mu_1(t) - \mu_2(t)]^2 \tag{3.7}$$

When the variance between the two segmented regions reaches its maximum, it is considered the optimal separation state of the two regions, thus determining the threshold T.

$$T = \max\left[\sigma_B^2(t)\right] \tag{3.8}$$

Determining the threshold by maximum variance does not require manually setting other parameters, making it an automatic method for selecting the threshold. However, the implementation of Otsu's method is relatively complex, and in practical applications, a simple iterative method is often used for automatic threshold selection. The method is as follows: First, choose an approximate threshold as the initial estimate, and then continuously improve this estimate. For example, use the initial threshold to generate sub-images, and select a new threshold based on the characteristics of the sub-images, then use the new threshold to segment the image. This approach yields better results than using the initial threshold to segment the image. The strategy for improving the threshold is key to this method. For example, one method is as follows:

① Choose the mean pixel value of the image as the initial threshold T;
② Use the threshold T to divide the image into two data sets R_1 and R_2;
③ Calculate the means u_1 and u_2 of regions R_1 and R_2;
④ Select a new threshold $T = (u_1 + u_2)/2$;

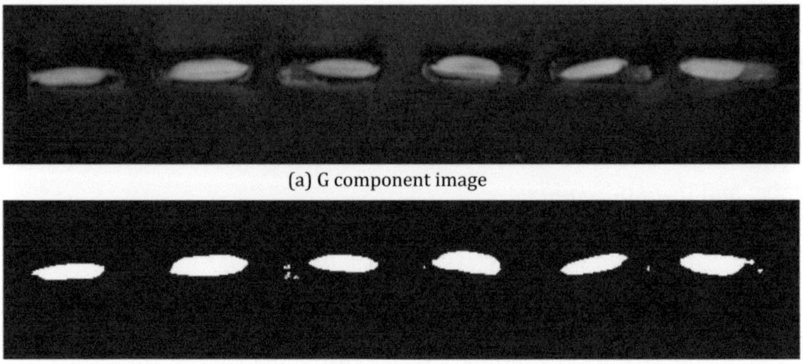

(a) G component image

(b) Binarized image

Fig. 3.4 Binary image of Otsu method

⑤ Repeat steps 2–4 until u_1 and u_2 no longer change.

Figure 3.4 shows the result of binarisation using the above Otsu's method on the *G* component image. For this image, the segmentation threshold calculated by Otsu's method is 52.

3.3 Colour-Based Target Extraction

3.3.1 Hue, Brightness, Saturation, and Others [1]

In Sect. 2.3, it was introduced that a colour image is composed of grayscale images of three components: red (*R*), green (*G*), and blue (*B*). When photographing green grass, the *G* component is stronger compared to the *R* and *B* components; for blue skies, the *B* component is stronger compared to the R and G components. Depending on the different values of the *R*, *G*, and *B* components, various colours can be seen. In colour image processing, it is necessary to consider not only position and grayscale information but also colour information.

For the same colour, different people may imagine different shades in their minds. To quantitatively represent colour, it can be divided into three characteristics: the first characteristic is hue (H), which is used to indicate the type of colour. The second characteristic is value V or brightness Y or intensity I, which is used to indicate the lightness or darkness of an image. The third characteristic is saturation or chroma (S), which is used to indicate the vividness of the colour. These three characteristics are known as the three basic attributes of colour. These three basic attributes of colour can be represented by an idealised double-cone HSI model, as shown in Fig. 3.5, which displays the coloured double-cone HSI model. The axis of the double cone represents the brightness value. The plane perpendicular to the axis represents hue and saturation, expressed in polar coordinates, where the angle represents hue and the radial distance represents saturation at a certain hue.

Simulated colour television signals also transform the *R*, *G*, *B* signals into a brightness signal *Y* and chrominance signals C_1, C_2. The relationship is as follows:

$$\begin{aligned} Y &= 0.3R + 0.59G + 0.11B \\ C_1 &= R - Y = 0.7R - 0.59G - 0.11B \\ C_2 &= B - Y = -0.3R - 0.59G + 0.89B \end{aligned} \tag{3.9}$$

Equation (3.10) illustrates the relationship between *R*, *G*, *B* signals and *Y*, C_1, C_2. Here, the brightness signal *Y* is equivalent to a grayscale image, and the chrominance signals C_1, C_2 are the parts remaining after removing the brightness signal. The formula for deriving *R*, *G*, *B* from the brightness and chrominance signals is as follows:

$$\begin{aligned} R &= Y + C_1 \\ G &= Y - \frac{0.3}{0.9}C_1 - \frac{0.11}{0.59}C_2 \\ B &= Y + C_2 \end{aligned} \tag{3.10}$$

The relationship between the aforementioned chrominance signals and hue, saturation is shown in Fig. 3.6. This figure is consistent with the projection plane perpendicular to the brightness axis shown in Fig. 3.5, i.e., the colour circle.

From Fig. 3.6, it can be seen that hue *H* represents the angle of rotation from the coordinate axis based on the chrominance signal *B-Y* (i.e., C_2), and saturation *S*

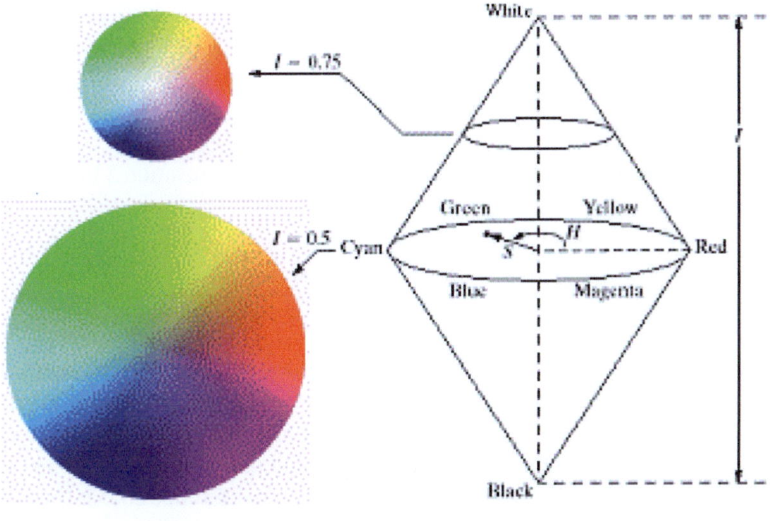

Fig. 3.5 Ideal model of colour

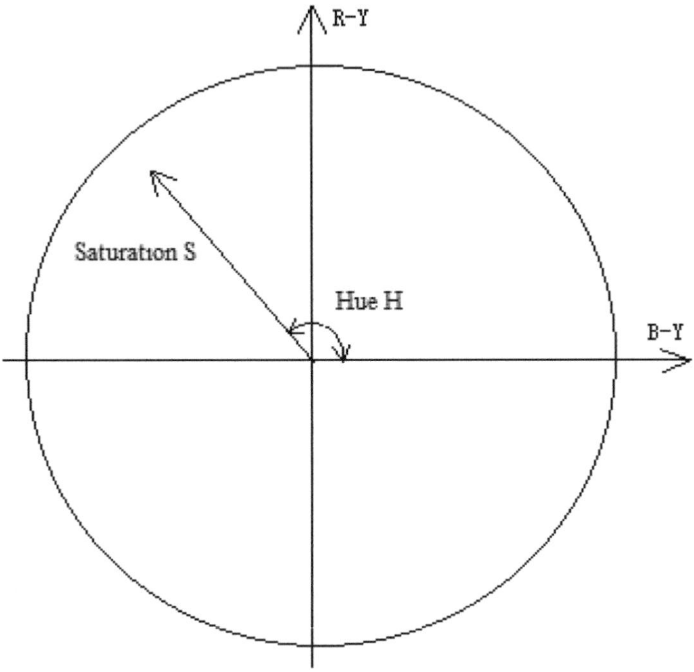

Fig. 3.6 Relationship between colour difference signal and hue, saturation and brightness

represents the distance from the origin. In formula terms, the relationship between hue H, saturation S, and chrominance is expressed as follows:

$$H = \arctan\left(\frac{C_1}{C_2}\right)$$
$$S = \sqrt{C_1^2 + C_2^2} \tag{3.11}$$

Conversely, the formula for transforming from hue H, saturation S to chrominance signals is as follows:

$$C_1 = \sin H$$
$$C_2 = S \cos H \tag{3.12}$$

Transforming the R, G, B of a colour image into images of brightness, hue, and saturation. The visualisation of the brightness signal image results in a grayscale image. Hue and saturation each use their differences as grayscale differences for image visualisation. The representation of hue is calculated from a reference colour, rotating between 0 and 180°; when it is the same as the reference colour (hue rotation angle is 0°), it is 255, and for the complementary colour in the opposite direction (hue rotation angle is 180°), it is 0, with 254 levels of grayscale in between. In the representation of hue, when saturation is 0 (i.e., no colour signal), hue is not calculated, and often a 0 grayscale level is assigned. The saturation image uses the minimum saturation value as the pixel's minimum value 0, and the maximum saturation value as the pixel's maximum value 255, converting the saturation data into image data proportionally.

The result of applying the aforementioned transformation to the actual image is shown in Fig. 3.7, where Fig. 3.7a is the original image, and Fig. 3.7b is the image of its brightness signal. The original image contains a pet rabbit with a significant red component. Since the hue signal is based on red, the hue signal image shown in Fig. 3.7c appears overall brighter. As the entire image is not very deeply coloured, the saturation signal in Fig. 3.7d appears darker, especially with the lowest saturation in the background floor tiles.

It can be observed that for this image, using the H or S signal image to perform binary extraction of the target object, the rabbit, should be somewhat easier. Therefore, converting RGB to HSI is sometimes more advantageous for extracting the target object, but compared to using RGB signals, it will require more the processing time.

In addition to RGB and HSI, there are many other models for describing colour, such as L*a*b*, UYV, XYZ, etc. These models can be applied to different purposes and scenarios as needed.

3.3.2 Colour Components and Their Combined Processing

For object extraction in natural environments, the colour characteristics of the target can be utilized by leveraging the R, G, and B components and their differential combinations. This approach effectively minimizes the impact of natural light variations and enables rapid and efficient target extraction. The following provides an example of target extraction based on colour differences.

(1) Extraction of Red Peaches on Fruit Trees

① Original Image. Figure 3.8 shows example images of colour photographs of peaches on fruit trees, representing various growth conditions such as single fruits, clusters of multiple fruits, fruits separated or in contact, as well as different lighting conditions and background scenarios. Figure 3.8a is taken under front lighting with strong illumination, showing a single fruit with leaf occlusion. The background primarily consists of leaves. Figure 3.8b under strong illumination, showing fruits in contact with leaf occlusion. The background primarily consists of branches and leaves. Figure 3.8c is taken under backlighting, showing both single fruits and fruits in contact, with partial occlusion by leaves. The background includes branches, leaves, and direct sunlight. Figure 3.8d is taken under low lighting with automatic camera flash, showing fruits in contact without occlusion. The background primarily consists of leaves. Figure 3.8e is taken under

Fig. 3.7 Three basic properties of colour

front lighting, showing both single fruits and fruits in contact, along with branch interference. Figure 3.8f is taken under strong illumination, showing both single fruits and fruits with mutual occlusion, as well as interference from branches and occlusion by leaves.

② Extraction of the Red Area of Peaches. Since ripe peaches generally have a red hue, the red area of peaches in the colour original image is first extracted using the red-green difference information, and then the complete area of the peaches is obtained by matching and expanding with the original image.

For the pixel points in the image (xi, yi), where xi and yi are the x and y coordinates of the pixel point ($0 \leq i < n$, and n is the total number of pixel points in the image), let the pixel values of the red (R) component and green (G) component be R(xi, yi) and G(xi, yi), respectively. Their difference is calculated as $\beta i = R(xi, yi) - G(xi, yi)$,

(a) Occlusion of leaves (b) Leaves obscure many fruits (c) Direct light

(d) Low light (e) Branch interference (f) Multiple fruit contact

Fig. 3.8 Colour original image

resulting in a grayscale image (RG image). If $\beta_i > 0$, set the pixel value at that point on the grayscale image to β_i, otherwise set it to 0 (black). Then calculate the mean of all non-zero pixel points in the RG image, α (as the threshold for binarization). Scan the RG image pixel by pixel; if $\beta_i > \alpha$, set the pixel value at that point to 255 (white), otherwise set it to 0 (black), obtaining a binary image, and perform hole filling and noise removal for areas smaller than 200 pixels (see Chap. 5).

Figure 3.9 shows the binary images extracted using the R-G colour difference mean as the threshold to extract the red regions of the peach from Fig. 3.8. From the extraction results in Fig. 3.9, it can be seen that this method can effectively extract the red regions of the peach under various lighting conditions and different background situations in Fig. 3.8.

For the binary image in Fig. 3.9, further steps such as boundary tracking, matching expansion, calculation of the centre point group, grouping of centre point groups, and calculation of the centre and radius are performed to obtain the detection results of the peach centre and radius as shown in Fig. 3.10. Since the other processing steps are beyond the scope of this chapter, they are not detailed here.

(2) Extraction of Green Wheat Seedlings [3]

From emergence to grain filling, wheat requires numerous field management operations, including soil loosening, fertilisation, weeding, spraying, irrigation, and growth detection. Different management operations have different targets. For example, in spraying, irrigation, and growth detection operations, the target is the wheat row (seedling row); in soil loosening and weeding operations, the target is the area between wheat rows (inter-row). Regardless of the operation, the wheat seedlings must first be extracted. Although the colour of wheat seedlings varies in different

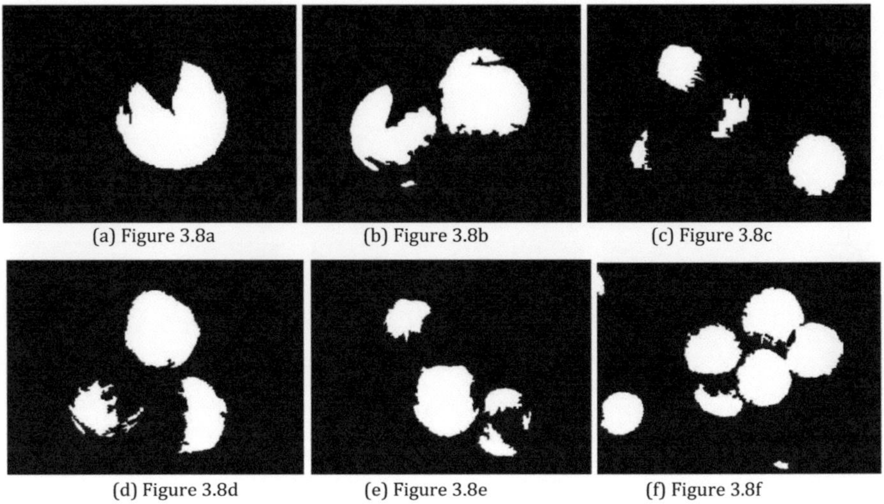

Fig. 3.9 Extract the binary image of the peach red region in Fig. 3.8

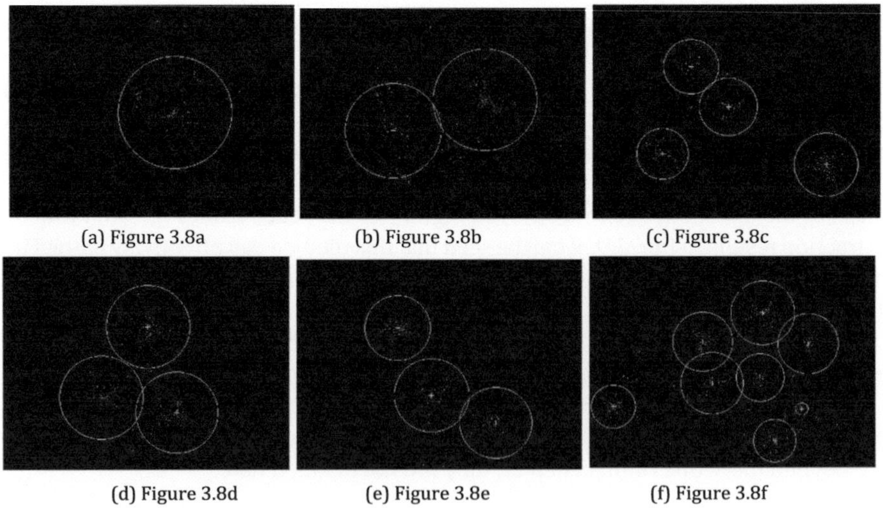

Fig. 3.10 Contour extraction and fitting results

seasons, they are always green. As shown in Fig. 3.11, (a) is an image from November (autumn) at the early growth stage of wheat on a cloudy day, with relatively moist soil; (b) is an image from February (winter) on a sunny day, with dry, cracked soil; (c) is an image from March (spring) during the wheat regreening period on a cloudy day, with relatively loose soil; (d) ~ (f) are images from different growth stages and

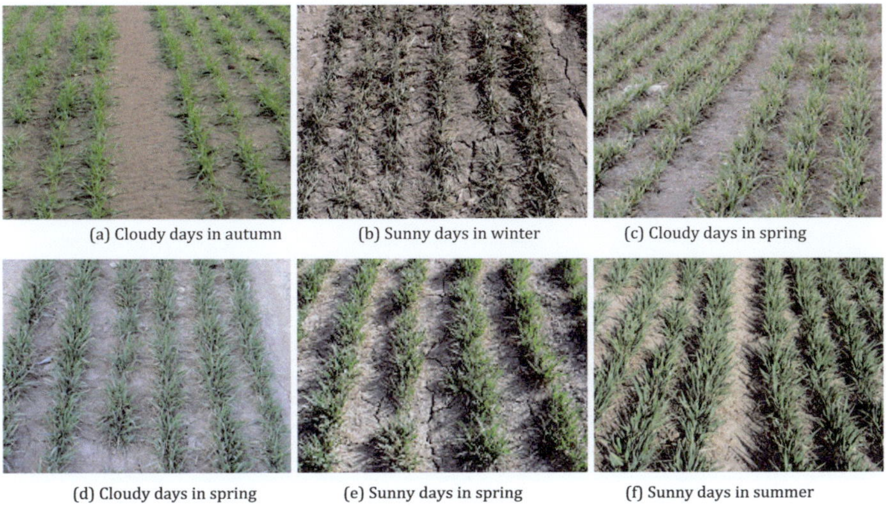

Fig. 3.11 Examples of original images of wheat fields at different growth stages

weather conditions thereafter. These six images represent different growth stages and weather conditions of wheat.

Since the green component of wheat seedlings is greater than the other two colour components, to extract the green wheat seedlings, the method of emphasising the green component and suppressing the other components can be used to convert the colour image of the wheat field into a grayscale image. The specific method is shown in formula (3.13).

$$pixel(x, y) = \begin{cases} 0 & 2G - R - B \leq 0 \\ 2G - R - B & Other \end{cases} \tag{3.13}$$

where G, R, B represent the green, red, and blue colour values of point (x, y) in the colour image, and pixel (x, y) represents the pixel value of point (x, y) in the processed grayscale image. Figure 3.12 is the grayscale image obtained through the above processing.

For the grayscale image in Fig. 3.12, use the Otsu method to determine the segmentation threshold for binarization, with the specific steps as follows:

① Calculate the average grayscale value of the grayscale image as the initial threshold t_0.
② Use t_0 to divide the grayscale image into two regions, Q_1 and Q_2, i.e., pixels with values less than t_0 are assigned to region Q_1, and those greater than t_0 are assigned to region Q_2.
③ Calculate Q_1 and Q_2 separately. The average grey values in two regions, t_1 and t_2, are considered, with the average of t_1 and t_2 set as the new threshold t_d, i.e., $t_d = (t_1 + t_2)/2$.

Fig. 3.12 Grayscale image of 2G-R-B

④ Determine whether t_0 is equal to t_d.

a. If they are equal, set the final threshold $T = t_d$.
b. If they are not equal, let $t_0 = t_d$, return to step ②, and repeat the process until the final threshold T is obtained.

Use T as the segmentation threshold to perform binarisation on the greyscale image, setting pixels with values greater than T as white (255) representing the seedling row, and pixels with values less than T as black (0) representing the inter-row. The processing result is shown in Fig. 3.13, where the white thin line on the binary image is the navigation line detected in subsequent processing. The binarisation result indicates that this adaptive threshold method is not affected by natural conditions such as lighting and background, effectively extracting the wheat seedlings without the need for noise elimination, filtering, or other auxiliary processing. Since the threshold determination does not require manual setting and is automatically determined based on the pixel value information of the image itself, it greatly improves processing accuracy.

3.4 Target Extraction Based on Differencing

Target extraction based on differencing is generally used for target extraction in motion images, with two methods: inter-frame differencing and background differencing. The following explains these two differencing target extraction methods using engineering practice projects.

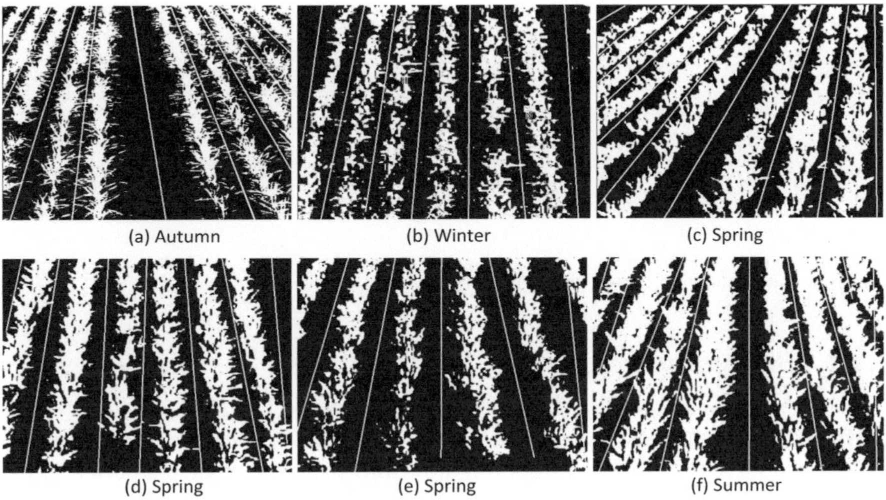

Fig. 3.13 Binarization processing results of Otsu method

3.4.1 Inter-frame Differencing [4]

Inter-frame differencing involves subtracting the pixel value of each point in the previous frame image from the corresponding point in the subsequent frame image (or vice versa). If the result exceeds a set threshold, it is set as a white pixel in the output image; otherwise, it is set as a black pixel. This can be expressed by the following formula.

$$f(x, y) = |f_1(x, y) - f_2(x, y)| = \begin{cases} 255 \geq thr \\ 0 \quad other \end{cases} \tag{3.14}$$

where $f_1(x, y)$, $f_2(x, y)$, and $f(x, y)$ represent the pixel values at point (x, y) in sequence image 1, sequence image 2, and the result image, respectively; || denotes the absolute value; *thr* is the set threshold.

This book explains the method of extracting badminton targets through inter-frame difference, using the statistical items of badminton techniques and tactics (specifically referenced in Sect. 11.3.1).

Figure 3.14 shows two adjacent frames from a video and the binarized image after differencing, with a threshold set at 5. The white pixels on the binary image represent the detected moving parts of the badminton shuttlecock and the players. Since the camera is stationary, the pixel values of the fixed parts in the sequence frames are essentially the same, resulting in a difference close to zero, while the moving areas such as the shuttlecock and players will have larger difference values, thus extracting the moving areas.

Fig. 3.14 Inter-frame difference and binarization results

3.4.2 Background Difference [5]

Traffic flow detection is an important topic in intelligent transportation systems (ITS). Traditional methods of collecting traffic flow information include the buried induction coil method, ultrasonic detector method, and infrared detection method, which have high equipment costs and are difficult to set up and maintain. With the rapid development of machine vision technology, visual detection technology for traffic flow is gradually replacing traditional methods due to its simple installation, easy operation, and convenient maintenance.

The goal of this project is to develop a road traffic image detection algorithm that is not affected by weather conditions, shadows, etc. The technical points are as follows:

① Real-time acquisition of background images;
② Extraction of vehicles in each frame of the image;
③ Elimination of the influence of vehicle shadows;
④ Differentiation of different vehicles in each frame;
⑤ Determination of the identity of vehicles in consecutive frames to achieve vehicle counting.

This project uses a laptop connected to a digital camera via an IEEE1394 interface for video image acquisition and storage, with an image size of 640 × 480 pixels and an image acquisition frame rate of 30 frames per second. The camera is installed at a height of approximately 6.6 m from the ground, with a depression angle of about 60°. The collected video images are colour images, with the red component R as the processing object.

The project first requires calculating a background image without vehicles, and due to the day-night transition of weather, the background image needs to be continuously calculated and updated regularly. This section does not introduce the calculation, updating, and other related algorithms of the background image, but focuses on the method of extracting target vehicles based on background difference. If the background image is known, the current image can be differenced with the background to extract moving vehicles.

(a) Background image (b) Images of the scene (c) Vehicle extraction results

Fig. 3.15 Vehicle extraction based on background differencing

Using the inter-frame difference formula (3.14), substitute f_1 for the current image, f_2 for the background image, set the threshold as the standard deviation of the background image pixel values, and perform noise removal processing on the resulting image f (see Chap. 5) to obtain ideal vehicle extraction results. Figure 3.15 is an example of a set of background difference images. Among them, image (a) is the background image of the highway, image (b) is a live image at a certain moment, and image (c) is the result of threshold segmentation and noise removal processing on the difference image of (a) and (b). The background image is calculated from a segment of actual images.

References

1. Chen Bingqi. Practical Digital Image Processing and Analysis [M]. 2nd Edition. Beijing: China Agricultural University Press, 2014.
2. Y. Liu, B. Chen, J. Qiao. Development of a Machine Vision Algorithm for Recognition of Peach Fruit in Natural Scene [J]. Transaction of the ASABE, 2011, 54(2):694–702.
3. H. Zhang, B. Chen, L. Zhang. Detection Algorithm for Crop Multi-centerlines Based on Machine Vision [J]. Transaction of the ASABE, 2008, 51(3):1089–1097.
4. Bingqi Chen, Zhiqiang Wang: A Statistical Method for Technical Data of a Badminton Match based on 2D Seriate Images [J]. Tsinghua Science and Technology, 2007, 12(5): 594–601.
5. Chen Wang, Chen Bingqi. Research on Highway Traffic Flow Statistics Method Based on Image Processing [J]. Computer Engineering and Applications, 2007, 43(6), 236–239.

Chapter 4
Edge Detection

4.1 Edge and Image Processing

In image processing, edges (or contours) refer not only to lines representing the boundaries of objects but also include line elements that depict image features, similar to the lines in a sketch. Of course, besides lines, colour and brightness are also important factors in images. However, many illustrations, charts, drawings, portraits, and comics are expressed using the method of depicting the object's edge lines. Although somewhat monotonous, we can still clearly comprehend what has been depicted there. Therefore, it seems somewhat incredible that simple edge lines can enable us to comprehend the object being expressed. For image processing, edge detection is also one of the important basic operations. The extracted edges can be used to identify specific objects, measure the area and perimeter of objects, find corresponding points between two images, and the processing of edge detection and extraction can further be used as a key pre-processing step for more complex image recognition and understanding.

Since the boundary between objects in an image or between an object and the background is the edge, it can be assumed that areas where the grayscale and colour of the image change abruptly can be considered edges. Since changes in colour in natural images are inevitably accompanied by changes in grayscale, for edge detection, focusing on grayscale alone is sufficient.

Figure 4.1 models typical examples of grayscale changes in images. Figure 4.1a shows the grayscale change of a step edge, a typical pattern that can be clearly seen as an edge, also known as a contour. Such step-like grayscale changes occur at the boundary between an object and the background. Figure 4.1b shows the grayscale change of the line itself, which can also be clearly regarded as an edge. Line-shaped objects and different lighting conditions that cast shadows on objects can produce line-shaped edges. Figure 4.1c has grayscale changes, but they are gradual, and the edge is not obvious. Figure 4.1d shows a zigzag grayscale change, which is not as obvious as Fig. 4.1b, but the sharp angle changes still reveal the edge.

© Chemical Industry Press 2026

B. Chen and S. Chen, *Machine Vision Technology*, Advanced and Intelligent Manufacturing in China, https://doi.org/10.1007/978-981-95-1184-6_4

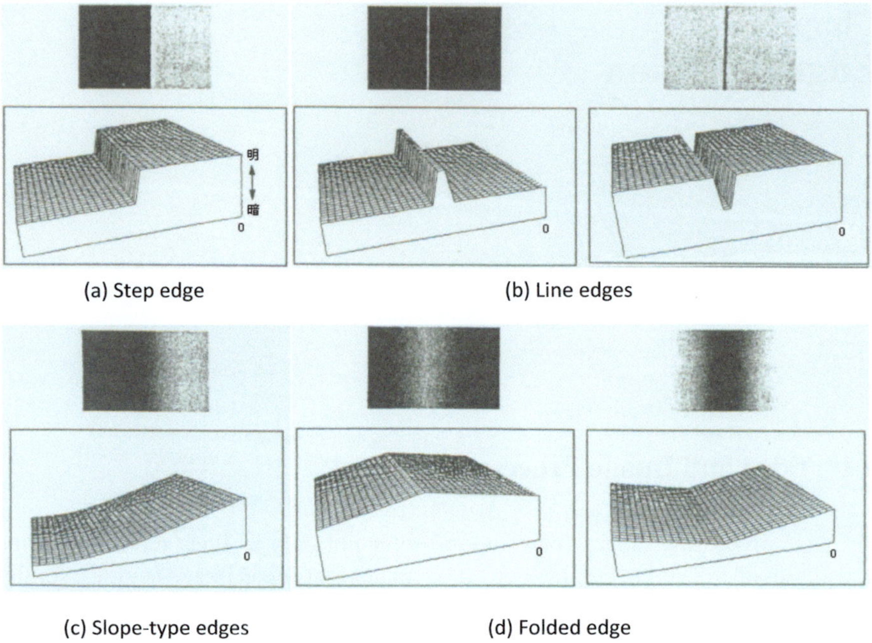

Fig. 4.1 Gray scale change model of the edge

Figure 4.2 shows the grayscale distribution of the contour part of a portrait photo. The edges, although quite clear, are not step-like but have become blunted, presenting a slope. Even for the same object's edge, the grayscale change varies at different locations, and blurred parts of the edge can be observed. Since most sensor elements have low-frequency characteristics, it is inevitable that step-like edges become slope-like edges and line-shaped edges become zigzag edges.

Therefore, in actual images (images produced by computer graphics are another matter), even if the eye can clearly identify them as edges, they will more or less become blunted, and the amount of grayscale change will decrease, making it unexpectedly difficult to extract clear edges. Consequently, various algorithms have been proposed.

4.2 Edge Detection Based on Differentiation

Since edges are parts where the grayscale value changes sharply, it is evident that differentiation, as an operation to extract parts of function changes, can be utilised in edge detection and extraction. In differentiation operations, there are first-order differential calculus (also known as gradient operations) and second-order differential

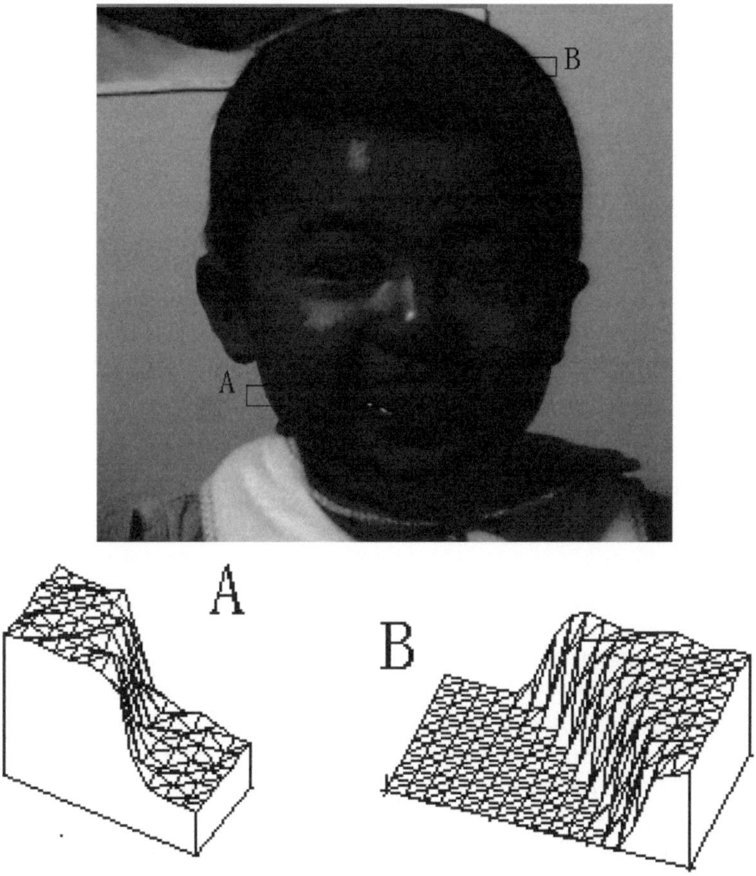

Fig. 4.2 Grayscale variation of the actual image

calculus (also known as Laplacian operations), both of which can be applied in edge detection and extraction.

(1) First-order Differential (Gradient Operation)

The first-order derivative of the grayscale slope at the coordinate point (*x*, *y*) can be represented by a vector G(*x*, *y*) = (f_x, f_y), which indicates both magnitude and direction. Here, f_x is the differential in the *x*-direction, and f_y is the differential in the *y*-direction.

The values of f_x and f_y in digital images are calculated using the following formulas:

$$\begin{aligned}&\text{Differential in the } x \text{ direction} \quad f_x = f(x+1, y) - f(x, y)\\&\text{Differential in the } y \text{ direction} \quad f_y = f(x, y+1) - f(x, y)\end{aligned} \tag{4.1}$$

Differential values of f_x and f_y are determined, and the edge's intensity and direction can be calculated using the following formulas.

$$\left[\text{Intensity}\right]:\ G=\sqrt{f_x^2+f_y^2} \tag{4.2}$$

$$[\text{Direction}]:\ \theta=\arctan(f_x/f_y)\ \text{Orientation of vector}\ \left(f_x,f_y\right) \tag{4.3}$$

The direction of the edge refers to the direction in which the greyscale changes from dark to light. It can be said that gradient operators are more suitable for detecting edges (step-like greyscale changes).

(2) Second-order differentiation (Laplacian operation)

Second-order differentiation $L\,(x, y)$ involves differentiating the gradient once more, used only for detecting the edge's intensity (not direction), and is represented in digital images by the following formula:

$$\begin{aligned}L(x,y)=4\cdot f(x,y)-f(x+1,y)-f(x-1,y)-f(x,y+1)\\ -f(x,y-1)\end{aligned} \tag{4.4}$$

Since data in digital images is arranged at discrete intervals, true differentiation is not feasible. Therefore, using the difference between adjacent pixels as in formula (4.1) or formula (4.4) is actually calculus of finite differences, conveniently referred to as differential calculus. The coefficient matrices used for pixel differentiation are called differential operators. The calculation formulas for f_x and f_y in gradient operations (4.1), as well as the Laplacian operation formula (4.4), are based on these differential operators. These differential operators, as shown in Tables 4.1 and 4.2, come in various types. The actual differential operation involves calculating the sum of the target pixel and its surrounding pixels each multiplied by the corresponding values in the differential operator matrix, and the result is used as the greyscale value of the target pixel after differentiation. Scanning the entire image and performing such differential operations on each pixel is called convolution.

Table 4.1 Differential operators for gradient calculation

Operator name	① General difference	② Roberts operator	③ Sobel operator
Template for f_x	0 0 0 0 1 1 0 0 0	0 0 0 0 1 0 0 0 1	−1 0 1 −2 0 2 −1 0 1
Template for f_y	0 0 0 0 1 0 0 1 0	0 0 0 0 0 1 0 1 0	−1 2 1 0 0 0 1 2 1

Table 4.2 Differential operators for Laplacian operation

Operator name	Laplace operator 1	Laplace operator 2	Laplace operator 3
Template	0 1 0 −1 4 1 0 1 0	−1 −1 −1 −1 8 −1 −1 −1 −1	1 −2 1 −2 4 −2 1 −2 1

4.3 Edge Detection Based on Template Matching

Template matching involves analyzing the consistency (or degree of match) between an image and a template. For this purpose, several standard patterns representing edges are prepared and compared with a part of the image, selecting the most similar part as the result image. As shown in Fig. 4.3, the Prewitt operator has 8 masks corresponding to 8 edge directions. Figure 4.4 demonstrates how these masks are compared with the actual image. Similar to differential operations, the target pixel and its surrounding (3×3 neighbourhood) pixels are each multiplied by the corresponding mask coefficients, and the sums of the products are calculated. Calculations are performed for each of the 8 masks, and the direction of the mask with the largest result is the edge direction, with the result being the edge intensity.

Figure 4.5 shows the results of processing a frame of image using different differential operators. Different differential operators produce varying results. In practical applications, the choice of differential operator can be tailored to the specific requirements of each situation. If the processing effects are similar, operators with less computation should be chosen to improve processing speed. For example, in Fig. 4.5, the differential effects of (b) and (d) are similar, but the Sobel operator in (b) requires much less computation than the Prewitt operator in (d).

Additionally, when the orientation of the target object is known, the template matching operator can be optimized by selecting only templates that match the target object's orientation for calculation. This approach can achieve good detection results while significantly reducing the computational load. For example, when detecting

	(1)	(2)	(3)	(4)	(5)	(6)	(7)	(8)
Mask mode	1 1 1 1 -2 1 -1 -1 -1	1 1 1 1 -2 -1 1 -1 -1	1 1 -1 1 -2 -1 1 1 -1	1 -1 -1 1 -2 -1 1 1 1	-1 -1 -1 1 -2 1 1 1 1	-1 -1 1 -1 -2 1 1 1 1	-1 1 1 -1 -2 1 -1 1 1	1 1 1 -1 -2 1 -1 -1 1
Corresponding edge								

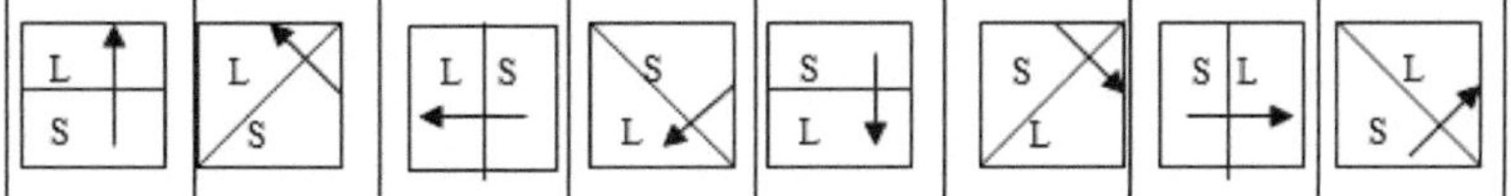

L: Light S: Shade

Fig. 4.3 Various mask modes for template matching (Prewitt operator)

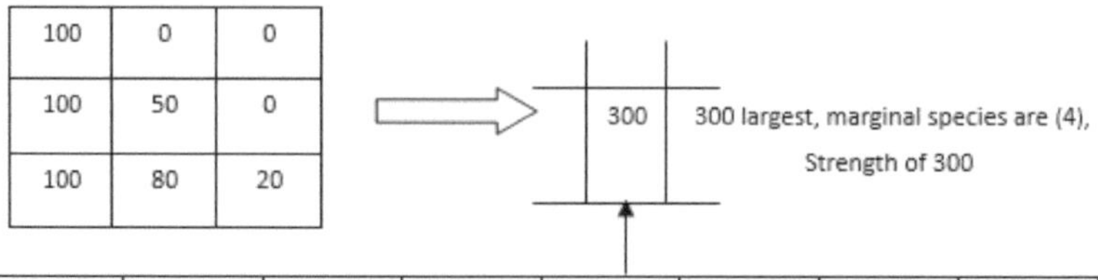

Mask	(1)	(2)	(3)	(4)	(5)	(6)	(7)	(8)
Value	-100	100	260	300	100	-100	-300	-260

(For the 8 neighborhoods of the current pixel, calculate the consistency of each mask)

For example, mask (1): 1×100 + 1×0 + 1×0 + 1×100 + (-2) ×50 + 1×0 + (-1)×100 + (-1)×80 + (-1)×20 = -100

Fig. 4.4 Calculation example of template matching

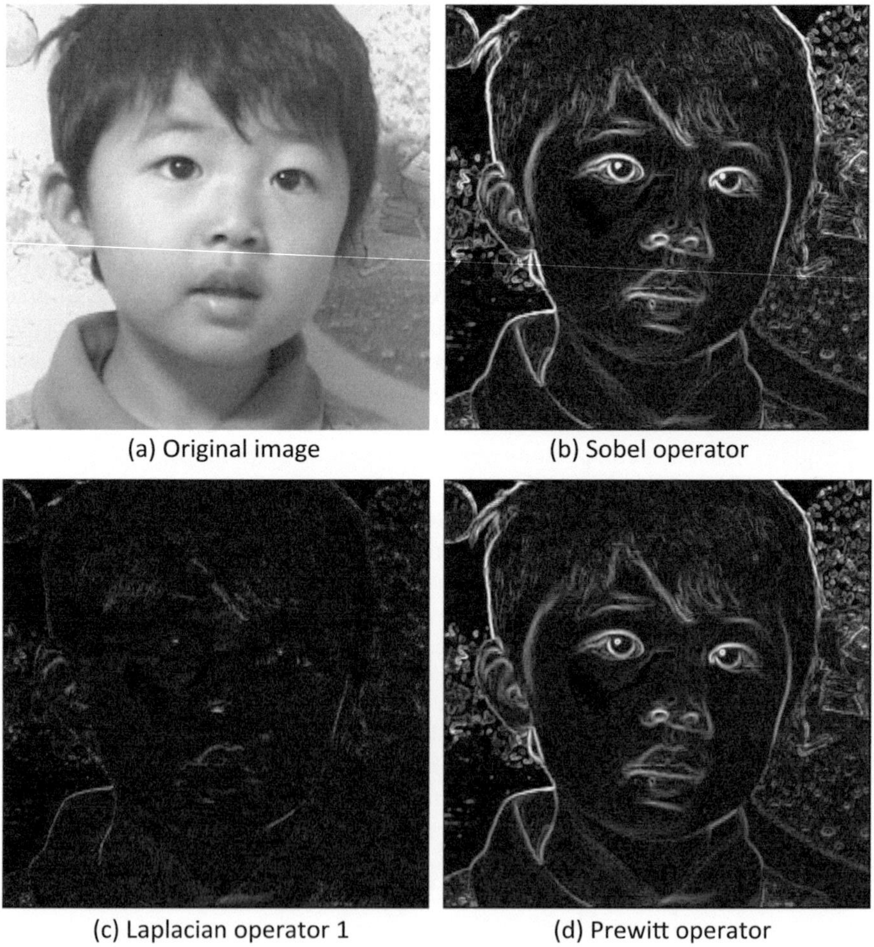

Fig. 4.5 Differential images of different operators

M1	M2	M3	M4	M5	M6	M7	M8
5 5 5	-3 5 5	-3 -3 5	-3 -3 -3	-3 -3 -3	-3 -3 -3	5 -3 -3	5 5 -3
-3 0 -3	-3 0 5	-3 0 5	-3 0 5	-3 0 -3	5 0 -3	5 0 -3	5 0 -3
-3 -3 -3	-3 -3 -3	-3 -3 5	-3 5 5	5 5 5	5 5 -3	5 -3 -3	-3 -3 -3

Fig. 4.6 Kirsch operator

M1	M2	M3	M4	M5	M6	M7	M8
1 2 1	2 1 0	1 0 -1	0 -1 -2	-1 -2 -1	-2 -1 0	-1 0 1	0 1 2
0 0 0	1 0 -1	2 0 -2	1 0 -1	0 0 0	-1 0 1	-2 0 2	-1 0 1
-1 -2 -1	0 -1 -2	1 0 -1	2 1 0	1 2 1	0 1 2	-1 0 1	-2 -1 0

Fig. 4.7 Robinson operator

lane lines on a highway, since the lane lines are vertically forward, only the left and right edges need to be detected, if the Prewitt operator is used, the calculation can be limited to the detection of the left and right edges, specifically ③ and ⑦. This can reduce the computational load to one-quarter of using all operators. Reducing the processing load is of great importance for real-time processing.

Moreover, in template matching, commonly used operators include the Kirsch operator shown in Fig. 4.6 and the Robinson operator shown in Fig. 4.7.

4.4 Binarisation of Edge Images

The image after differential processing is still a grayscale image, and generally requires further binarisation. For the binarisation of differential images, the ppp-parameter method introduced in Chap. 3 is applied. By setting the threshold at the position corresponding to the top 5% (brightest portion) of the histogram, a reliable and stable processing effect is achieved. Figure 4.8 shows the binarisation effect of the differential image in Fig. 4.5 using this method.

4.5 Thinning Process

Thinning is the process of organising edge lines with uneven line widths into lines of uniform width (generally 1 pixel wide) on the binarised image after threshold processing.

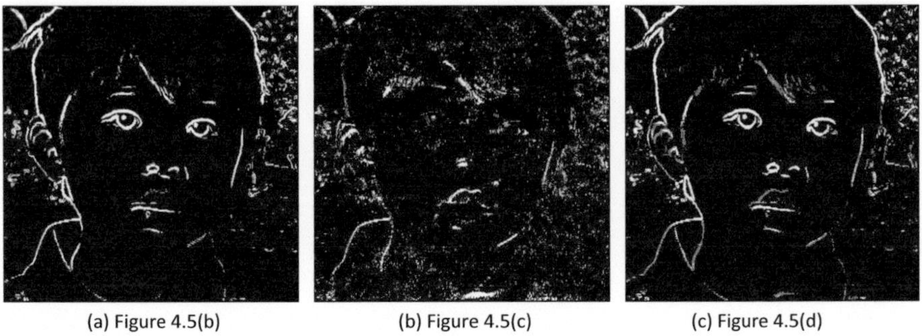

Fig. 4.8 Extraction results of the top 5% pixels in the differential image in Fig. 4.5

As illustrated in Fig. 4.9, the thinning process involves progressively removing pixel layers from the outer edges of a thick edge line until the line is reduced to a width of one pixel. This process must ensure the continuity of the lines, following specific pixel removal rules, which makes the algorithm relatively complex. The result of applying this thinning process to the input binarized edge image is shown in Fig. 4.10, producing an edge image with a consistent line width of one pixel.

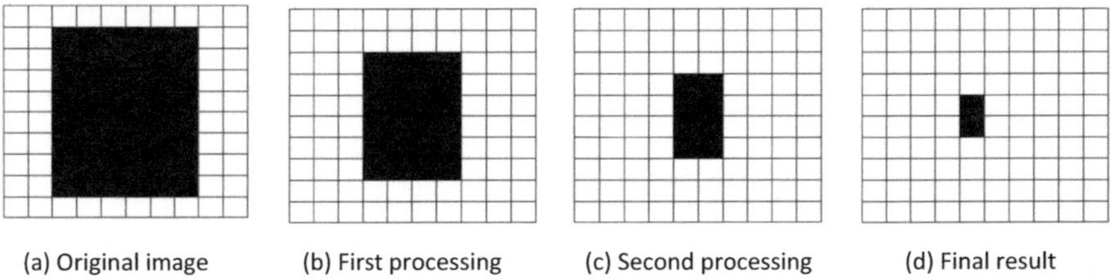

Fig. 4.9 Thinning process

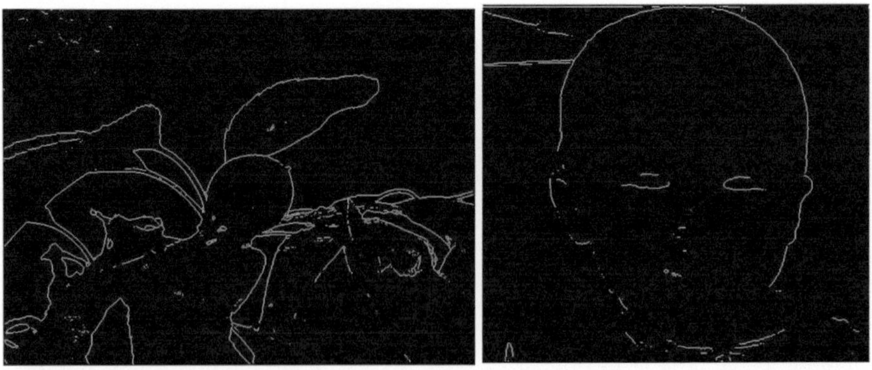

Fig. 4.10 Fineness processing results

4.6 Canny Algorithm

The Canny algorithm is a multi-stage edge detection algorithm developed by John F. Canny in 1986. Although the Canny algorithm is quite old, it can be said to be a classic algorithm for edge detection and is therefore widely used.

Canny's goal was to find an optimal edge detection algorithm, which means:

① Optimal detection. The algorithm should identify as many real edges in the image as possible, with the probability of missing real edges and falsely detecting non-edges being as low as possible.
② Optimal localisation criterion. The detected edge points should be as close as possible to the actual edge points, or the degree to which the detected edges deviate from the true edges of the object due to noise should be minimal.
③ One-to-one correspondence between detected points and edge points. The edge points detected by the operator should correspond one-to-one with the actual edge points.

To meet these requirements, Canny used the calculus of variations, a method for finding the optimised function of a specific function. Optimal detection is represented by four exponential function terms, which are very similar to the first derivative of the Gaussian function.

The Canny edge detection algorithm can be divided into the following five steps:

① Apply Gaussian filtering to smooth the image, aiming to remove noise.
② Find the intensity gradients of the image.
③ Apply non-maximum suppression to eliminate false edges (those that are detected but should not be).
④ Apply a double threshold method to determine potential edges.
⑤ Use hysteresis to track the edges.

The following explains each step in detail.

(1) Image Smoothing (Denoising)

Denoising is an important step in target extraction, and Chap. 5 specifically introduces various image denoising methods. The first step of the Canny algorithm is to convolve the original data with a Gaussian filter (details in Chap. 5), resulting in a slightly blurred image compared to the original. This way, individual pixel noise becomes almost negligible on the Gaussian-smoothed image. Below is a 5×5 Gaussian filter (standard deviation of the Gaussian kernel $\sigma = 2$), where A is the original image, and B is the smoothed image. Figure 4.11 shows the original image and the Gaussian smoothed result image.

$$B = \frac{1}{84} \times \begin{bmatrix} 1\,2\,3\,2\,1 \\ 2\,5\,6\,5\,2 \\ 3\,6\,8\,6\,3 \\ 2\,5\,6\,5\,2 \\ 1\,2\,3\,2\,1 \end{bmatrix} \times A$$

(a) Original image (b) Gaussian smoothing

Fig. 4.11 Original image and Gaussian smooth image

(2) Finding the intensity gradient in the image (gradient operation)

The basic idea of the Canny algorithm is to find the positions in an image where the change in grey intensity is the strongest, i.e., the gradient direction. The gradient of each pixel in the smoothed image can be obtained using operators such as the Sobel operator from Sect. 4.2. Taking the Sobel operator as an example, first calculate the gradients in the horizontal (x) and vertical (y) directions f_x and f_y; then use Eq. (4.2) to determine the gradient intensity value G for each pixel. By replacing each point in the smoothed image with G, the gradient intensity image of Fig. 4.12 can be obtained. As seen in Fig. 4.12, in areas with drastic changes (at the boundaries), a larger gradient intensity value G is obtained, corresponding to a larger pixel value. However, since these boundaries are usually very broad, making it difficult to determine the true position of the boundary, it is also necessary to store the gradient direction [Eq. (4.3)] and perform non-maximum suppression. This step requires storing two pieces of data: one is the gradient intensity information, and the other is the gradient direction information.

(3) Non-maximum suppression

The purpose of this step is to make the blurred boundaries sharp. Simply put, it retains the maximum gradient intensity value at each pixel point and removes other values.

For the gradient intensity image, perform the following operations on each pixel:

a. Approximate its gradient direction to one of 0, 45, 90, 135, 180, 225, 270, 315, i.e., up, down, left, right, and 45° directions.
b. Compare the pixel value of this point with the pixel values in its gradient direction (positive and negative).
c. If the pixel value of this point is the largest, retain it; otherwise, suppress it (i.e., set it to 0).

Fig. 4.12 Gradient intensity image

To better explain this concept, refer to Fig. 4.13.

The numbers in the figure indicate the gradient intensity of each pixel, while the arrows represent the gradient direction. For example, consider the third pixel in the second row. Its gradient direction is upward, so its intensity (7) is compared with the intensities of the two neighboring pixels above (5) and below (4). Since this pixel has the highest intensity among the three, it is retained. The result of non-maximum suppression on Fig. 4.12 is shown in Fig. 4.14.

(4) Double thresholding

There are still many noise points in the image after non-maximum suppression. The Canny algorithm applies a technique called double thresholding, which sets a high threshold and a low threshold. If a pixel point in the image is greater than the high threshold, it is considered definitely a boundary (called a strong edge), and its pixel

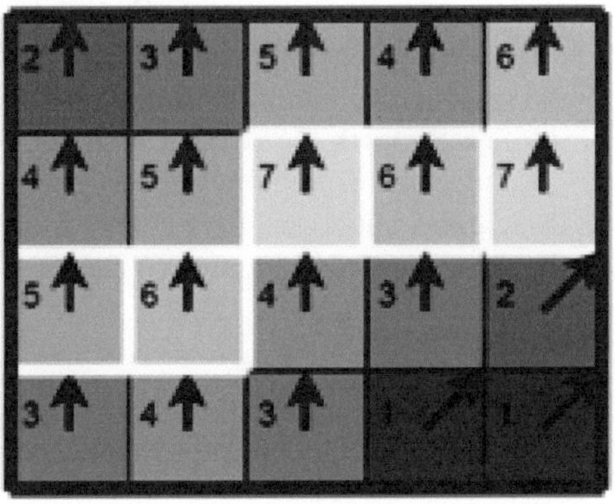

Fig. 4.13 Strength diagram

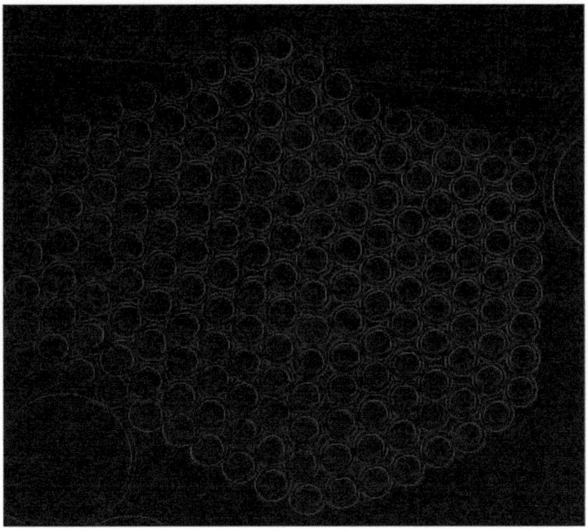

Figure. 4.14 Results of non-maximum suppression processing on Fig. 4.12

value is set to 255; if it is less than the low threshold, it is considered definitely not a boundary, and its pixel value is set to 0. Pixels in between are considered boundary candidates (called weak edges), and their pixel values remain unchanged.

The setting of high and low thresholds uses the p parameter method introduced in Chapter 3, generally setting the positions at 80% and 40% of the histogram's lower part (dark part) for better processing results. The threshold values can also be adjusted to see if the processing effect improves.

In actual execution, this step only calculates the specific values of the high and low thresholds through proportion and does not change the pixel values of Fig. 4.14.

(5) Hysteresis edge tracking

Perform hysteresis edge tracking on the image after double thresholding, with the specific steps as follows:

① Simultaneously scan the non-maximum suppression image I (Fig. 4.14) and the gradient intensity image G (Fig. 4.12). When a pixel $p\ (x, y)$ is not 0 in the I image and is greater than the high threshold but less than 255 in the G image, set $p\ (x, y)$ to 255 (white) in both the I and G images, and designate this pixel point as $q\ (x, y)$, track the contour line starting from $q\ (x, y)$.

② Examine the 8-neighbourhood of $q\ (x, y)$ in image I and image G. If a certain 8-neighbourhood pixel $s\ (x, y)$ is greater than zero in image I and greater than the low threshold in image G, then set this pixel to white in image G and set this pixel as $q\ (x, y)$.

③ Repeat step ② until there are no pixels meeting the conditions.

④ Repeat steps ①–③ until the entire image has been scanned.

⑤ Scan image G and set non-white pixels to 0.

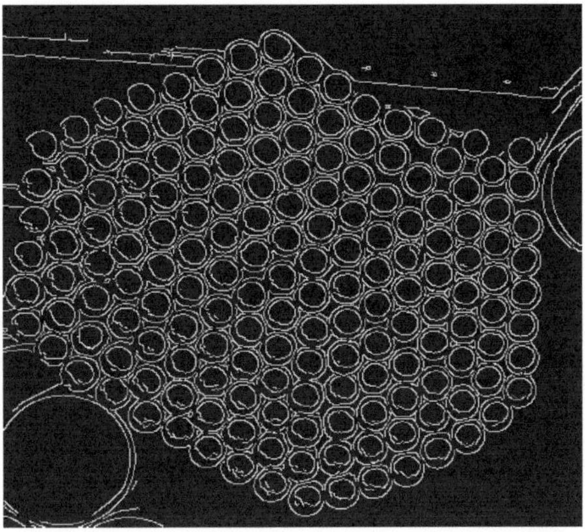

Figure. 4.15 Canny differential processing results

At this point, the edge detection of the Canny algorithm is completed. Figure 4.15 shows the result of hysteresis edge tracking.

The Canny algorithm includes many adjustable parameters that will affect the computation time and effectiveness of the algorithm.

Size of the Gaussian filter: All smoothing filters in the first step will directly affect the result of the Canny algorithm. Smaller filters produce less blurring, allowing for the detection of smaller, more distinct fine lines. Larger filters produce more blurring, colouring a larger image area with the colour value of a specific point. This result is more useful for detecting larger, smoother edges, such as the edge of a rainbow.

Double threshold: Using two thresholds is more flexible than using one, but it still has the common problem of thresholds. If the set threshold is too high, important information may be missed; if too low, trivial information will be considered important. It is difficult to provide a universal threshold applicable to all images. Currently, there is no verified implementation method.

Reference

1. Chen Bingqi. Practical Digital Image Processing and Analysis [M]. 2nd Edition. Beijing: China Agricultural University Press, 2014.

Chapter 5
Image Smoothing Processing

5.1 Image Noise and Common Smoothing Methods [1]

Images are subjected to various noise interferences during acquisition and transmission, which degrades image quality. To suppress noise and improve image quality, smoothing processing is required. The term noise, simply put, refers to obstacles. Image noise can be understood as obstacles on the image. For example, when a television image becomes chaotic and difficult to watch due to poor antenna conditions, this state is referred to as image degradation. This image degradation can be roughly divided into two categories: one is salt & pepper noise, which has a similar amplitude but appears in very random locations; the other is random noise, which has a random distribution of both position and amplitude.

Figure 5.1 is an image with noise, where it can be seen that there is a sharp grey level difference between the noise and its surrounding grey levels, and it is precisely these sharp grey level differences that cause viewing obstacles. The method of eliminating such noise in images is called image smoothing or simply smoothing. However, the edge parts of the target image also have sharp grey level differences. In image smoothing processing, the ability to distinguish between edge parts and noise parts is crucial for effective noise reduction.

Image smoothing processing involves suppressing image noise while preserving as much detail as possible. The method of eliminating noise varies according to the nature of the noise. Below are some commonly used methods for noise elimination (filtering).

Spatial domain filtering: Directly perform spatial transformations on image data to achieve filtering purposes.

Frequency filtering: First, transform the spatial domain image to the frequency domain for processing, and then transform it back to the spatial domain image (Fourier transform, wavelet transform, etc.).

Linear filtering: The output pixel is a linear combination of the input pixel's neighbouring pixels (moving average, Gaussian filtering, etc.).

© Chemical Industry Press 2026

B. Chen and S. Chen, *Machine Vision Technology*, Advanced and Intelligent Manufacturing in China, https://doi.org/10.1007/978-981-95-1184-6_5

Fig. 5.1 Image with random noise

Non-linear filtering: The output pixel is a non-linear combination of the input pixel's neighbouring pixels (median filtering, edge-preserving filtering, etc.).

The following introduces several commonly used image filtering methods.

5.2 Moving Average [1]

The moving average method (also known as the averaging filter) is the simplest approach to eliminating noise. As shown in Fig. 5.2, this method replaces the pixel value with the average value of the 3 × 3 pixel range surrounding a certain pixel. Its principle is to blur the image to the extent that small noise is not visible. However, this method blurs both noise and edges indiscriminately, resulting in the target image being blurred while the noise is eliminated.

The best result for noise elimination should be that the noise is removed while the edges are well preserved. The most famous method to achieve this effect is median filtering.

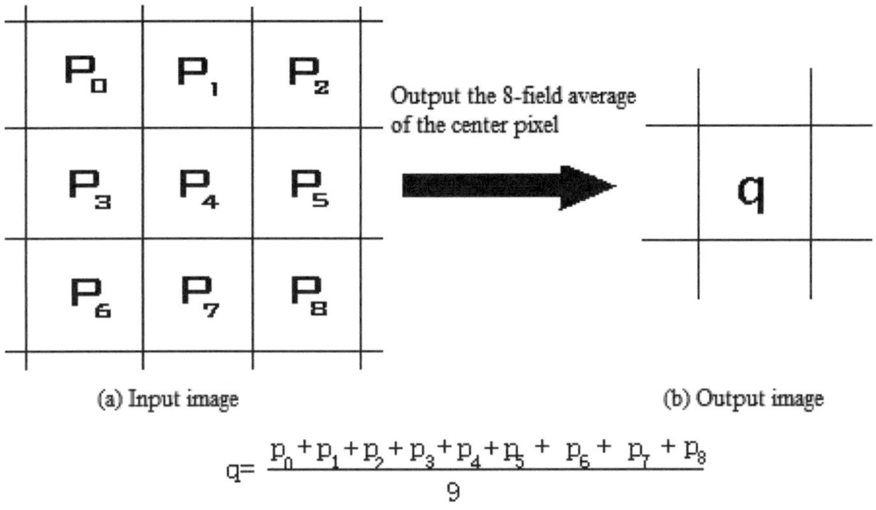

Fig. 5.2 Moving average method

5.3 Median Filtering [1]

As shown in the grayscale image data in Fig. 5.3, to determine the pixel value surrounded by ○, examine the 9 pixels' grayscale within the 3 × 3 neighbourhood (the area enclosed by the black frame), and arrange them in ascending order, as shown below.

2 2 3 3 ④ 4 4 5 10

The median value (also known as the median) at this time is the grayscale value of the 5th pixel out of the 9 pixels after sorting, which is 4. The pixel with a grayscale value of 10 was intentionally introduced as noise and successfully removed through

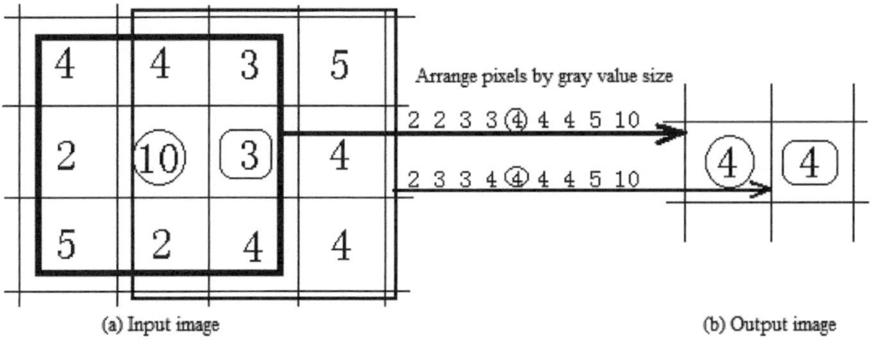

Fig. 5.3 Median filtering

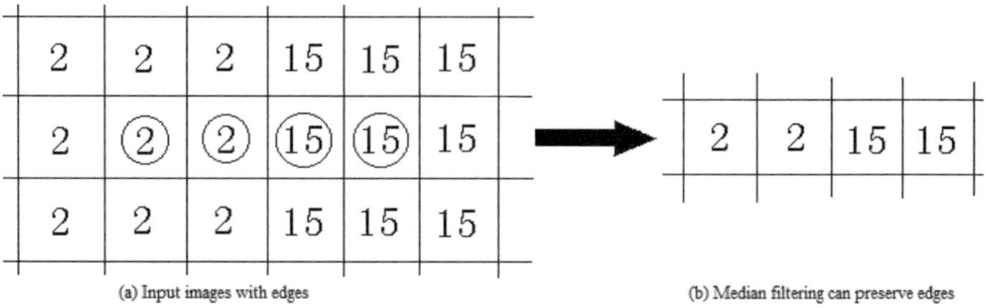

Fig. 5.4 Median filtering of images with edges

median filtering. This occurs because the grayscale value of the noise significantly differs from that of the surrounding pixels. When the values are sorted, the noise pixel appears at either the extreme left or right of the sorted sequence and is therefore not chosen as the median.

So, what about the pixel on its right (the pixel surrounded by □)? Check the pixels within the neighbourhood enclosed by the thin frame.

2 3 3 4 ④ 4 4 5 10

The median is 4, but in reality, it was 3 that became 4, which is the damage caused by the processing. However, it is not visually noticeable.

The question is whether the edge part is preserved. Figure 5.4a is an image with edges, and the result obtained for the pixel surrounded by ○ is shown in Fig. 5.4b, where the edges are completely preserved.

In the moving average method, because the noise component is included in the average calculation, the output is affected by the noise. However, in median filtering, since the noise component is unlikely to be chosen, it has minimal impact on the output. Therefore, when comparing the same 3×3 area, the noise removal capability of median filtering is superior.

Figure 5.5 shows the results of removing noise using median filtering and the moving average method, clearly demonstrating that median filtering is an excellent method both in eliminating noise and preserving edges. However, the computation time for median filtering is many times that of the moving average method.

5.4 Gaussian Filtering

The Gaussian filter is a linear smoothing filter that selects weights based on the shape of the Gaussian function. The Gaussian smoothing filter is very effective in removing noise that follows a normal distribution. Equations (5.1) and (5.2) are the formulas for the one-dimensional zero-mean Gaussian function and the two-dimensional zero-mean Gaussian function, respectively. Figure 5.6a and b are schematic diagrams

(a) Original image (b) Median filter (c) Moving average method

Fig. 5.5 Comparison between median filtering and moving average method

of the distribution of the one-dimensional zero-mean Gaussian function and the two-dimensional Gaussian zero-mean function, respectively.

$$G(x) = \frac{1}{\sqrt{2\pi}\delta} e^{-\frac{x^2}{2\delta^2}} \tag{5.1}$$

$$G(x, y) = \frac{1}{2\pi\delta^2} e^{-\frac{x^2+y^2}{2\delta^2}} \tag{5.2}$$

where, σ is the standard deviation of the normal distribution, which determines the width of the Gaussian function.

The Gaussian function has the following five important characteristics.

① The two-dimensional Gaussian function has rotational symmetry, meaning the degree of smoothing by the filter is the same in all directions. Generally, the edge direction of an image is not known in advance, so it is impossible to determine before filtering that more smoothing is needed in one direction than another. Rotational symmetry means that the Gaussian smoothing filter will not bias any direction in subsequent edge detection.

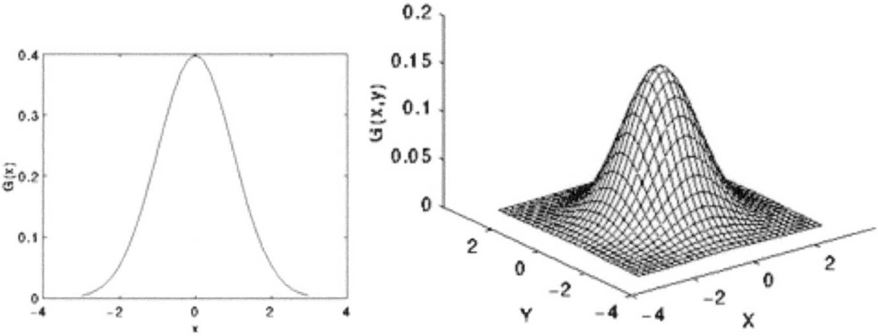

(a) One-dimensional Gaussian function (b) Two-dimensional Gaussian function

Fig. 5.6 Distribution of Gaussian function

② The Gaussian function is a single-valued function. This indicates that the Gaussian filter replaces the pixel value at a point with the weighted mean of the pixel neighbourhood, and the weight of each neighbourhood pixel point monotonically increases or decreases with the distance from that point to the centre point. This property is important because an edge is a local feature of an image, and if the smoothing operation still has a significant effect on pixel points far from the operator centre, the smoothing operation will distort the image.

③ The Fourier transform spectrum of the Gaussian function is single-lobed. This property indicates that the Fourier transform of the Gaussian function is equal to the Gaussian function itself. Images are often contaminated by unwanted high-frequency signals (noise and fine textures), while desired image features (such as edges) contain both low-frequency and high-frequency components. The single-lobed nature of the Gaussian function's Fourier transform means that smoothing the image will not be contaminated by unwanted high-frequency signals while retaining most of the desired signals.

④ The width of the Gaussian filter (which determines the degree of smoothing) is determined by the parameter σ, and the relationship between σ and the degree of smoothing is very simple. A larger σ produces a wider Gaussian kernel in the spatial domain but a narrower bandwidth in the frequency domain, thereby increasing the smoothing strength by suppressing more high-frequency components. By adjusting the parameter σ, the degree of smoothing of the image can be effectively adjusted. σ is also known as the smoothing scale.

⑤ The separability of the Gaussian function allows for efficient implementation of filtering with large-sized Gaussian filters. The convolution of a two-dimensional Gaussian function can be performed in two steps: first, convolve the image with a one-dimensional Gaussian function, and then convolve the result with the same one-dimensional Gaussian function in the perpendicular direction. Therefore, the computational load of two-dimensional Gaussian filtering increases linearly with the width of the filter template rather than quadratically.

These characteristics demonstrate that the Gaussian smoothing filter is a highly effective low-pass filter in both spatial and frequency domains and has been effectively used by engineers in practical image processing.

For image processing, a commonly used two-dimensional zero-mean discrete Gaussian function is used as a smoothing filter. When designing a Gaussian filter, for ease of calculation, it is generally desired that the filter weights are integers. A value is taken at a corner of the template, and a K is chosen so that the value at the corner is 1. This coefficient allows the filter to be converted into an integer form. However, since the sum of the weights in the integerized template does not equal 1, the filter template must be normalized to ensure that uniform gray areas in the image remain unaffected.

Below are several Gaussian filter templates.

$\sigma = 1$, 3×3 template

$$\frac{1}{16} \times \begin{bmatrix} 1 & 2 & 1 \\ 2 & 4 & 2 \\ 1 & 2 & 1 \end{bmatrix}$$

$\sigma = 2$, 5×5 template

$$\frac{1}{84} \times \begin{bmatrix} 1 & 2 & 3 & 2 & 1 \\ 2 & 5 & 6 & 5 & 2 \\ 3 & 6 & 8 & 6 & 3 \\ 2 & 5 & 6 & 5 & 2 \\ 1 & 2 & 3 & 2 & 1 \end{bmatrix}$$

$\sigma = 3$, 7×7 template

$$\frac{1}{365} \begin{bmatrix} 1 & 2 & 4 & 5 & 4 & 2 & 1 \\ 2 & 6 & 9 & 11 & 9 & 6 & 2 \\ 4 & 9 & 15 & 18 & 15 & 9 & 4 \\ 5 & 11 & 18 & 21 & 18 & 11 & 5 \\ 4 & 9 & 15 & 18 & 15 & 9 & 4 \\ 2 & 6 & 9 & 11 & 9 & 6 & 2 \\ 1 & 2 & 4 & 5 & 4 & 2 & 1 \end{bmatrix}$$

After obtaining the Gaussian filter template, similar to the differentiation operation (see Chap. 4), convolve the image to obtain a smoothed image.

Figure 5.7 shows the original colour image of 640 × 480 pixels and the images filtered using the aforementioned three Gaussian filter templates. It can be observed that the filtered images appear cleaner and clearer than the original image. As the smoothing scale increases, especially with the increase in template size, larger noise can be removed, but the image also becomes more blurred, and the smoothing time is longer. In practical applications, different template sizes and smoothing scales σ should be used according to the actual noise size. By comparison, it is found that Gaussian filtering is quite effective in removing random noise and Gaussian noise (especially noise following a normal distribution).

5.5 Clarification of Blurred Images

Images are a very useful source of information, so they are required to be clear. A clear image is one where the subtle differences in brightness and colour of the object are distinctly captured. However, the images obtained through a camera are not necessarily clear. For example, animals photographed in the dark or locusts in grass may blend into a background with similar brightness or colour, making such images difficult to distinguish. Even for such images, it is possible to amplify

Fig. 5.7 Gaussian filter image

the slight differences in colour and brightness between the animals, locusts, and the background, making the posture of the animals and locusts in the background visible. Methods that amplify the brightness and colour information contained in the image, or transform this information into other forms, to obtain a clear image are called image enhancement. Image enhancement, depending on the information enhanced, includes methods such as edge enhancement, grey level enhancement, and colour saturation enhancement. The following introduces several methods that can be used to enhance images.

5.5.1 Contrast Enhancement [1]

The ratio of the grey levels of the bright and dark parts of an image is called contrast. In a high-contrast image, the contours of the illuminated objects are clearly visible, making it a clear image; conversely, in a low-contrast image, the contours of objects are blurred, making it an unclear image. For example, when looking at a photograph taken a long time ago, you may find it appears entirely white, and it is difficult to distinguish between black and white. For such low-contrast images, transformations that make the white parts whiter and the black parts blacker, i.e., contrast enhancement, can be used to obtain a clear image.

Let us explain the method of contrast enhancement.

Look at Fig. 5.8, where the entire image is very dark. By examining the grey level histogram (Fig. 5.9), it is found that the grey values of the image are overly concentrated at the low end of the grey level range. So, how can such an image be processed to make it a clear image? Simply disperse the overly concentrated grey values to expand the difference between the background and the object. One processing method is to multiply the grey values of the pixels in the image by n, i.e.:

$$g(x, y) = nf(x, y) \tag{5.3}$$

The grey value $f(x, y)$ at position (x, y) in the original image is multiplied by n, and the grey value of the processed image at (x, y) becomes $g(x, y)$. Since the range of image data is 0–255, if the calculated result exceeds 255, it is set to 255, with 255 as the maximum limit.

Figure 5.10 shows the results of changing the n value (2–5) for the image in Fig. 5.8. As the n value increases, the image becomes brighter and clearer. However, when the n value is too large, the image becomes entirely white and bright, making it difficult to distinguish. For this image, it can be seen that the image is clearest when $n = 3$. By examining its greyscale histogram (Fig. 5.11), it is evident that when the greyscale values are expanded threefold, the greyscale distribution almost covers the entire range from 0 to 255. This results in a clear distinction between light and dark areas in the image, enhancing its contrast. Therefore, one can sequentially increase the multiplier n to seek the optimal value for obtaining a clear image. So, is there

Fig. 5.8 Original image

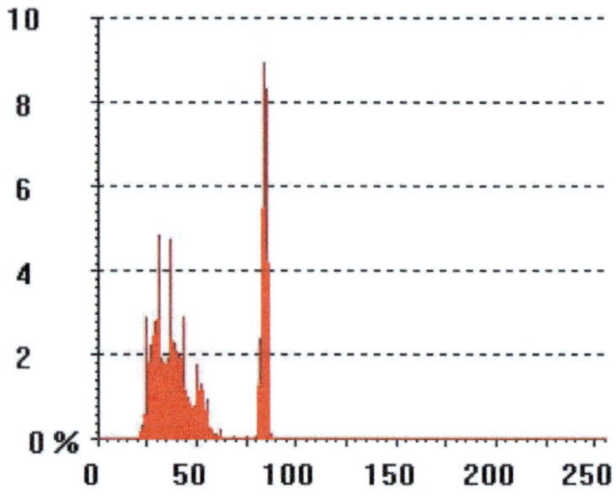

Fig. 5.9 Gray histogram

a method to automatically enhance contrast by automatically analysing the original image? (Fig. 5.9).

5.5.2 *Automatic Contrast Enhancement*

From the results obtained in the previous section, it is clear that if the greyscale range of the original image can fill the entire allowable greyscale range, a clear image can be automatically obtained.

For the greyscale histogram, the range can be transformed from $[a, b]$ shown on the left side of Fig. 5.12 to $[a', b']$ shown on the right side using Eq. (5.4).

$$z' = \frac{(b' - a')}{(b - a)} \times (z - a) + a' \tag{5.4}$$

According to this equation, the greyscale z $(a \leq z \leq b)$ of any pixel can be transformed into greyscale z'. This transformation form is easier to understand when expressed with a greyscale transformation curve. The greyscale transformation curve uses the pre-transformation image greyscale values as the horizontal axis and the post-transformation greyscale values as the vertical axis. The greyscale transformation curve of Eq. (5.4) is shown in Fig. 5.13. From this figure, it can be seen that the minimum and maximum greyscale values a and b of the pre-transformation image are transformed into a' and b', and any value z is transformed into z'. Therefore, if the variables a and b in Eq. (5.4) are the minimum and maximum greyscale values of the original image, and the variables a' and b' are the minimum (0) and maximum (255) greyscale values processed by memory, then an image with enhanced contrast from the original image will be automatically obtained. Figure 5.14 shows the processing result. Figure 5.15 shows its greyscale histogram, where the greyscale values span the entire range from 0 to 255. Compared with Fig. 5.8, the contrast is enhanced, and the image layers are clearly defined.

Fig. 5.10 Results after the gray value of the image is enlarged by n times

Fig. 5.11 Gray histogram when n = 3

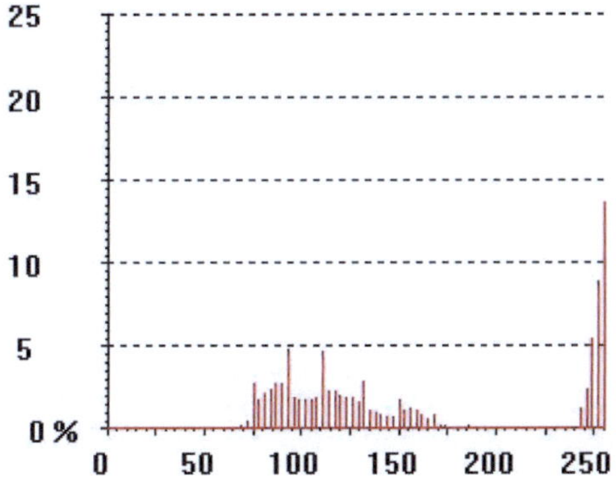

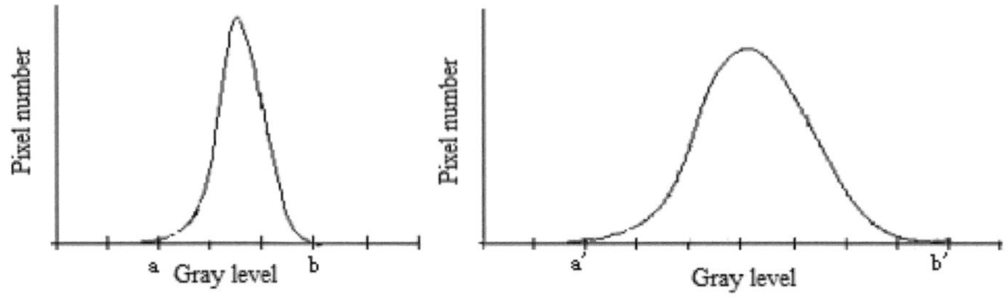

Fig. 5.12 Stretching of gray histogram

Fig. 5.13 Gray transform curve

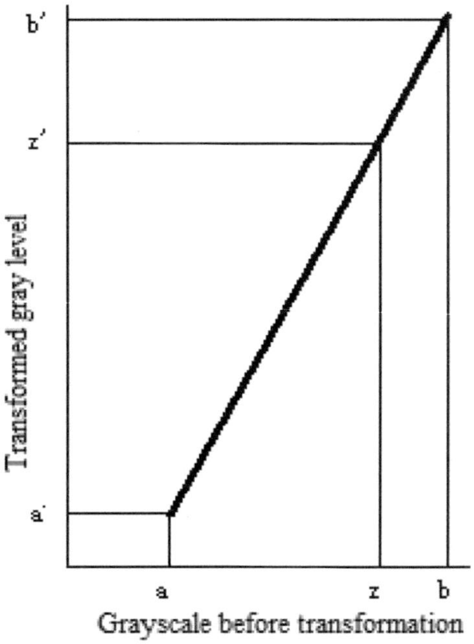

However, for the original image shown in Fig. 5.16, the apples and branches are relatively dark. The result after contrast enhancement is shown in Fig. 5.17. Compared with the original image, it is found that it has hardly become clearer. Why is this? Let us examine the greyscale histogram of its original image. As shown in Fig. 5.18, although there are few pixel points in most of the middle region on its greyscale histogram, there are quite a few greyscale levels at both the low and high ends, making it impossible to stretch the greyscale histogram, and thus, contrast enhancement cannot be performed.

For this situation, there are two methods for image contrast enhancement.

One method is to compress the greyscale levels with fewer pixels, extracting only the greyscale value range of the part to be enhanced and performing a greyscale range transformation (grey-scale transformation or grey-level transformation). In

Fig. 5.14 Gray scale transformation results

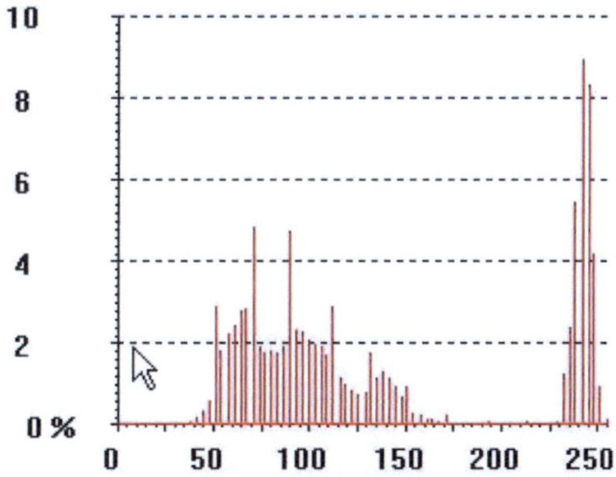

Fig. 5.15 Gray histogram

Fig. 5.16 Original image

Fig. 5.17 Gray scale transformation results

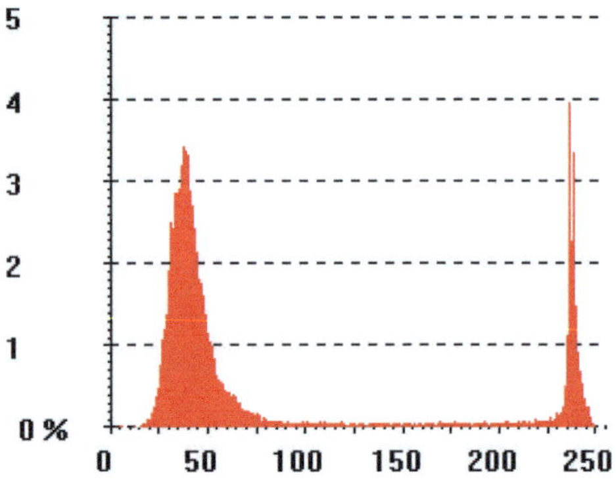

Fig. 5.18 Gray histogram

other words, in Eq. (5.4), instead of taking *a* and *b* as the minimum and maximum greyscale values, the part to be enhanced is taken as the minimum and maximum values.

Another method is to transform all greyscales on the greyscale histogram into a distribution form with the same number of pixels, a method known as histogram equalisation.

The former method requires knowing the greyscale range of the part to be enhanced, while the latter method can enhance contrast without examining the greyscale range. Below is a detailed explanation of histogram equalisation.

5.5.3 Histogram Equalisation [1]

Histogram equalisation involves compressing the parts of the original image with fewer pixels and stretching the parts with more pixels. If the pixels are concentrated within a certain grey level range, the contrast of the entire image is enhanced because

Gray level	7	6	5	4	3	2	1	0
Number of pixels in original image	0	4	9	11	5	7	4	0

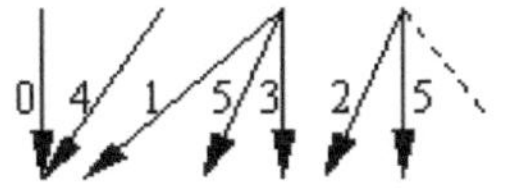

Number of equalized pixels	5	5	5	5	5	5	5	5

Fig. 5.19 Gray histogram equalization

the stretched parts have more pixels compared to the compressed parts, making the image clearer.

Below is a simple example to illustrate the histogram equalisation algorithm. The number of pixels corresponding to each grey level from 0 to 7 is shown in Fig. 5.19. After equalisation, the number of pixels allocated to each grey level should be the total number of pixels divided by the total number of grey levels, i.e., $40 \div 8 = 5$. Starting with the pixels of the original image with the highest grey value, allocate 5 pixels at a time, beginning with 7. For the image shown in Fig. 5.19, all the pixels originally at gray levels 7 and 6, along with 1 pixel from the 9 pixels at gray level 5, are reassigned to gray level 7. There are two possible algorithms for selecting this 1 pixel from gray level 5:

① Random selection.
② Sequential selection from the pixels with a higher average grey level of the surrounding pixels.

Algorithm ② is slightly more complex than algorithm ①, but the result obtained from algorithm ② has less noise than that from algorithm ①.

Algorithm ② is chosen here. For the next grey level, select 5 pixels from the remaining 8 pixels of grey level 5 in the original image using the previous method, as the number of pixels for grey level 6. This process is repeated for all pixels to reallocate grey levels.

Figures 5.20 and 5.21 show the results of histogram equalisation applied to Figs. 5.8 and 5.16, respectively. It is evident that both examples demonstrate that histogram equalisation is quite effective in improving contrast.

5.5.4 Dark Channel Prior Method for Dehazing [2]

Apart from dim lighting causing image blurring, another reason for image blurring is the effect of haze in the environment. The impact of hazy weather on the quality of digital images is mainly due to the scattering phenomenon caused by light interacting with suspended particles, which attenuates the light reflected by the target object. The basic principles of dehazing technology can be divided into image enhancement dehazing and inversion dehazing. The histogram equalisation introduced in the

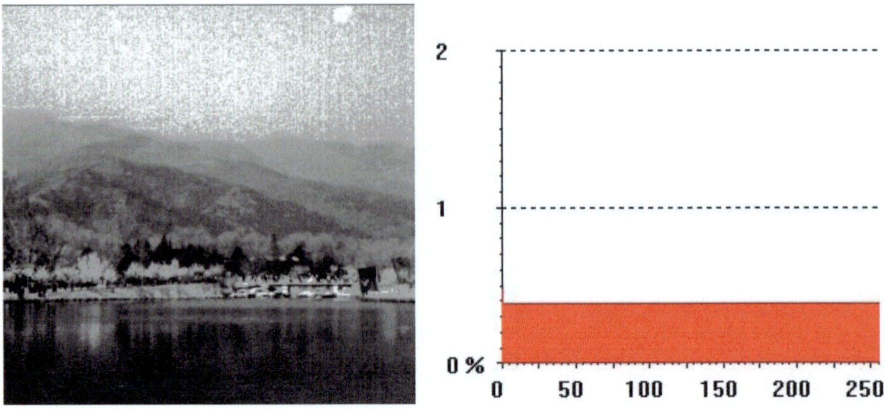

Fig. 5.20 Results of gray histogram equalization of Fig. 5.8

Fig. 5.21 Results of gray histogram equalization of Fig. 5.16

previous section is an image enhancement algorithm widely used for clarifying hazy images. Another method is to establish a mathematical model based on the physical principles of hazy image degradation and use mathematical derivation to effectively reverse the effects of haze and restore image clarity. The dark channel prior method introduced in this section belongs to this type.

(1) Degradation Model of Hazy Images

If we use I to represent the image obtained by the visual system in hazy weather, and J to represent the desired image, which is a clear image without haze, then the difference between these two images is the degraded image, which is related to the atmospheric light and the transmission rate of the air (or directly understood as haze). Let the atmospheric light coefficient be A and the air transmission rate coefficient is t, then the following mathematical model can be obtained:

$$I = Jt + A(1 - t) \tag{5.5}$$

In this model, Jt is the desired image multiplied by the transmission rate coefficient, representing the direct attenuation of the image; $A(1 - t)$ represents the atmospheric light component in the image. The goal of dehazing is to recover J from I.

Figure 5.22 shows a clear image and a hazy image along with their histograms, where Figure (a) represents the clear image condition, and Figure (b) represents the hazy environment condition. It can be seen that the pixel value distribution range of the hazy image's RGB is relatively narrow, as the image appears whitish, with its histogram distribution concentrated on the right. The RGB value distribution of the clear image is relatively uniform and mostly centred, with good contrast.

(2) Dark Channel Prior Dehazing Method

The dark channel prior is a statistical rule based on a large number of haze-free images, which states that except for continuously bright areas like the sky, most local image regions can find a colour channel with a very low pixel value, known as the dark pixel, and the channel with this dark pixel is called the dark channel. In

(a) Clear image and its histogram

(b) Atomization image and its histogram

Fig. 5.22 Clear image, atomization image and its histogram

most haze-free images, the intensity value of the dark channel is very small, even approaching zero. This statistical rule is known as the dark channel prior (DCP). The pixel value represents the intensity of the sensor's sensitivity to light. If this minimum value is defined as $J(x)$, where x represents the centre of this small block area, then $J(x)$ is the dark channel value. $J(\mathrm{x})$ can be expressed as:

$$J(x) = \min_{y\in\Omega(x)}[\min_{y\in\Omega(\mathrm{r,g,b})} J^c(y)] \to 0 \tag{5.6}$$

$\Omega(x)$ represents a neighbourhood region centred at x, c represents the R, G, B three channels, $J(y)$ represents traversing all pixel values of the three channels in $\Omega(x)$.

The goal of dehazing is to restore J from I, then by transforming Eq. (5.5), we obtain Eq. (5.7)

$$J = \frac{I - A(1-t)}{t} = \frac{I - A + At}{t} = \frac{I - A}{t} + A \tag{5.7}$$

In this equation, I is our existing image to be dehazed, J is the haze-free image to be restored. This equation has two unknowns, t and A, and without further information input, this equation cannot be solved. However, if the dark channel prior knowledge is incorporated, it can be transformed into a solvable equation. Assuming A is known, divide both sides of Eq. (5.5) by A to obtain

$$\frac{I}{A} = \frac{J}{A}t + 1 - t \tag{5.8}$$

By representing the colour channels together in Eq. (5.8), we obtain Eq. (5.9).

$$\frac{I^c(x)}{A^c} = \frac{J^c(x)}{A^c}t(x) + 1 - t(x) \tag{5.9}$$

The superscript c represents the R, G, B three channels. $t(x)$ is the transmission coefficient within each window. By performing the minimum operation twice on both sides of Eq. (5.9), we obtain the following equation:

$$\min_{y\in\Omega(x)}\left[\min_{c}\frac{I^c(x)}{A^c}\right] = t(x)\min_{y\in\Omega(x)}\left[\min_{c}\frac{J^c(x)}{A^c}\right] + 1 - t(x) \tag{5.10}$$

Equation (5.10) is the expression of Eq. (5.8) after adding channels and regions, where $\Omega(x)$ represents a small region centred at x, generally set to 15×15 pixels. Combining with Eq. (5.6), we can obtain Eq. (5.11).

$$\min_{y\in\Omega(x)}\left[\min_{c}\frac{J^c(x)}{A^c}\right] = 0 \tag{5.11}$$

Substituting Eq. (5.11) into Eq. (5.10), we obtain:

$$t(x) = 1 - \min_{y \in \Omega(x)} \left[\min_{c} \frac{J^c(x)}{A^c} \right] \tag{5.12}$$

In the above derivation, it is assumed that the atmospheric light coefficient A value is known. In actual calculations, the method to obtain the A value is to take the top 0.1% of pixel positions according to brightness from the dark channel map, and find the pixel value with the highest frequency at these positions in the original hazy image I as the A value. The t value is obtained from the A value using Eq. (5.12), and then the desired image J is obtained using Eq. (5.7).

For the apple image in Fig. 5.22, when calculating the atmospheric light coefficient A value, emphasis is placed on the red channel. The dark channel of the entire image is calculated, and the top 0.1% brightness area is taken. Among these pixels, the maximum number of brightness values in the red channel corresponding to the pixel points in the original hazy image are used as the A value. Using this method, the atmospheric light coefficient A for the hazy image in Fig. 5.22b is calculated to be 251.

5.6 Smoothing of Binary Images [1]

The noise in binary images, as shown in Fig. 5.23, is generally salt-and-pepper noise. Of course, this noise can be eliminated using median filtering, but since it is binary, dilation and erosion can also be used to eliminate it.

Dilation is a process where if there is at least one white pixel in the neighbourhood of a pixel, that pixel changes from black to white, while others remain unchanged. Erosion is a process where if there is at least one black pixel in the neighbourhood of a pixel, that pixel changes from white to black, while others remain unchanged. After dilation → erosion processing in Fig. 5.24, the dilation becomes thicker, and

Fig. 5.23 Salt and pepper noise

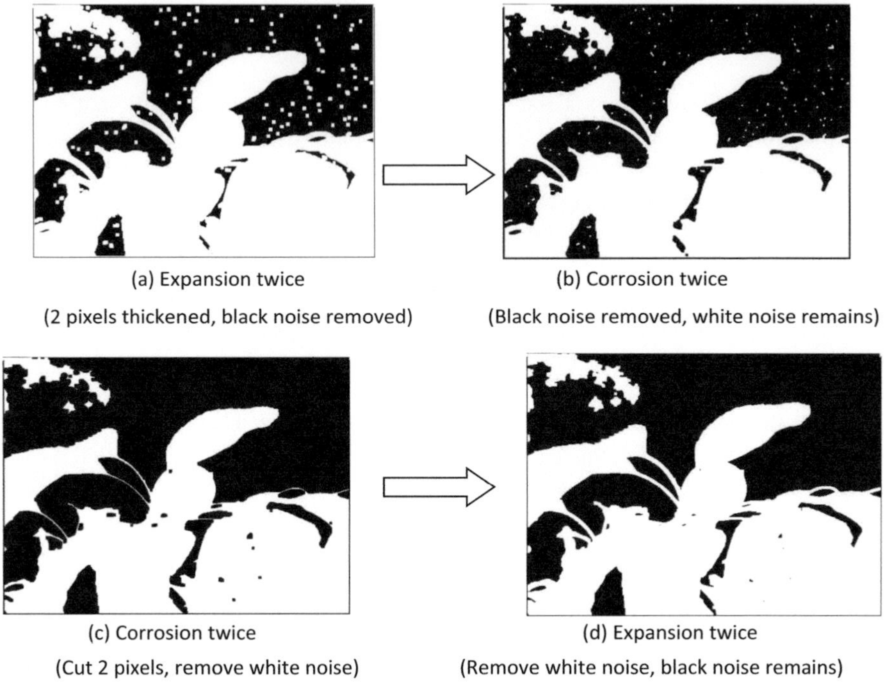

Fig. 5.24 Expansion and corrosion treatment of Fig. 5.23 (the order of expansion and corrosion is different, and the results are different)

the erosion becomes thinner, resulting in almost no change in the image. Conversely, after erosion → dilation processing, the white isolated point noise is eliminated during erosion.

In addition to dilation and erosion, noise can also be removed by calculating the area size. The size of the area is actually the number of pixels contained in the connected region, which will be introduced in Chap. 6 on geometric parameter detection. Figure 5.25 shows the binary image of rice field seedlings and the result image after denoising the 50-pixel white area. Compared with dilation and erosion, area denoising does not destroy the connectivity between regions.

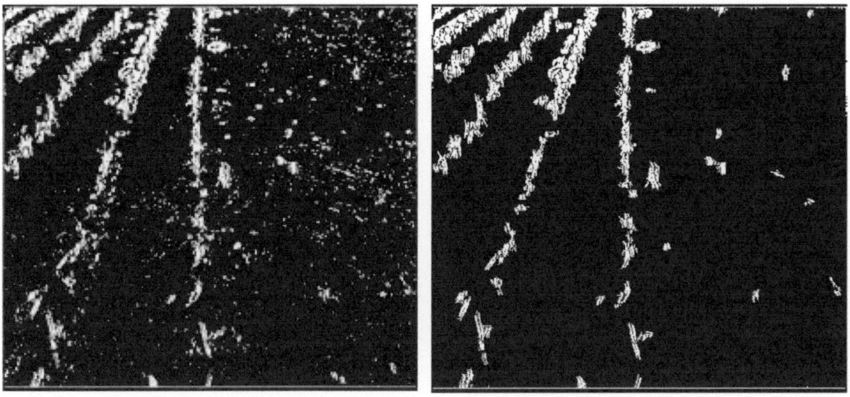

(a) Binary image (b) Image after denoising of the 50-pixel white area

Fig. 5.25 Denoising processing of the area of a binary image

References

1. Chen Bingqi. Practical Digital Image Processing and Analysis [M]. 2nd Edition. Beijing: China Agricultural University Press, 2014.
2. Zhu Deli, Chen Bingqi, et al. Dark Channel Prior Dehazing Method for Apple Picking Robot Vision System [J]. Journal of Agricultural Engineering, 2016, 32(16): 151–158.

Chapter 6
Geometric Parameter Detection

6.1 Automatic Recognition Based on Image Features

Currently, there are numerous examples of automatically distinguishing objects by analyzing image features using computers. For instance, coin recognition in vending machines, automatic product quality inspection in factories via cameras, automatic sorting of letters through postal code recognition, electronic keys based on fingerprint recognition, and more recently, facial recognition to prevent terrorism. This chapter explains these features, especially the feature selection of images [1].

To facilitate understanding, this chapter employs simple binary images as examples, introducing techniques to extract desired objects and eliminate unwanted noise by analyzing features like shape and size.

6.2 Feature Parameters of Binary Images

The so-called features of an image are the characteristics of the objects included in the image. If you aim to extract a banana from Fig. 6.1, what steps should you take?

Computers do not inherently comprehend the concept of a banana. People can only indicate to the computer the features of the object to be extracted, for example, that a banana is a slender object. In other words, it is essential to provide the computer with information about the object's characteristics in the image, including size and shape, and to highlight features like size, roundness, or angularity, round objects, or angular objects. Of course, this indication relies on parameters that describe the shape characteristics (shape representation and description) of the object.

Below, several representative feature parameters and calculation methods are explained. Table 6.1 lists several figures and their corresponding parameters.

© Chemical Industry Press 2026

B. Chen and S. Chen, *Machine Vision Technology*, Advanced and Intelligent Manufacturing in China, https://doi.org/10.1007/978-981-95-1184-6_6

Fig. 6.1 Original image

Table 6.1 Figures and their features

Type	Circle	Square	Regular triangle
Image	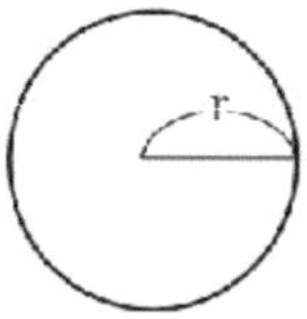	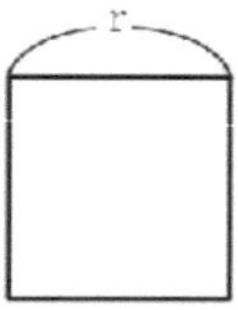	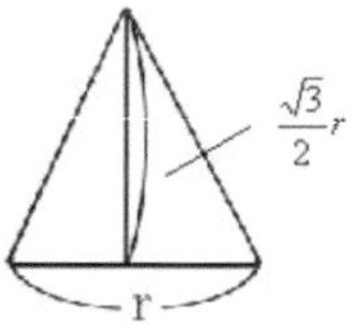
Area	πr^2	r^2	$\frac{\sqrt{3}}{4}r^2$
Perimeter	$2\pi r$	$4r$	$3r$
Circularity	1.0	$\frac{\pi}{4} = 0.79$	$\frac{\pi\sqrt{3}}{9} = 0.60$

[Area]

Calculate the number of pixels contained within the object (or region).

[Perimeter]

The perimeter of the object (or region) contour line refers to the sum of the distances between pixels on the contour line. There are two cases for the distance between pixels, as shown in Figure 6.2a and b. Figure 6.2a represents adjacent pixels, which can be in four directions: up, down, left, and right, with a distance of 1 pixel between these adjacent pixels. Figure 6.2b represents pixels connected in an oblique direction, with four directions: top-left, bottom-left, top-right, and bottom-right, with a distance of $\sqrt{2}$ pixels between these oblique direction pixels. When measuring the perimeter, it is necessary to calculate the distance according to the connection method between pixels. Figure 6.2c is an example of a perimeter measurement.

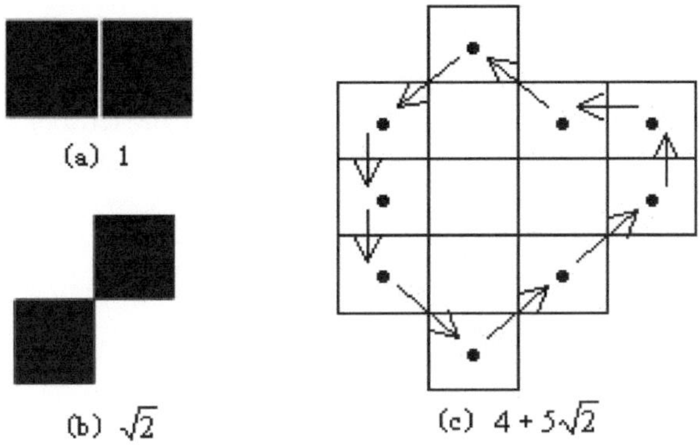

Fig. 6.2 Distance between pixels (pixels)

As shown in Fig. 6.3, extracting the contour line requires tracking the contour line according to the following steps.

① Scan the image, sequentially investigating the value of each pixel on the image, looking for boundary points without the scan mark a_0.
② If a_0 is surrounded by all black pixels (0), it indicates that a_0 is an isolated point, and tracking stops.
③ Otherwise, find the next boundary point in the order shown in Fig. 6.3. Use the same method to track each boundary point.
④ Upon reaching the next junction point a_0, it proves that the object has been circled once, and the scan is terminated.

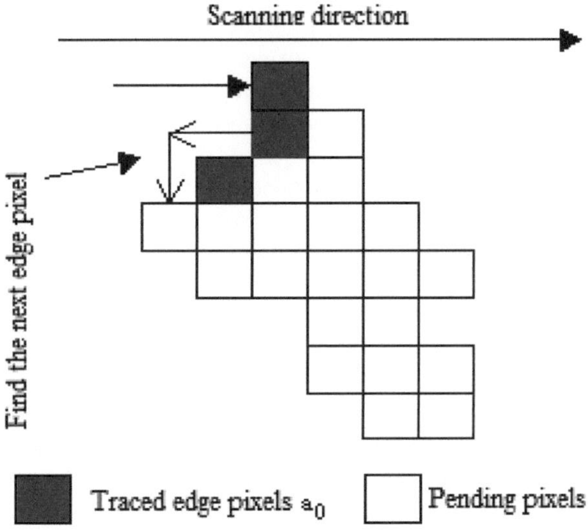

Fig. 6.3 Tracing of contour lines

[Compactness]

Compactness is a feature quantity that calculates the complexity of the shape of an object (or region) based on area and perimeter. For example, consider a circle and a pentagram. If the area of the pentagram is equal to the area of the circle, then its perimeter must be longer than that of the circle. Therefore, the following parameter can be considered:

$$e = \frac{4\pi \times Area}{(\text{Perimeter})^2} \tag{6.1}$$

e is the compactness. For a circle with radius r, the area is equal to πr^2, and the perimeter is equal to $2\pi r$, so the compactness e equals 1. As can be seen from Table 6.1, the closer the shape is to a circle, the larger e is, with a maximum of 1; the more complex the shape, the smaller e is, with e's value between 0 and 1.

[Center of Gravity or Centroid]

The center of gravity is the average value of the pixel coordinates within the object (or region). For example, if the coordinates of a certain white pixel are (x_i, y_i) $(i = 0, 1, 2, \ldots, n-1)$, its centroid coordinates (x_0, y_0) can be obtained by the following formula:

$$(x_0, y_0) = \left(\frac{1}{n}\sum_{i=0}^{n-1} x_i, \sum_{i=0}^{n-1} y_i\right) \tag{6.2}$$

In addition to the above parameters, there are many other feature parameters such as length and breadth, Euler's number, and moments that can view the length direction of the object, which will not be introduced one by one here.

Using the aforementioned parameters, it seems possible to distinguish bananas from other fruits. Bananas have the smallest roundness among those fruits. However, first, it is necessary to extract everything from the background, which can be achieved by extracting the bright parts using binarization. Figure 6.4 is the image from Fig. 6.1 after binarization (with a threshold above 40) and denoised through two median filters.

At this point, it is still not enough; each object must also be distinguished. To differentiate each object, it is necessary to investigate whether the pixels are connected, a process known as region labeling.

6.3 Region Labeling

Region labeling involves assigning the same label to groups of connected pixels (known as connected components) and distinct labels to separate connected components. Region labeling occupies a crucial role in binary image processing. Figure 6.5

Fig. 6.4 Binary image of Fig. 6.1

shows the image after region labeling, through which each connected component is distinguished, allowing the investigation of the shape characteristics of each connected component.

There are also many methods for region labeling; a simple method is introduced below, with steps as follows (refer to Fig. 6.6).

① Scan the image, and when encountering a target pixel (white pixel) P that has not been labeled, assign a new label.
② Assign the same label to pixels connected to P (i.e., the same connected component).
③ Further, assign the same label to all pixels connected to the labeled pixels.

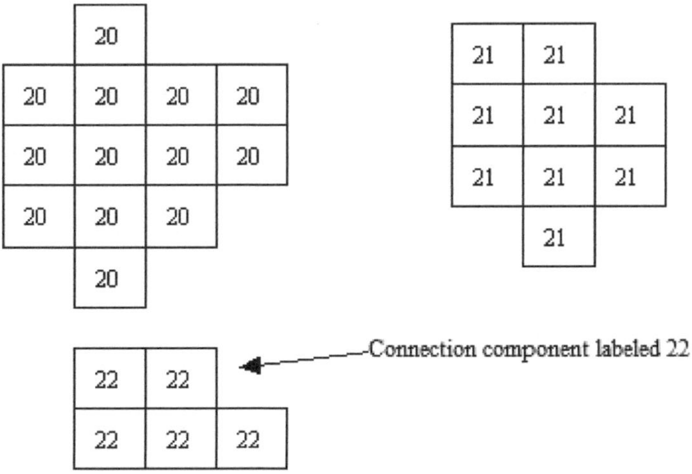

Fig. 6.5 Image after region labeling

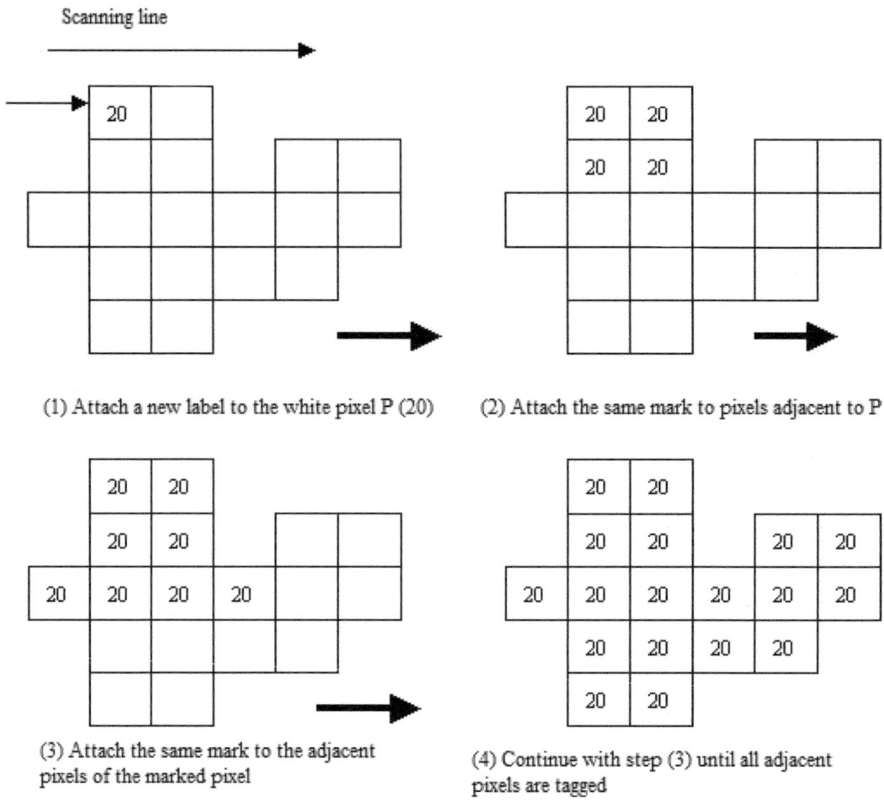

Fig. 6.6 Assigning label 20 to a connected component

④ Continue step 2 until all connected pixels are labeled. In this way, a connected component is assigned the same label.
⑤ Return to step 1, search for new unlabeled pixels, and repeat the above steps.
⑥ The process ends after the entire image is scanned.

6.4 Extracting Objects Based on Feature Parameters

After the above processing steps, the preparation for extracting bananas from Fig. 6.1 is finalized. The steps for investigating the characteristics of each object are shown in Fig. 6.7, and the processing results are shown in Table 6.2. Figure 6.8 shows the processed image, with the contour lines and centroid positions appearing brighter.

From Table 6.2, it can be seen that there are two objects with small roundness, which might be bananas. If bananas are to be extracted, follow the steps in Fig. 6.7, and then extract the connected components with a certain roundness. The image of the extracted connected components is shown in Fig. 6.9. This processing results in a

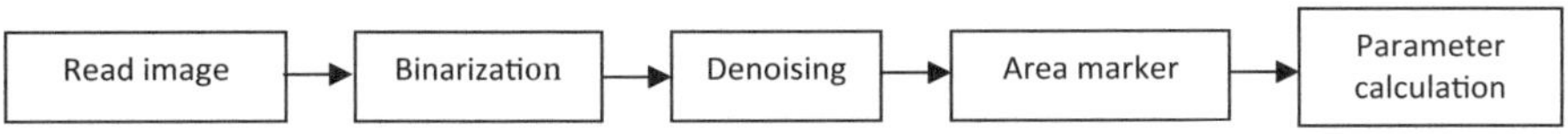

Fig. 6.7 Steps for investigating object characteristics

Table 6.2 Feature parameters of each object (unit: pixel)

Serial number	Area	Perimeter	Circularity	Gravity position
0	21718	894.63	0.3410	(307, 209)
1	22308	928.82	0.3249	(154, 188)
2	9460	367.85	0.8785	(401, 136)
3	14152	495.14	0.7454	(470, 274)
4	8570	352.98	0.8644	(206, 260)

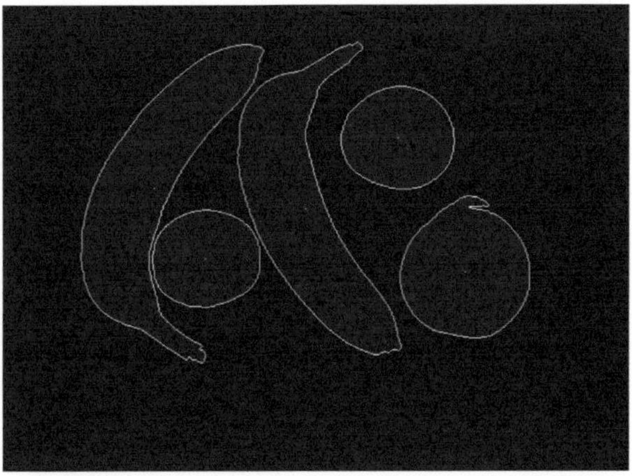

Fig. 6.8 Depicts the traced contour along with the center of gravity

mask image, which can be used to extract bananas from the original image (Fig. 6.1). The extraction result is shown in Fig. 6.10.

6.5 Noise Removal Based on Feature Parameters

Up to now, the previous discussion has been about processing aimed at extracting objects, which can also be used to remove unnecessary items. For example, it can be used for noise removal. Noise in binary images can be eliminated using area, as briefly explained in Chap. 5. After distinguishing each connected component through region labeling, small-area connected components can be removed. The processing flow is shown in Fig. 6.11, and the processing result is shown in Fig. 6.12 (using a green pepper sample as an example). The image obtained from differential processing

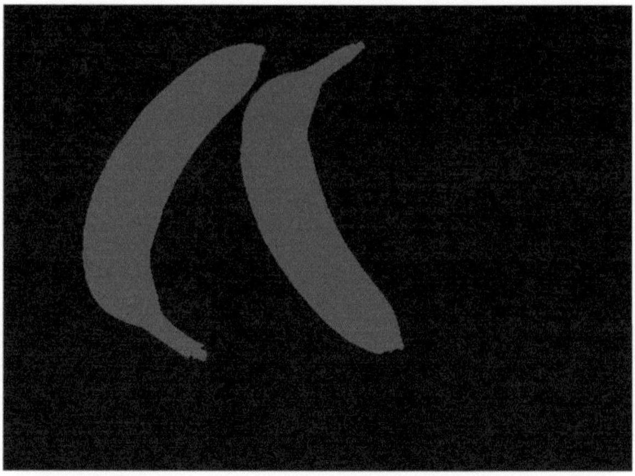

Fig. 6.9 Extraction results of objects with roundness less than 0.5 in Fig. 6.8

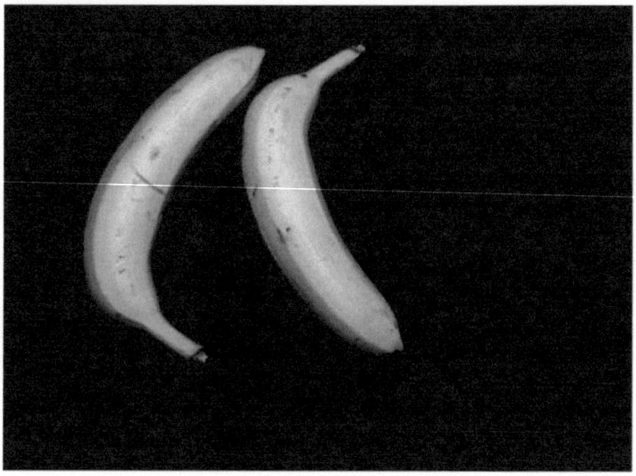

Fig. 6.10 Extracts bananas from Fig. 6.1 using Fig. 6.9

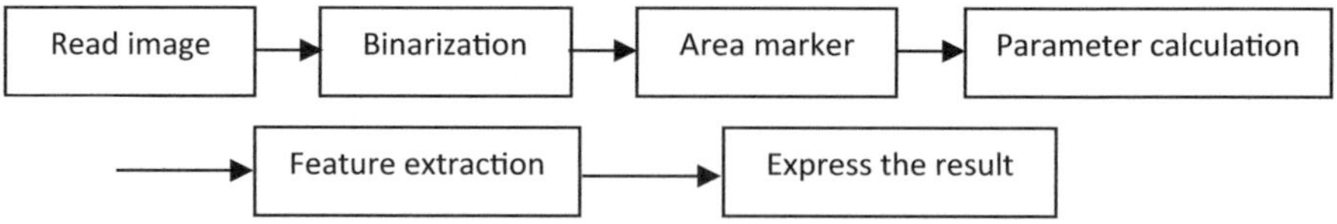

Fig. 6.11 Steps for removing noise from characteristic parameters

(Prewitt operator) (Fig. 6.12c) is used as the input image, and the result after noise removal is shown in Fig. 6.12d. The removed noise is connected components with an area smaller than 80 pixels, and it can be seen that the dot-like noise has completely disappeared.

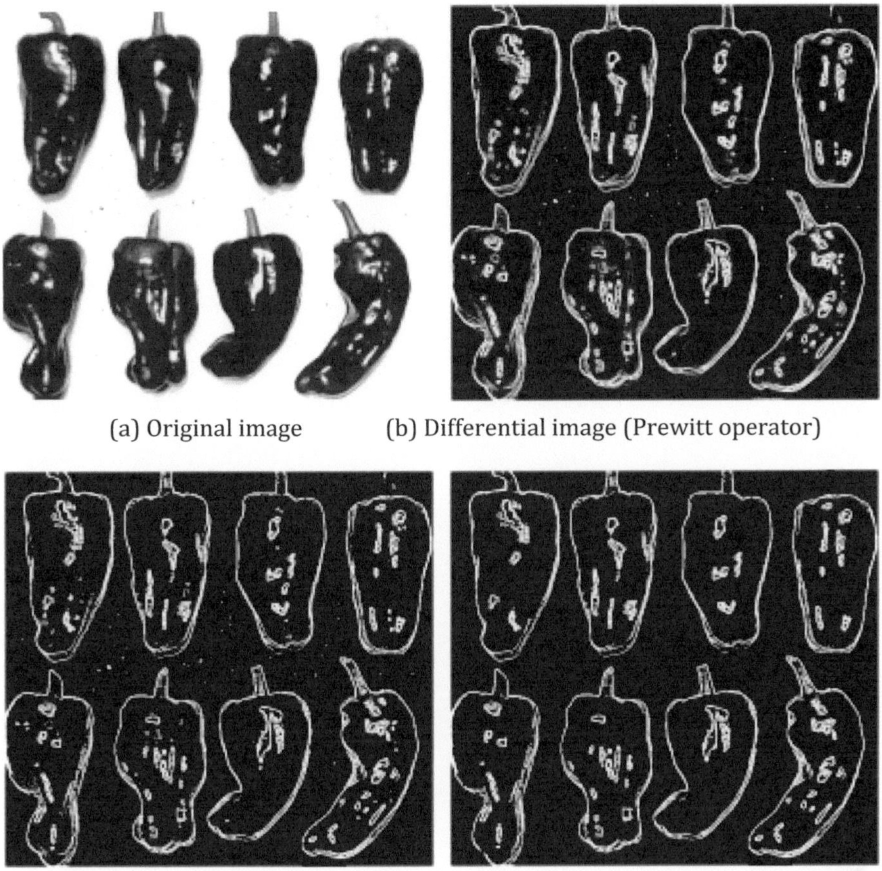

(a) Original image (b) Differential image (Prewitt operator)

(c) Binary image (threshold 100) (d) Image with an area of 80 pixels denoised

Fig. 6.12 Example of noise cancellation using area parameters

Reference

1. Chen Bingqi. Practical Digital Image Processing and Analysis [M]. 2nd Edition. Beijing: China Agricultural University Press, 2014.

Chapter 7
Hough Transform

The Hough Transform is an effective technique for edge detection. Its fundamental concept is to convert a point in the measurement space into a curve or surface in the parameter space, where points sharing the same parameter characteristics intersect. By judging the accumulation at the intersection, the detection of feature curves is completed. Based on different parameter properties, Hough Transform can detect lines, circles, ellipses, hyperbolas, etc. This chapter will mainly introduce the method of detecting lines using the Hough Transform.

7.1 Traditional Hough Transform for Line Detection [1]

Paul Hough proposed the Hough Transform method in 1962 and applied for a patent. This method transforms the detection challenge in image space into parameter space, accomplishing the detection task via simple accumulation statistics within the parameter space, and describes the region boundary curve of the image using a certain parameter form that most boundary points satisfy. This method has good fault tolerance for images with noisy interference or discontinuous region boundaries. Initially, the Hough Transform was primarily utilized for detecting lines in image space, particularly through early line transforms conducted in two Cartesian coordinate systems, which presented challenges for identifying lines with infinite slopes. In 1972, Duda transformed the form of the transform, converting points in data space into curves in ρ - θ parameter space, improving its performance in line detection. This method has been continuously researched and developed, and has found very wide applications in fields such as image analysis, computer vision, and pattern recognition, becoming an important tool in pattern recognition.

The equation of a line can be represented by Eq. (7.1).

$$y = kx + b \tag{7.1}$$

© Chemical Industry Press 2026

B. Chen and S. Chen, *Machine Vision Technology*, Advanced and Intelligent Manufacturing in China, https://doi.org/10.1007/978-981-95-1184-6_7

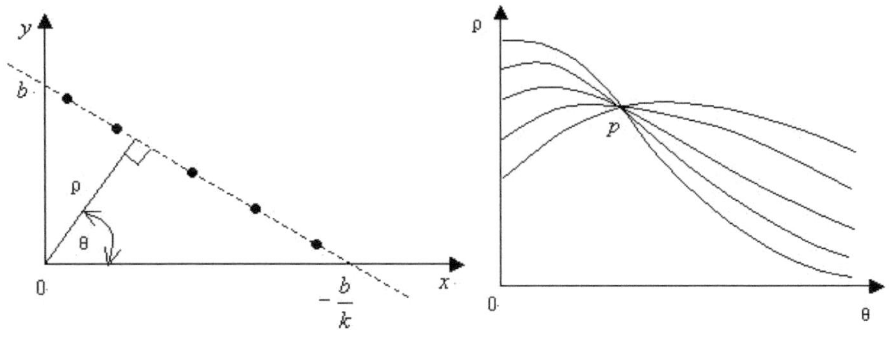

Fig. 7.1 Schematic diagram of the duality of Hough transform

where k and b are the slope and intercept, respectively. All lines passing through a certain point (x_0, y_0) on the x - y plane satisfy the equation $y_0 = kx_0 + b$. That is, a family of lines passing through the point (x_0, y_0) on the x - y plane corresponds to a line in the k-b parameter plane.

Since the line equation in the form of Eq. (7.1) cannot represent lines in the form $x = c$ (c is a constant) (where the slope of the line is infinite), in practical applications, the polar coordinate parameter equation in the form of Eq. (7.2) is generally used.

$$\rho = x\cos\theta + y\sin\theta \tag{7.2}$$

where ρ is the perpendicular distance from the origin to the line, and θ is the angle between ρ and the x axis (as shown in Fig. 7.1).

According to Eq. (7.2), different points on the line are transformed into a family of sine curves intersecting at point p in parameter space, thus allowing the detection of lines in the x–y coordinate system by detecting the local maximum point p in parameter space.

The general steps of the Hough Transform are as follows.

① Quantise the parameter space into $m \times n$ (m is the number of equal divisions of θ, n is the number of equal divisions of ρ) units, and set up an accumulator matrix $Q\,[\,m \times n]$;
② Assign an accumulator $Q(\theta_i, p_j)$ $(0 < i < m\text{-}\ 1,\ 0 < j < n\text{-}\ 1)$ to each unit in parameter space, and initialise the accumulator's value to zero;
③ Transform each point (x_k, y_k)($k = 1,2,\ldots, s$, where s is the number of points in the Cartesian coordinate system) substitute into Eq. (7.2), then substitute θ_0 ~ $\theta_{m\text{-}1}$ into it as well, and calculate the corresponding values p_j respectively;
④ In the parameter space, find the unit corresponding to each (θ_i, p_j) and increment the accumulator of that unit by 1, i.e., $Q(\theta_i, p_j) = Q(\theta_i, p_j) + 1$, and perform a vote for that unit;
⑤ After all points in the x - y coordinate system have been processed, check the accumulators in the parameter space. There must be one that reaches a maximum

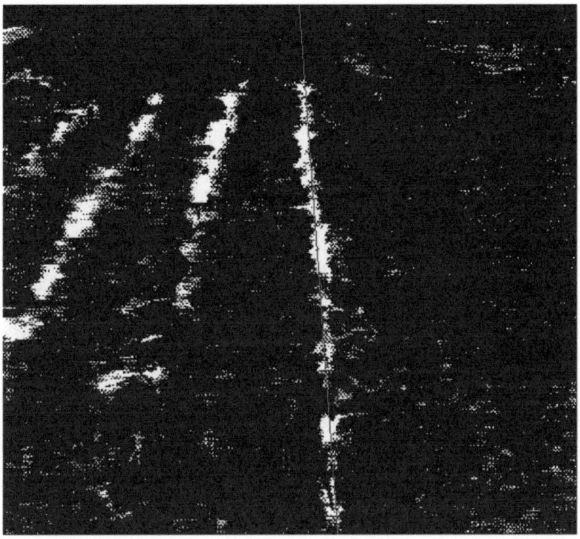

Fig. 7.2 Processing result of binary image through traditional Hough transform

value, and the parameter values of the unit corresponding to this accumulator are output as the parameters of the desired line.

From the above steps, it can be seen that the specific implementation of the Hough transform uses a voting method, where each point on the curve can vote for several parameter combinations, and the parameters that win the majority vote are the winners. The peak value of the accumulator array represents the parameters of a line. This basic strategy of the Hough transform can also be extended to the detection of planar curves.

Figure 7.2 shows the result of line detection on a binary image using the traditional Hough transform. The image size is 512 $\times$ 480 pixels, and the computation time is 652ms (CPU speed is 1GHz).

The Hough transform is a global detection method with excellent anti-interference capability, effectively suppressing interference present in concentrated data points, and can fit multiple lines to a set of data points. However, the precision of the Hough transform is difficult to control, making it unsuitable for practical problems requiring high precision in line fitting. Additionally, the large computational demand results in slow processing speed, limiting its application in fields with high real-time requirements.

7.2 Line Detection Through Known Points Using Hough Transform [2]

The Hough transform line detection method introduced above is an exhaustive search with high computational and spatial complexity, making it difficult to apply in fields with high real-time requirements. To address this issue, many scholars have been dedicated to researching the acceleration of the Hough transform algorithm over the years. For example, combining random processes, fuzzy theory, etc., with the Hough transform, or introducing hierarchical iteration and cascading concepts into the Hough transform process, has greatly improved the efficiency of the Hough transform. This section introduces a fast line detection method using an improved Hough transform through known points as an example.

The enhanced Hough transform method leveraging known points is founded on the fundamental principles of the Hough transform, transforming the voting process from each point across the entire parameter space into a rapid line detection approach that only votes within a "known point" parameter space. The basic idea is: first, find a point belonging to the line, define the coordinates of this known point p_0 as (x_0, y_0), and define the slope of the line passing through p_0 as m. The relationship between the coordinates and the slope can be expressed by the following equation:

$$(y - y_0) = m(x - x_0) \tag{7.3}$$

Define the coordinates of the target pixel p_{i} within the region as (x_{i}, y_{i}), ($0 \leq i < n$, n For the total number of target pixels within the region, the slope m_i of the line connecting point p_i and point p_0 can be expressed by the following formula:

$$m_i = (y_i - y_0)/(x_i - x_0) \tag{7.4}$$

The slope values are mapped onto a set of accumulators. Each time a slope is calculated, the corresponding accumulator's value is incremented by 1. Since the slope calculated for points on the same line is consistent, when there is a linear component in the target area, the corresponding accumulator will show a local maximum. The slope corresponding to this value is taken as the slope of the desired line.

When $x_i = x_0$, m_i becomes infinite, and formula (7.4) does not hold. To avoid this phenomenon, when $x_i = x_0$, let $m_i = 2$. When $m_i > 1$ or $m_i <$ -1, use the calculated value from formula (7.5) to replace m_i, thus limiting the infinite domain of m_i to a finite range of (-1, 3). In practical operations, set the slope interval to [-2, 4].

$$m_i\prime = 1/m_i + 2 \tag{7.5}$$

The specific steps of the Hough transform through a known point are as follows:

① Divide the set slope interval into 10 sub-intervals, i.e., the width of each sub-interval is 1/10 of the set slope interval width.

② Set an accumulator n_j ($1 \leq j) \leq 10$) for each sub-interval.

③ Initialise the value of each accumulator to 0, i.e., $n_j = 0$.
④ Scan the image point by point from top to bottom, left to right. When a target pixel is encountered, calculate its slope m with the known point p_0 using formulas (7.4) and (7.5). The value of m determines which sub-interval's accumulator is incremented by 1.
⑤ After scanning the entire processing area, take the sub-interval with the maximum accumulator value and its two adjacent sub-intervals (a total of 3 sub-intervals) as the slope interval for the next vote. Repeat steps 1 to 4 until the width of the slope interval is smaller than the predefined slope detection precision, e.g., $m = 0.05$. At this point, the middle value of the sub-interval with the highest accumulated value is taken as the slope of the desired line, following the inverse transformation based on the condition set in Eq. (7.5).

The line detection process through a known point Hough transform is shown in Fig. 7.3.

Figure 7.4 shows the line detection result through a known point Hough transform. The " + " on the detected line indicates the position of the known point, and the processing time is 35ms. In other words, under the same conditions, the processing speed of the Hough transform through a known point is nearly 20 times faster than the general Hough transform.

The crucial aspect when employing the line detection method using the Hough transform through a known point lies in selecting the known point accurately.

In practical operations, feature points that are easy to obtain are generally chosen as known points, such as the centre of pixel distribution within a certain area. In practical applications, the processing speed of the Hough transform is often improved by analysing the characteristics of the detection object to obtain a small number of target pixels, thereby reducing the processing object. The characteristics of the detection object generally use brightness or colour features. For example, when detecting road lane lines, by analysing the brightness or a certain colour component of the lane lines, the distribution centre points of the lane lines on each horizontal scan line can be identified first, and then the Hough transform is applied only to these centre points, which can greatly improve processing speed. When extracting feature points, some feature points may have errors, but due to the statistical nature of the Hough transform, some errors will not affect the final detection result.

7.3 Curve Detection with Hough Transform

The Hough transform can detect not only straight lines but also curves, such as arcs, ellipses, parabolas, etc. However, as the complexity of the curve escalates, so do the parameters required to describe it, resulting in an increase in the dimensionality of the parameter space during the Hough transform. Since the essence of the Hough transform is to cluster pixels in the image space that have a certain relationship, and to find the accumulation of corresponding points in the parameter space that can relate

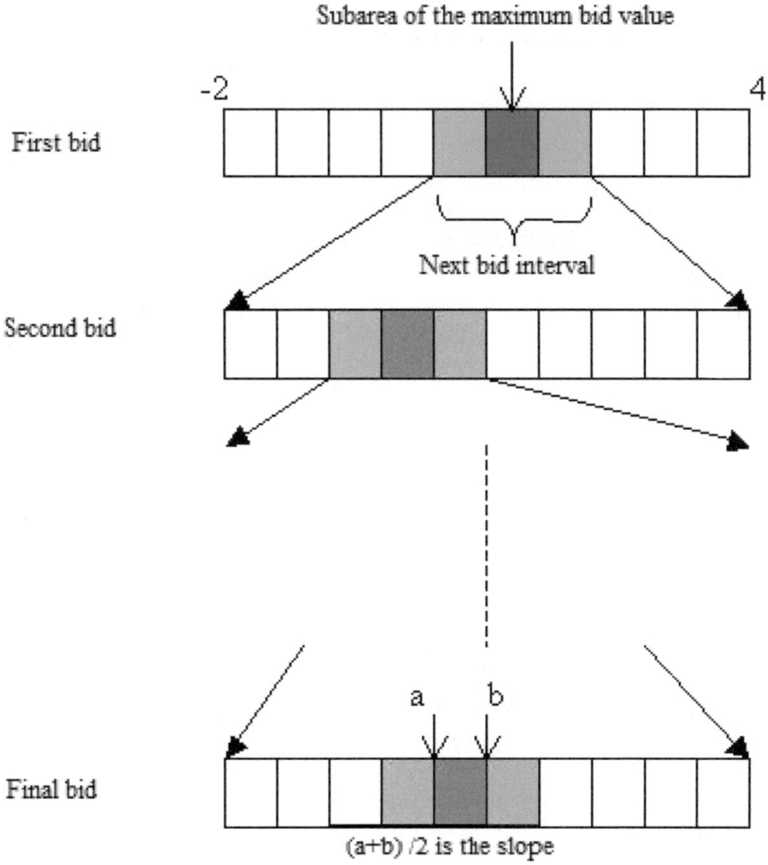

Fig. 7.3 Linear detection process through a known point Hough transform

these pixels with a certain analytical expression, this transformation has ideal effects when the parameter space does not exceed two dimensions. However, when it exceeds two dimensions, the time consumption and required storage space increase sharply, making it theoretically feasible but almost impossible to implement in practical applications. At this time, it is often necessary to find characteristics from specific application situations, such as using some prior knowledge of the detected image to try to reduce the dimensionality of the parameter space to reduce the transformation process time.

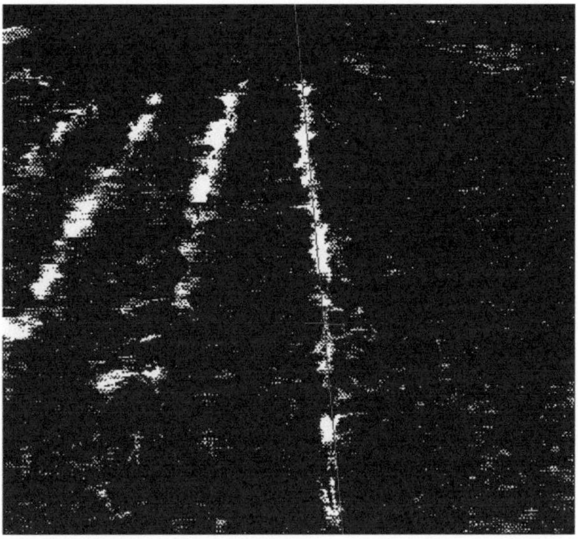

Fig. 7.4 The line detection result through a known point Hough transform

References

1. Chen Bingqi. Practical Digital Image Processing and Analysis [D]. 2nd Edition. Beijing: China Agricultural University Press, 2014.
2. Chen Bingqi, Watanabe Kenji, Higashijo Kiyohide. Research on the Visual Part of Rice Planting Robots (2nd Report) [J]. Journal of the Japanese Society of Agricultural Machinery, 1997, 59(3): 23-28.

Chapter 8
Geometric Transformations

8.1 About Geometric Transformation

The deformed image shown in Fig. 8.1 is referred to as a geometric transformation in the field of image processing. Figure 8.1 shows the result obtained after performing a perspective transformation on an image of a pet rabbit. Geometric transformations are applied in numerous situations. For example, the cloud images seen in weather forecasts are images obtained after geometric transformations. Since images taken from artificial satellites contain distortions caused by the lens, geometric transformations are needed to correct them to obtain distortion-free images [1].

So, what kind of processing is a geometric transformation? Geometric transformation is achieved by changing the position of pixels. In contrast, processing outside this chapter involves changing the grey value. Geometric transformations include dilation, translation, rotation, and other types of processing, which will be explained with simple examples below.

First, an explanation of the coordinate system used in this chapter. Usually, the coordinate system for image processing uses the top-left corner as the origin, with the right and downward directions as positive. However, if the image is enlarged with the origin as the centre, the range of the image only moves outward in the lower right direction, as shown in Fig. 8.2a. Enlarging the image with its center as the origin allows it to expand outward equally in all directions, creating a more natural effect. Therefore, the coordinate system with the centre of the image as the origin, as shown in Fig. 8.2b, is more convenient and is the coordinate system adopted in this chapter.

© Chemical Industry Press 2026

B. Chen and S. Chen, *Machine Vision Technology*, Advanced and Intelligent Manufacturing in China, https://doi.org/10.1007/978-981-95-1184-6_8

Fig. 8.1 Perspective transformation of pet rabbits

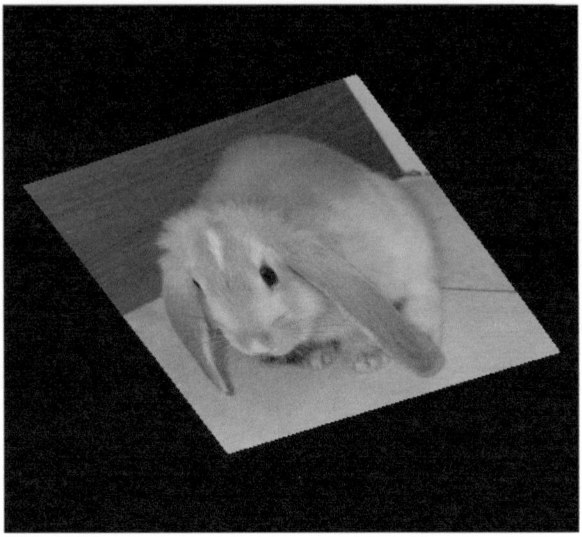

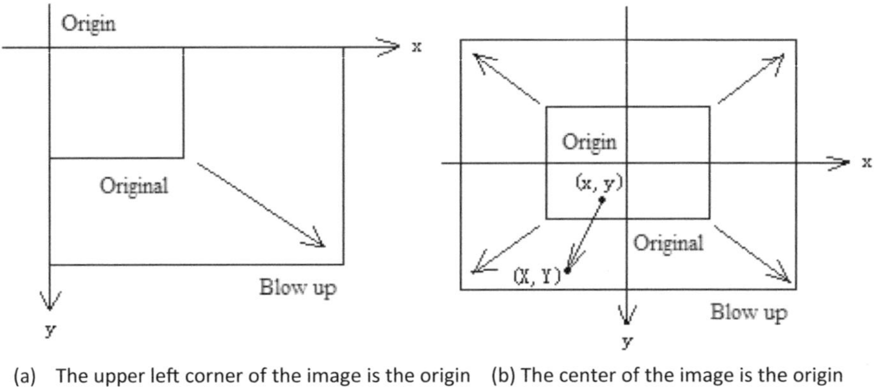

Fig. 8.2 Coordinate system

8.2 Dilation and Contraction

First, consider changing the size of the image. As illustrated in Fig. 8.2b, a given point (x, y) shifts to (X, Y) following dilation or contraction, and the relationship between them is defined as:

$$\begin{aligned} X &= \mathrm{a}x \\ Y &= \mathrm{b}y \end{aligned} \tag{8.1}$$

where a and b are the x-direction and y-direction respectively. The magnification rate in the direction. When a and b are greater than 1, the image is enlarged; when less than 1, it is reduced. For all pixel points (x, y), calculate and substitute the grayscale value of the point (x, y) on the input image into the point (X, Y) on the output image, thus enlarging or reducing the image.

Using the aforementioned method, an example where the image is reduced by half ($a = b = 1/2$) and enlarged by a factor of 2 ($a = b = 2$) is shown in Fig. 8.3b and c, respectively. The image reduced by 1/2 seems fine, but the image enlarged by 2 times appears somewhat odd. What is the reason for this?

Let's look at Fig. 8.4. When the pixel p of the input image corresponds to p' of the output image, and the neighbouring point q and the next neighbouring point r on the input image correspond to q' and r' on the output image, q' and r' either approach

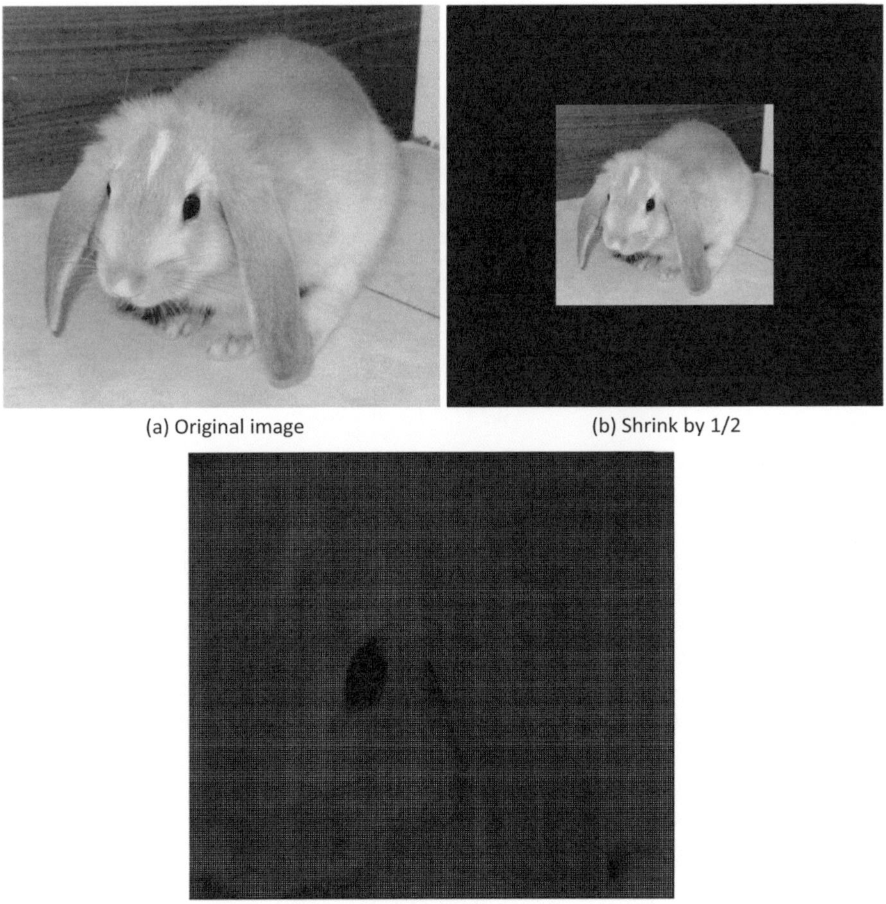

(a) Original image (b) Shrink by 1/2

(c) Magnified twice

Fig. 8.3 Processing example of intuitive zooming and shrinking

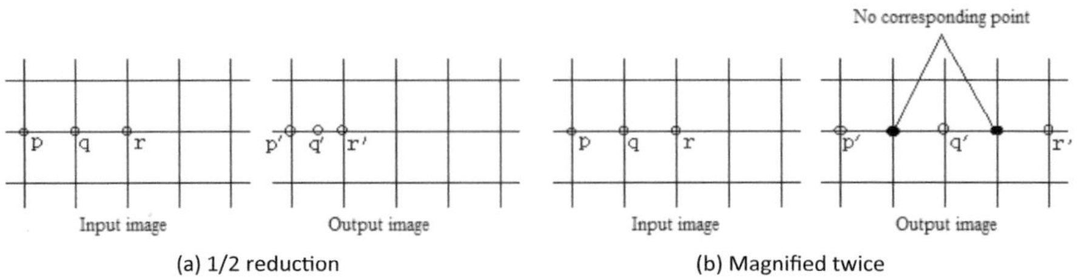

Fig. 8.4 Diagram of zooming in and out processing

or move away from the point p' according to the magnification rate. When reduced by 1/2, as shown in Fig. 8.4a, the position of the point q' corresponding to q is not at a pixel position, so it is automatically cancelled on the output image, making p' and r' neighbouring points. On the other hand, in the case of enlarging by 2 times, as shown in Fig. 8.4b, q' and r' are arranged with a pixel gap, meaning nothing is written between the neighbouring points of p' and q' on the output image. This is why the pixels appear discontinuous in Fig. 8.3c.

The above method is based on finding the corresponding points on the output image from the input image, resulting in some positions on the output image having no corresponding pixel values when enlarging. If the output image is used as the basis, finding the corresponding pixel for each pixel on the output image in the input image can avoid the above phenomenon. For this, consider the inverse operation of Eq. (8.1), namely:

$$\begin{aligned} x &= X/\mathrm{a} \\ y &= Y/\mathrm{b} \end{aligned} \tag{8.2}$$

By utilizing Eq. (8.2) to calculate and locate the corresponding pixel (x, y) on the input image for each pixel (X, Y) on the output image, and then assigning the grayscale value of this pixel, the artifact observed in Fig. 8.3c can be avoided. Examples of reducing by 1/2 and enlarging by 2 times in this way are shown in Fig. 8.5, which appear more normal.

Equation (8.2) involves real number calculations, and x and y include decimal places. However, the pixel addresses of the input image must be integers, so it is necessary to adopt some form of integerisation for address calculation. Here, the commonly used method is rounding. In terms of the image, as shown in Fig. 8.6, it involves selecting the point on the grid closest to the coordinate point (x, y), thus called the nearest neighbour approach, also known as zero-order interpolation. For this method, from the enlarged image in Fig. 8.5c, it can be seen that the image appears mosaic-like, and this phenomenon becomes more pronounced as the magnification rate increases.

To improve accuracy, a method called bilinear interpolation can be used. This method involves, when the calculated address is not on the grid, finding the ratio of distances to the four points on the grid, and using this ratio to perform grayscale

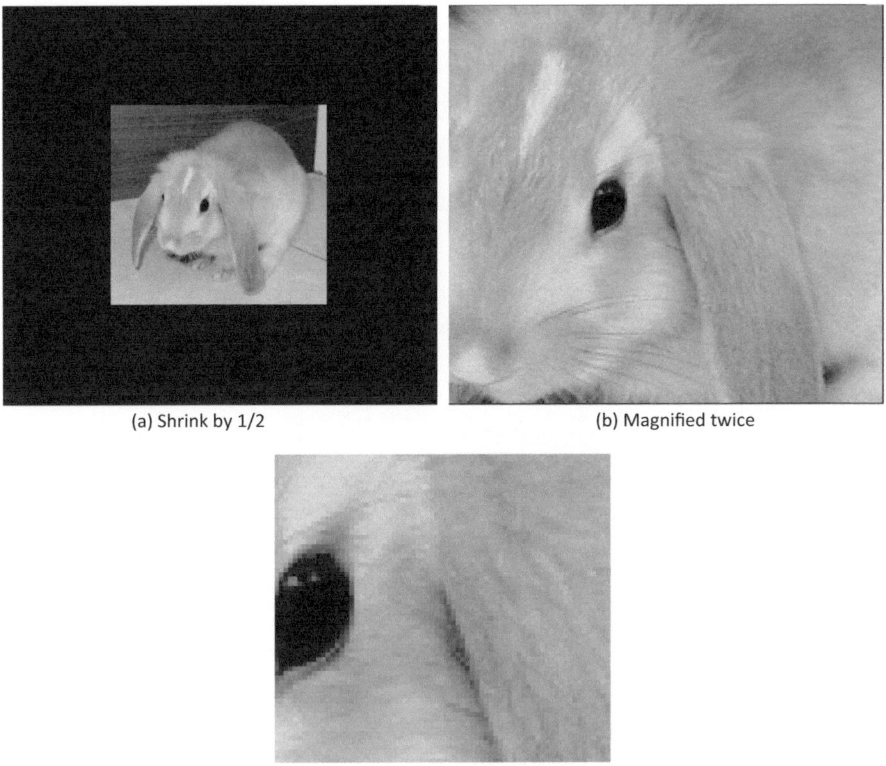

(a) Shrink by 1/2 (b) Magnified twice

(c) Enlarged image center position

Fig. 8.5 Example of zooming in and out (nearest neighbor method)

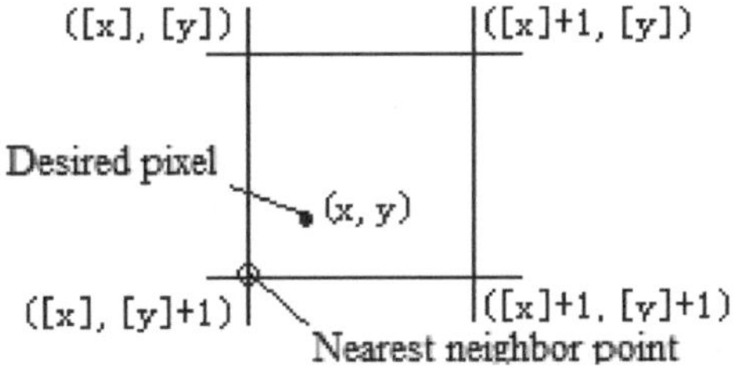

Fig. 8.6 Nearest neighbor method

interpolation with the grayscale values of the four nearest neighbours, as shown in Fig. 8.7.

The formula for calculating this grayscale value is as follows: (x, y) represents the greyscale value at the coordinates (x, y), where $[x]$ and $[y]$ are integer values not exceeding x and y respectively. An example of processing using bilinear interpolation is shown in Fig. 8.8. The enlarged image in Fig. 8.8c does not appear mosaic-like but rather very smooth. This bilinear interpolation method can not only use the

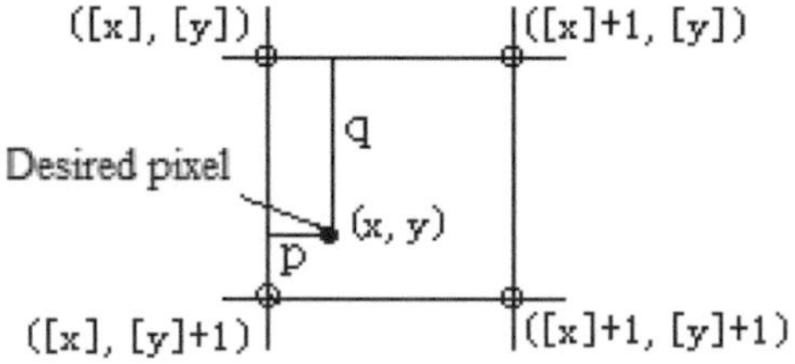

Fig. 8.7 Bilinear interpolation

aforementioned 4-neighbour points but also 8-neighbour points, 16-neighbour points, 24-neighbour points, etc., for higher-order interpolation.

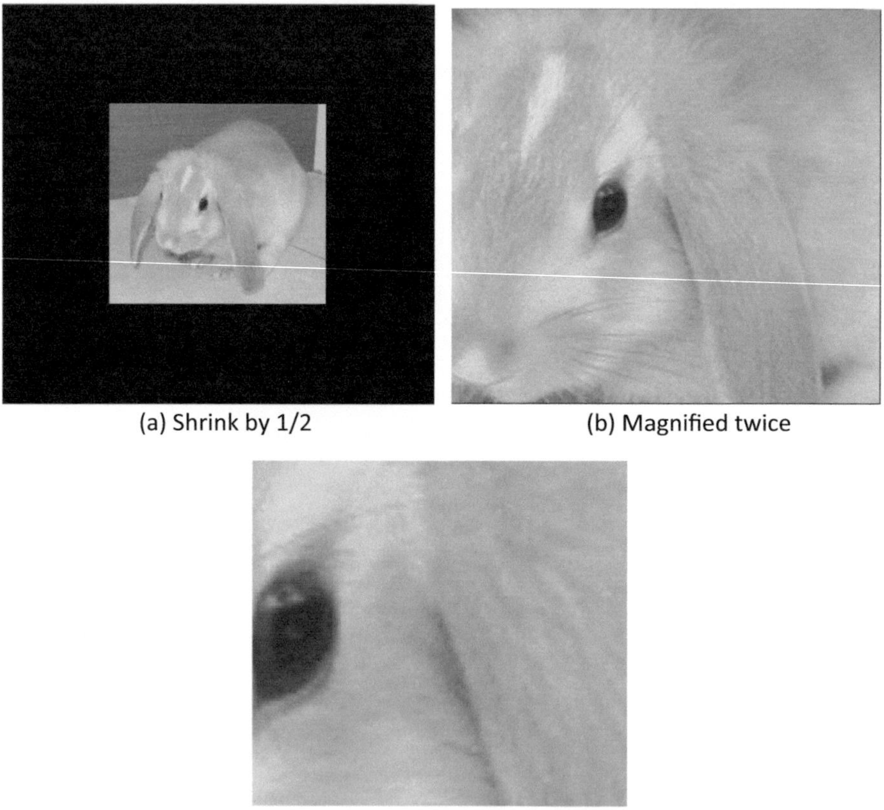

(a) Shrink by 1/2

(b) Magnified twice

(c) Enlarged image center position

Fig. 8.8 Example of zooming in and out (Bilinear interpolation)

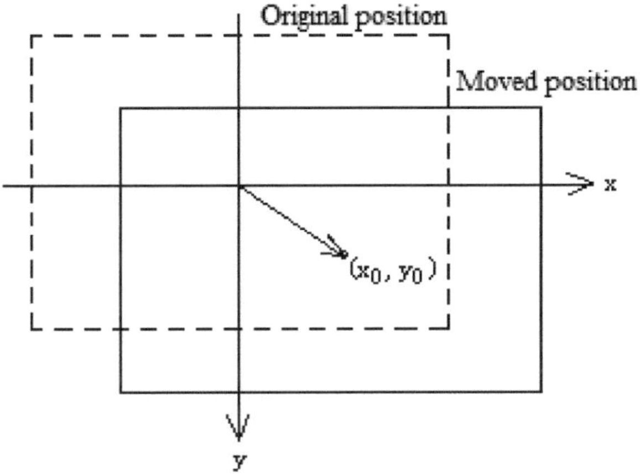

Fig. 8.9 Translation

8.3 Translation

Let us now analyse the movement of image positions. As shown in Fig. 8.9, in order to translate the image downwards to the right along the x and y coordinates by x_0 and y_0 respectively, the following translation transformation formula is required:

$$\begin{aligned} X &= x + x_0 \\ Y &= y + y_0 \end{aligned} \tag{8.4}$$

The inverse transformation formula is as follows:

$$\begin{aligned} x &= X - x_0 \\ y &= Y - y_0 \end{aligned} \tag{8.5}$$

An example illustrating the application of translation transformation is depicted in Fig. 8.10.

8.4 Rotation

Let us now consider rotating an image. As shown in Fig. 8.11, to rotate the image counterclockwise by θ °, the following transformation formula is required:

$$\begin{aligned} X &= x\cos\theta + y\sin\theta \\ Y &= -x\sin\theta + y\cos\theta \end{aligned} \tag{8.6}$$

The inverse transformation formula is as follows:

Fig. 8.10 Translation processing example

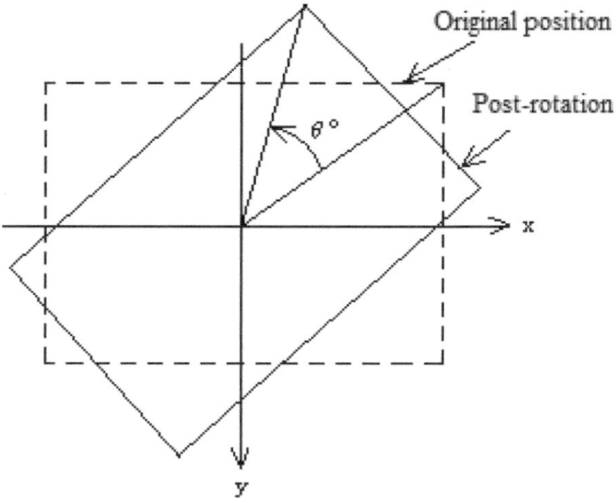

Fig. 8.11 Rotation

$$
\begin{aligned}
x &= X\cos\theta - Y\sin\theta \\
y &= X\sin\theta + Y\cos\theta
\end{aligned} \tag{8.7}
$$

An example demonstrating the application of rotation transformation is presented in Fig. 8.12.

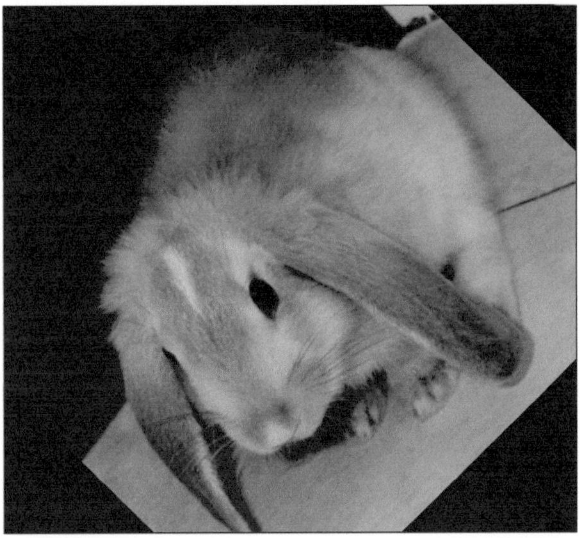

Fig. 8.12 Processing example of rotation

8.5 Complex Deformation

By combining the aforementioned scaling, translation, and rotation, various deformations can be achieved. So far, the methods described have been centred on the origin, but it is also possible to rotate, scale, and translate centred on any point. For example, to rotate centred on (x_0, y_0), as shown in Fig. 8.13, first translate $(-x_0, -y_0)$ to bring (x_0, y_0) back to the origin, then rotate by $\theta°$, and finally translate back by (x_0, y_0).

This method necessitates continuous address calculation and pixel access to determine grayscale values during processing, resulting in significant time consumption. To save time, addresses can be calculated in advance using formula (8.8):

$$\begin{aligned} X &= (x - x_0)\cos\theta + (y - y_0)\sin\theta + x_0 \\ Y &= -(x - x_0)\sin\theta + (y - y_0)\cos\theta + y_0 \end{aligned} \tag{8.8}$$

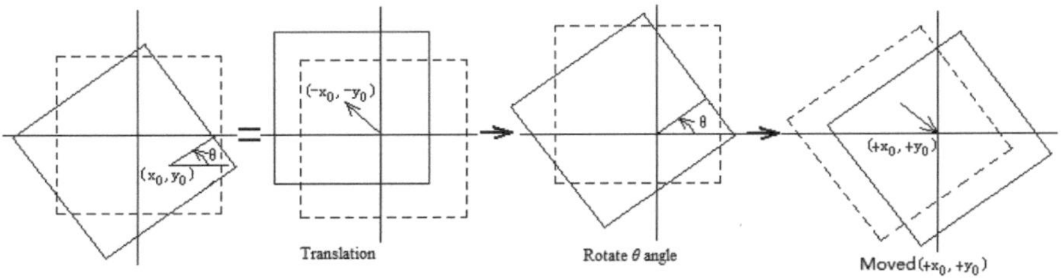

Fig. 8.13 Revolves around (x_0, y_0)

The inverse transformation formula is as follows:

$$\begin{aligned} x &= (X - x_0)\cos\theta - (Y - y_0)\sin\theta + x_0 \\ y &= (X - x_0)\sin\theta + (Y - y_0)\cos\theta + y_0 \end{aligned} \tag{8.9}$$

After calculating the address centrally, reading a pixel once allows the computation of the transformed result's grayscale value. This type of geometric transformation is called a two-dimensional affine transformation. The general representation formula for a two-dimensional affine transformation is as follows:

$$\begin{aligned} X &= ax + by + c \\ Y &= dx + ey + f \end{aligned} \tag{8.10}$$

The inverse transformation formula is as follows:

$$\begin{aligned} x &= AX + BY + C \\ y &= DX + EY + F \end{aligned} \tag{8.11}$$

Although the parameters differ, the forms are the same. The previously mentioned scaling formula (8.2), translation formulas (8.4) and (8.5), and rotation formulas (8.6) and (8.7) are all included in formulas (8.10) and (8.11).

Formulas (8.10) and (8.11) are first-degree polynomials. If they are made into higher-degree polynomials, more complex geometric transformations will result.

The image shown in Fig. 8.14 is an example of processing called perspective transform. When drawing, distant objects are depicted smaller, and perspective transform can also generate similar effects. As shown in Fig. 8.14, when viewing an object from a point (viewpoint), the projected image of the object on the imaging plane is the perspective transform image. This perspective transform is expressed by the following two formulas:

$$\begin{aligned} X &= (ax + by + c)/(px + qy + r) \\ Y &= (dx + ey + f)/(px + qy + r) \end{aligned} \tag{8.12}$$

The inverse transformation formula is shown as follows:

$$\begin{aligned} x &= (AX + BY + C)/(PX + QY + R) \\ y &= (DX + EY + F)/(PX + QY + R) \end{aligned} \tag{8.13}$$

The forms of the forward and inverse transformations are the same. Here, a, b, c and A, B, C, etc., are transformation coefficients, determined by the position of the viewpoint, the position of the imaging plane, and the size of the object. These coefficients can be simply calculated using the matrix form of homogeneous coordinates, with a processing example shown in Fig. 8.15.

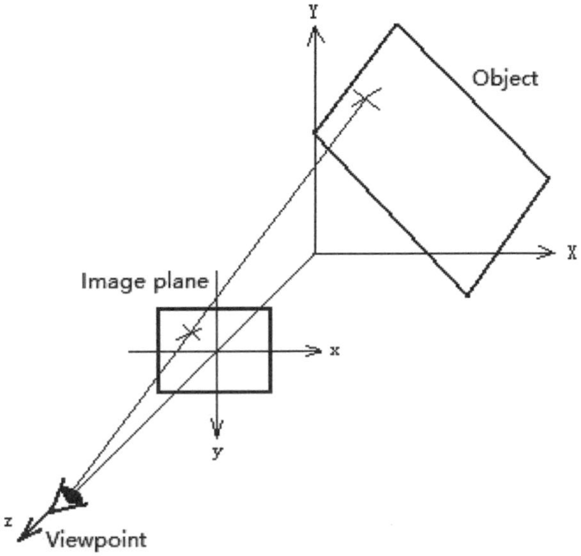

Fig. 8.14 Perspective transformation

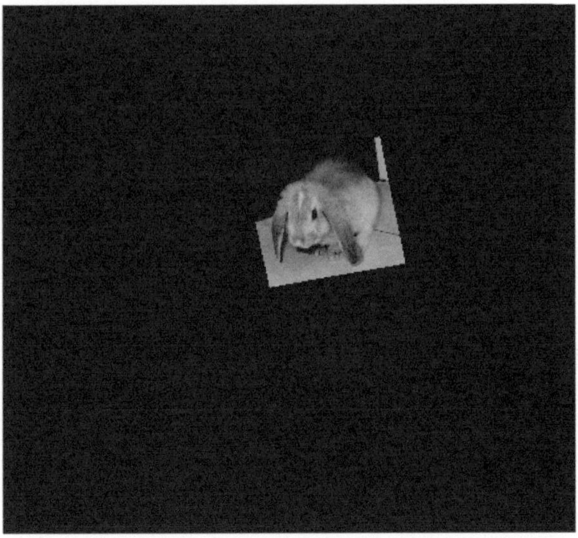

Fig. 8.15 Processing example of perspective transformation

8.6 Homogeneous Coordinate Representation

Geometric transformations are more conveniently handled using matrices. Geometric transformations of a 2D plane (x, y) can be represented using a 2D vector $[x, y]$ and a 2×2 matrix, but cannot represent translation. Therefore, to handle translation similarly, a virtual dimension 1 is added, typically using a 3D vector $[x, y, 1]^T$ and a 3

× 3 matrix. This 3D space coordinate $(x, y, 1)$ is called the homogeneous coordinate of (x, y).

Based on this homogeneous coordinate, the affine transformation can be expressed as:

$$\begin{bmatrix} X \\ Y \\ 1 \end{bmatrix} = \begin{bmatrix} a\,b\,c \\ d\,e\,f \\ 0\,0\,1 \end{bmatrix} \begin{bmatrix} x \\ y \\ 1 \end{bmatrix} \tag{8.14}$$

The above formula is consistent with formula (8.10). Additionally, scaling is represented as:

$$\begin{bmatrix} X \\ Y \\ 1 \end{bmatrix} = \begin{bmatrix} 1\,0\,x_0 \\ 0\,1\,y_0 \\ 0\,0\,1 \end{bmatrix} \begin{bmatrix} x \\ y \\ 1 \end{bmatrix} \tag{8.15}$$

The homogeneous coordinate representation of translation is:

$$\begin{bmatrix} X \\ Y \\ 1 \end{bmatrix} = \begin{bmatrix} a\,0\,0 \\ 0\,b\,0 \\ 0\,0\,1 \end{bmatrix} \begin{bmatrix} x \\ y \\ 1 \end{bmatrix} \tag{8.16}$$

The homogeneous coordinates for rotation are expressed as:

$$\begin{bmatrix} X \\ Y \\ 1 \end{bmatrix} = \begin{bmatrix} \cos\theta\,\sin\theta\,0 \\ -\sin\theta\,\cos\theta\,0 \\ 0\,0\,1 \end{bmatrix} \begin{bmatrix} x \\ y \\ 1 \end{bmatrix} \tag{8.17}$$

Equations (8.15) to (8.17) are consistent with the aforementioned Eq. (8.1), Eq. (8.4), and Eq. (8.6) respectively. By combining these matrices, various affine transformations can be represented. For example, rotation centred at (x_0, y_0) can be expressed in the form of a product of translation and scaling matrices as shown in Eq. (8.18):

$$\begin{bmatrix} X \\ Y \\ 1 \end{bmatrix} = \begin{bmatrix} 1 & 0 & x_0 \\ 0 & 1 & y_0 \\ 0 & 0 & 1 \end{bmatrix} \begin{bmatrix} \cos\theta & \sin\theta & 0 \\ -\sin\theta & \cos\theta & 0 \\ 0 & 0 & 1 \end{bmatrix} \begin{bmatrix} 1 & 0 & -x_0 \\ 0 & 1 & -y_0 \\ 0 & 0 & 1 \end{bmatrix} \begin{bmatrix} x \\ y \\ 1 \end{bmatrix} \tag{8.18}$$

After expanding Eq. (8.18), it is consistent with Eq. (8.8).

Perspective transformations and others are transformations in 3-dimensional space, represented using 4-dimensional vectors and 4 × 4 matrices. If a point in space has coordinates(X, Y, Z) and (x, y, z) in two different coordinate systems, the coordinate transformation formula can be described using a rotation matrix $\boldsymbol{R}$ and a translation matrix t as:

$$\begin{bmatrix} X \\ Y \\ Z \\ 1 \end{bmatrix} = \begin{bmatrix} R & t \\ 0^T & 1 \end{bmatrix} \begin{bmatrix} x \\ y \\ z \\ 1 \end{bmatrix} \tag{8.19}$$

where $\boldsymbol{R}$ is a 3×3 rotation matrix; t is a 3-dimensional translation vector; $\mathbf{0}^{\mathbf{T}} = (0,0,0)$. This type of perspective transformation is often applied in fields such as computer graphics.

Reference

1. Chen Bingqi. Practical Digital Image Processing and Analysis [M]. 2nd Edition. Beijing: China Agricultural University Press, 2014.

Chapter 9
Monocular Vision Measurement

This chapter introduces the hardware composition, fundamental principles, implementation methodologies, and measurement precision of monocular measurement, leveraging a portable monocular measurement system [1, 2].

9.1 Hardware Composition

Monocular measurement pertains to the utilization of a single camera to capture an image for measurement purposes. Its advantages lie in its simple structure, ease of portability and calibration, and relatively high measurement accuracy. In recent years, monocular measurement technology has attracted attention and has been widely applied in fields such as indoor scene measurement of buildings and traffic accident investigations. In a monocular vision system, it is necessary to place both the calibration template and the object to be measured in the same scene for shooting. This means that a single image must simultaneously contain both calibration information and measurement information, resulting in a large amount of data. Therefore, there are relatively high requirements for the imaging quality of the image acquisition equipment, especially the image resolution.

The components of a monocular vision system, as depicted in Fig. 9.1, are as follows.

① Calibration board. Different calibration templates are selected according to different calibration methods. Since monocular vision technology is only applicable to two-dimensional planes in space, it is essential to ensure that the calibration template and the object to be measured are placed on the same plane.
② Camera. Used for image acquisition.
③ Tripod. Used to adjust the height and viewing angle of the camera. It is also possible to shoot handheld without a tripod.
④ Computer. Used for storing, processing, and saving images.

© Chemical Industry Press 2026

B. Chen and S. Chen, *Machine Vision Technology*, Advanced and Intelligent Manufacturing in China, https://doi.org/10.1007/978-981-95-1184-6_9

Fig. 9.1 Composition of monocular visual system

Nowadays, a tablet (or mobile phone) combined with a calibration board can replace the aforementioned hardware devices, making monocular measurement more convenient. The monocular measurement system discussed in this chapter is a portable system based on a tablet, as shown in Fig. 9.2, with a camera resolution of 4096 × 3072 pixels. The accompanying calibration ruler and marker points are shown in Figs. 9.3 and 9.4, respectively.

To facilitate the automatic detection of the calibration ruler and marker points, this system incorporate a calibration ruler featuring a combination of blue, yellow,

Fig. 9.2 Portable monocular measurement system

Fig. 9.3 Calibration scale

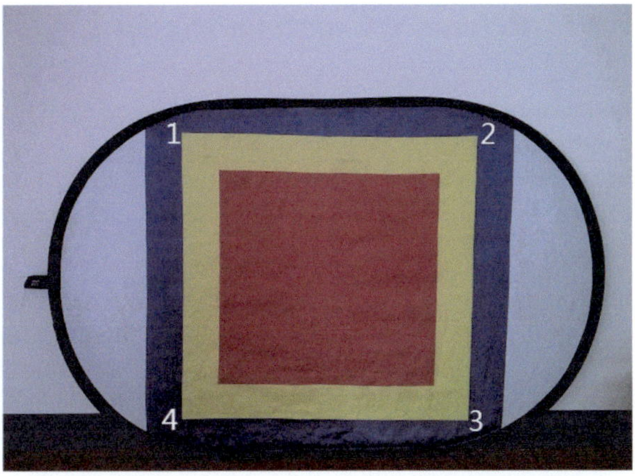

Fig. 9.4 Marker point

and red colours (Fig. 9.3) and marker points as detection targets (Fig. 9.4). The blue-yellow boundary of the calibration ruler is a square with a length and width of 80 cm each, and the four corners of this square are detected as calibration data.

The marker point is a square with a side length of 20 cm, and the blue-yellow boundary of the marker point points to the intersection below, serving as the target point for detection. During measurement, the downward-facing yellow corner point is positioned at the target location. By detecting its position in the image, the system can accurately determine the target's location.

9.2 Camera Model

In machine vision, the projection of an object's three-dimensional spatial position onto the imaging plane within the world coordinate system is represented by a geometric model. This geometric model links the 2D coordinates of the image with the 3D coordinates in real space, which is commonly referred to as the camera model.

9.2.1 Reference Coordinate Systems

In the camera model, four types of coordinate systems are generally involved: world coordinate system, camera coordinate system, image physical coordinate system, and image pixel coordinate system. Comprehending the significance and interrelationships of these four coordinate systems is crucial for image restoration and information reconstruction.

① Image pixel coordinate system: Digital images are represented in computers in the form of discrete pixel points, and the brightness or greyscale value of each pixel point in the image is stored in the computer in the form of an array. A plane rectangular coordinate system with the pixel point at the top-left corner of the image as the origin is established, forming the image pixel coordinate system. The coordinate value of each pixel point in this coordinate system indicates the relative position of that point in the image plane with respect to the top-left corner pixel point.

② Image physical coordinate system: A plane rectangular coordinate system established in the image with the intersection of the camera's optical axis and the image plane (usually located at the centre of the image) as the origin, expressed in physical units (such as millimetres), as shown in the coordinate system XO_1Y in Fig. 9.5. The coordinate value of a pixel point in this coordinate system can reflect the physical position of that point in the image.

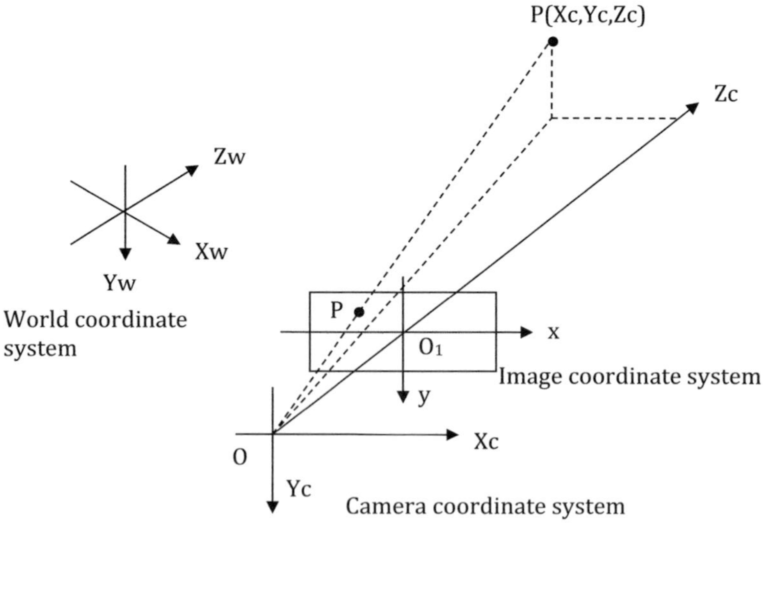

Fig. 9.5 Reference coordinate system

③ Camera coordinate system: In Fig. 9.5, the three-dimensional coordinate system composed of the origin *Oc* and the *Xc*, *Yc*, *Zc* axes is the camera coordinate system, where *Oc* is the optical centre of the camera, and the *Xc* and *Yc* axes are parallel to the *X* and *Y* axes of the image coordinate system. $_c$ axis is the optical axis of the camera, perpendicular to the image plane.

④ World Coordinate System: A three-dimensional coordinate system chosen based on the real environment, where the actual position coordinates of the camera and the scene are relative to this coordinate system. The world coordinate system is generally described using the O_w point and X_w axis, Y_w axis, Z_w axis, which can be selected arbitrarily according to the actual situation.

Among the four coordinate systems involved in the above camera model, the most concerned are the world coordinate system and the image pixel coordinate system.

9.2.2 *Camera Model Analysis*

The camera imaging model is generally divided into linear camera models and nonlinear camera models. The linear camera model, also known as the pinhole model, is the most commonly used imaging model in perspective projection. This model is an ideal state imaging model and does not consider the impact of camera lens distortion on imaging. Therefore, in situations where lens distortion is significant, the nonlinear model can more accurately describe the camera imaging process. However, with the improvement of camera lens manufacturing technology, the distortion of many modern camera lenses can be almost negligible. In such cases, the difference between the linear model and the nonlinear model is not significant, and the linear model is simple to solve and easy to use, thus having a wider application in visual measurement. This chapter introduces the mapping relationship between spatial points and their image points based on the linear camera model.

In the linear model, the object point, camera optical centre, and image point are collinear, as shown in Fig. 9.5. The intersection of the line connecting the spatial point and the optical centre with the imaging plane is its corresponding image point. An object point has a unique image point corresponding to it on the image plane. The relationship between the image pixel coordinates of any point *P* in the scene and the world coordinates can be expressed in the form of homogeneous coordinates and matrices as Eq. (9.1).

$$Z_c \begin{bmatrix} u \\ v \\ 1 \end{bmatrix} = \begin{bmatrix} \frac{1}{dx} & 0 & u_0 \\ 0 & \frac{1}{dy} & v_0 \\ 0 & 0 & 1 \end{bmatrix} \begin{bmatrix} f & 0 & 0 & 0 \\ 0 & f & 0 & 0 \\ 0 & 0 & 1 & 0 \end{bmatrix} \begin{bmatrix} R & T \\ 0^T & 1 \end{bmatrix} \begin{bmatrix} X_w \\ Y_w \\ Z_w \\ 1 \end{bmatrix}$$

$$
\begin{aligned}
&= \begin{bmatrix} f_x & 0 & u_0 & 0 \\ 0 & f_y & v_0 & 0 \\ 0 & 0 & 1 & 0 \end{bmatrix} \begin{bmatrix} R & T \\ 0^T & 1 \end{bmatrix} \begin{bmatrix} X_w \\ Y_w \\ Z_w \\ 1 \end{bmatrix} \\
&= M_1 M_2 X_w = M X_w
\end{aligned}
\tag{9.1}
$$

where, (u, v) are the image pixel coordinates of the projection point of P on the image plane, $X_w = [X_w, Y_w, Z_w, 1]^T$, describing its world coordinates. $f_x = f/dx$, is the focal length of the camera in the x direction; $f_y = f/dy$, is the focal length of the camera in the y direction. The parameters f_x, f_y, u_0, v_0 in M_1 are related to the internal structure of the camera itself, and therefore, they are called internal parameters. M_1 is the intrinsic matrix. The rotation matrix R and the translation vector T in M_2 represent the position of the camera relative to the world coordinate system, and thus are called external parameters. M_2 is the extrinsic matrix. M is the product of M_1 and M_2, forming a 3×4 matrix known as the projection matrix, which can reflect the relationship between the image pixel coordinates and the world coordinates of any spatial point.

Through Eq. (9.1), it is known that if the projection matrix M and the world coordinates of the spatial point X_w are known, the image coordinates (u, v) of the spatial point can be obtained. Therefore, in a linear model, a physical point corresponds to a unique image point on the imaging plane. Conversely, if the image point coordinates (u, v) and the projection matrix M are known, substituting into Eq. (9.1) only yields two linear equations about X_w. These two linear equations represent the line connecting the image point and the optical centre, meaning all points on this line correspond to the image point.

To obtain the distance parameters of the target to be measured, one of the key steps is to restore the coordinate information of the target in the three-dimensional scene from the two-dimensional image. As discussed above, in a linear model, the physical point corresponding to an image point is not unique. Therefore, it is unrealistic to perform three-dimensional reconstruction of the image scene using only one image. However, in many scenarios, the target to be measured can be approximately considered to be located on the same plane. In this case, establishing the correspondence between the plane where the target is located (hereinafter referred to as the "world plane") and the image plane can achieve three-dimensional reconstruction of the target. The linear camera model can also be simplified to a planar camera model, as shown in Fig. 9.6.

In Fig. 9.6, C is the optical centre of the camera, i.e., the pinhole in pinhole imaging. The corresponding point of the spatial point X on the image plane is the image point x. Let $X = [X, Y, Z, 1]$ and $x = [x, y, 1]$ represent the homogeneous coordinates of the spatial point in the world coordinate system and the image pixel coordinate system, respectively. Then, according to the transformation of Eq. (9.1), the following relational expression can be obtained:

$$
\lambda x = PX \tag{9.2}
$$

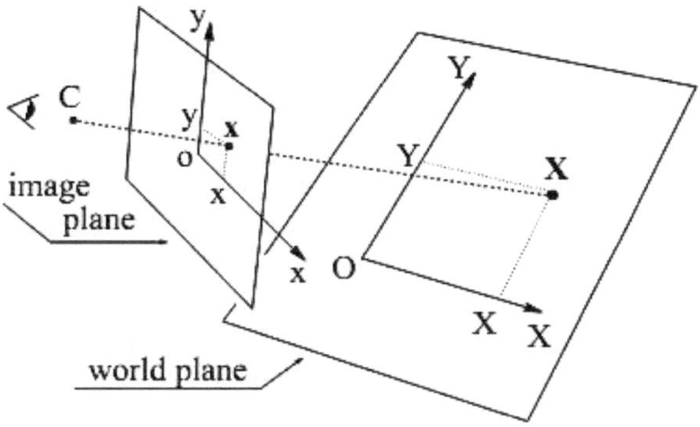

Fig. 9.6 Flat camera model

In Eq. (9.2), P is a 3×4 matrix, and $\lambda \in R$ is a scale factor related to the homogeneous world coordinates X. By setting the origin of the world coordinate system, the X axis, and the Y axis on the plane to be measured, the Z axis is perpendicular to the plane to be measured. The homogeneous coordinates of X can be simplified to $[X, Y, 0, 1]$. Substituting into Eq. (9.2) gives:

$$\lambda \begin{bmatrix} x \\ y \\ 1 \end{bmatrix} = [P_1, P_2, P_3, P_4] \begin{bmatrix} X \\ Y \\ 0 \\ 1 \end{bmatrix} = [P_1, P_2, P_4] \begin{bmatrix} X \\ Y \\ 1 \end{bmatrix}$$

$$= \begin{bmatrix} H_{11} & H_{12} & H_{13} \\ H_{21} & H_{22} & H_{23} \\ H_{31} & H_{32} & H_{33} \end{bmatrix} \begin{bmatrix} X \\ Y \\ 1 \end{bmatrix} \tag{9.3}$$

From the above equation, it can be understood that the relationship between points on a three-dimensional space plane and points on the image plane can be described by a 3×3 homogeneous matrix $\boldsymbol{H} = [P_1, P_2, P_3]$, where $\boldsymbol{H}$ is the homography matrix. The world coordinates can be converted into image pixel coordinates using Eq. (9.3), and conversely, image pixel coordinates can be converted into world coordinates using Eq. (9.4).

$$sX = H^{-1}x \tag{9.4}$$

9.3 Camera Calibration

The purpose of camera calibration is to establish a mapping relationship between the three-dimensional object points in the world coordinate system and the two-dimensional image points in the image coordinate system. The mutual relationship between the three-dimensional geometric position of a point on the surface of a spatial object and its corresponding point in the image is determined by the geometric model of camera imaging. In the linear model, the projection relationship between the three-dimensional object points and the corresponding image points is related to the internal and external parameters of the camera, described by a 3×4 projection matrix $\boldsymbol{M}$. The process of camera calibration is the process of solving the internal and external parameters of the camera, i.e., the process of obtaining the projection matrix $\boldsymbol{M}$.

This chapter introduces a special linear camera model—the planar camera model. In this model, the projection relationship between the world coordinate system and the pixel coordinate system is described using the homography matrix $\boldsymbol{H}$. When the target to be measured is on the same plane, the relationship between the measured plane and the image plane can be represented by the homography matrix $\boldsymbol{H}(\boldsymbol{H}^{-1})$. Once $\boldsymbol{H}^{-1}$ is obtained, the pixel coordinates of the target to be measured can be converted into world coordinates on the measured plane, and further parameters such as distance can be calculated. The process of obtaining the homography matrix $\boldsymbol{H}^{-1}$ is the camera calibration process.

The primary algorithms for deriving the homography matrix encompass point correspondence algorithms, line correspondence algorithms, and algorithms that employ the homography relationship between two images as a constraint. The following introduces the point correspondence algorithm.

Assuming in the planar camera model, there are N pairs of corresponding points, with known world coordinates and image coordinates. Suppose the world coordinates and image coordinates of a certain point are $[X_i, Y_i, 1]^{\mathrm{T}}$ and $[x_i, y_i, 1]^{\mathrm{T}}$ respectively, then according to Eq. (9.4), two linear equations as shown in Eq. (9.5) can be obtained. Here, $\boldsymbol{h} = (h_0, h_1, h_2, h_3, h_4, h_5, h_6, h_7, h_8)^{\mathrm{T}}$ is the vector form of the matrix $\boldsymbol{H}^{-1}$.

$$\begin{gathered}\left(x_i\ y_i\ 1\ 0\ 0\ 0\ -x_iX_i\ -y_iX_i\ -X_i\right)\boldsymbol{h} = 0 \\ \left(0\ 0\ 0\ x_i\ y_i\ 1\ -x_iY_i\ -y_iY_i\ -Y_i\right)\boldsymbol{h} = 0\end{gathered} \tag{9.5}$$

Thus, N pairs of corresponding points can yield $2N$ linear equations about h, since $\boldsymbol{H}^{-1}$ is a homogeneous matrix, it has 9 elements but only 8 are independent. In other words, although it has 9 parameters, there are actually only 8 unknowns. Therefore, when $N \geq 4$, there are enough equations to estimate the homography matrix $\boldsymbol{H}^{-1}$ and complete the camera calibration.

9.4 Calibration Scale Detection

The main purpose of calibration scale detection (scale detection) is to automatically extract the image coordinates of the calibration points. The generalizability, speed, and accuracy of scale detection directly influence the overall performance of the

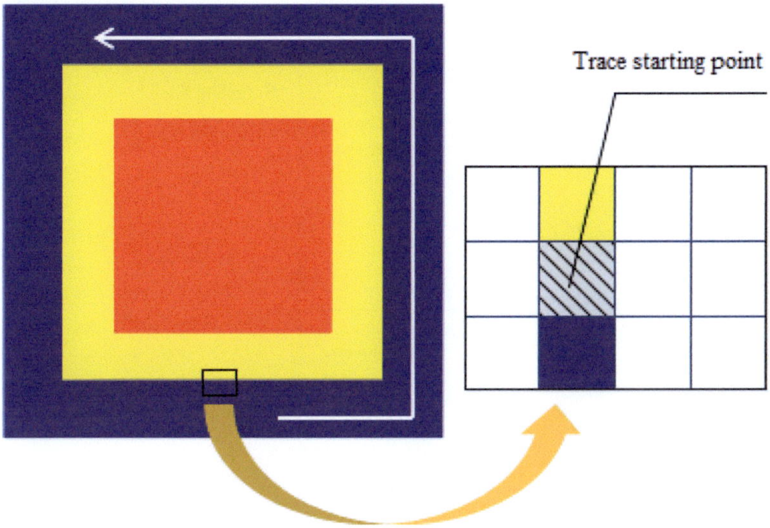

Fig. 9.7 Tracking the outer boundary points of the yellow area

measurement system. This system has developed the following scale detection algorithm. First, the colour scale image is read into the system memory, and a fixed step size is used to scan the entire image from bottom to top. When a junction of the blue and yellow areas at the bottom of the scale is detected, the scan stops, and this point is taken as the starting point for tracking. Then, a local scanning method is used to track all the outer boundary points of the yellow area counter clockwise, as shown in Fig. 9.7. The coordinates of the tracked boundary points are stored in a linked list. After tracking is completed, four corner point coordinates are extracted from the linked list, and the corner point positioning accuracy is improved through Hough transform and pixel value fine positioning, finally confirming the corner point coordinates.

9.4.1 Locating the Tracking Starting Point

The method used to locate the tracking starting point is to perform a line scan of the entire image with a fixed step size. The image is divided equally into 10 parts horizontally, with the dividing lines being $x = xsize/10$, $xsize/5$, $3\ xsize/10$, …, $9\ xsize/10$, where $xsize$ is the image width. Using these dividing lines as targets, the line scanning operation is performed from $x = xsize/2$, starting from the centre of the image and proceeding to both sides, as shown in Fig. 9.8, where the dashed lines represent the scan line positions.

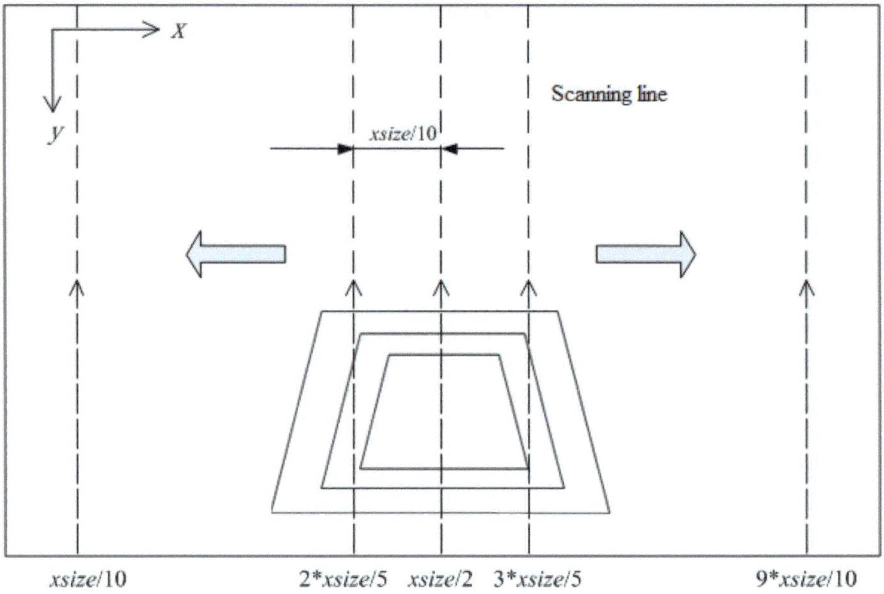

Fig. 9.8 Schematic diagram of line scanning

The specific steps of the line scan are as follows ($step = 20$).

① From the top to the bottom of the image, read the red (R), green (G), and blue (B) components of the current scan line pixel points, and store them in the arrays *buff_r* [*ysize*], *buff_g* [*ysize*], *buff_b* [*ysize*], where *ysize* is the image height.

② Define an integer *num* to record the number of consecutive pixel points in the scale area obtained by scanning, with an initial value of 0. Starting from $j = ysize - 1$ (j is the y coordinate of the current scan point, and $step \le j < ysize$), scan the arrays *buff_r*, *buff_g*, *buff_b* element by element, i.e., scan each point on the scan line from the bottom to the top of the image, and determine whether their *RGB* components satisfy Eq. (9.6), which describes the *RGB* value relationship of the blue area pixel points of the scale.

$$\begin{cases} B > 100 \\ B - R > 30 \\ B - G > 30 \end{cases} \text{or} \begin{cases} B \le 100 \\ B - R > 10 \\ B - G > 10 \end{cases} \text{or} \begin{cases} B > 200 \\ B > R \\ B > G \end{cases} \text{or} \begin{cases} \frac{B}{R} > 1.1 \\ \frac{B}{G} > 1.1 \end{cases} \tag{9.6}$$

③ When the target point P_1 is scanned, and its *RGB* values satisfy Eq. (9.6), it indicates that this point carries the colour information of the blue area of the scale. Consider P_1 as a candidate point for the blue area of the scale, and evaluate the target point P_2 located *step* pixels above, with *RGB* components being *buff_r* [j − *step*], *buff_g* [j − *step*], *buff_b* [j − *step*]. If it satisfies Eq. (9.7) or Eq. (9.8) [both Eqs. (9.7) and (9.8) represent the *RGB* value relationship of the yellow area of the scale, the former under normal lighting conditions, and the latter under strong reflection conditions], then point P_2 is considered a point in the yellow area of the scale, and temporarily consider the candidate point P_1 as a point in the blue area of the scale, increment the value of *num* by 1. Set $j = j - 1$, and continue scanning upwards. If the pixel above is also a point in the blue area of the scale, then increment the *num* value by 1 again; otherwise, reset the *num* value. When *num* > *step* /4, it is considered that there is scale information on this scan line, stop scanning, and record the current j value.

$$\begin{cases} R < 100 \\ R - B > 10 \\ G - B > 10 \end{cases} \text{or} \begin{cases} R \geq 100 \\ R - B > 50 \\ G - B > 50 \end{cases} \text{or} \begin{cases} R \geq 100 \\ R > B \\ G > 1.2B \end{cases} \quad (9.7)$$

$$\begin{cases} R > 200 \\ G > 200 \\ R > B \\ G > B \end{cases} \text{or} \begin{cases} R = 255 \\ G = 255 \end{cases} \quad (9.8)$$

④ Set $Y = start = j - step/4$ as the starting point, and perform an upward local scan on the pixels within the interval $[start - step, start]$ on the scan line. When the *RGB* values of the target point satisfy Eq. (9.9), it indicates that the boundary of the yellow-blue area has been scanned, at which point the scan is stopped, and the current target point (x, y) is recorded and marked with the marking colour F_c, with red, green, and blue components being 250, 0, 0 respectively. This point is used as the starting point for tracking, to trace the outer contour points of the yellow area of the scale.

$$R > B \text{ or } G > B \text{ or } R = G = B = 255 \quad (9.9)$$

⑤ In step ②, if after the current column scan ends, no target point meeting the conditions is found, or in step ③, the number of continuous pixel points in the scanned ruler area is not greater than the threshold value step/4, it is considered that there is no ruler information on this scan line, and steps ① ~ ⑤ are repeated to scan the next column.

⑥ If after all columns of the image have been scanned, no scan line with ruler information is found, it is considered that there is no ruler target in the current image, and no further detection is performed.

9.4.2 *Blue-Yellow Boundary Detection*

Using the boundary point $P_S(x_0, y_0)$ extracted in the previous section as the starting point for tracking, perform a local scan of the image and track all outer boundary points counterclockwise. During the tracking process, use a linked list to store the

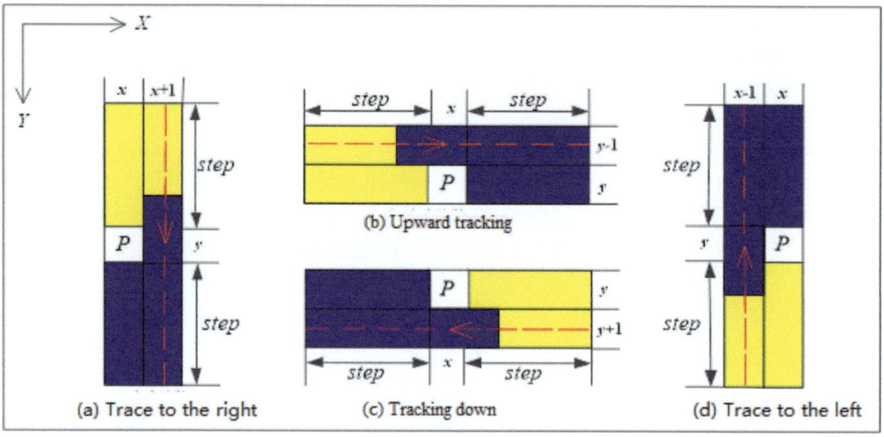

Fig. 9.9 Trace diagram

coordinate information of the detected boundary points. The specific process is as follows:

(1) Track to the right

First, track to the right. Take P_S as the tracked point $P(x, y)$, and use $X = x + 1$ as the scan line to perform a top-down scan operation in the interval $Y = sy = y - step$ to $Y = ey = \mathrm{y} + step$, as shown in Fig. 9.9a. During the scan, the following three situations may occur:

① When tracking to the right, it is first necessary to determine whether the tracking process has completed a full cycle. If the current scan line satisfies Eq. (9.10), it indicates that after a full cycle, the tracking returns to the starting point, at which point scanning and tracking stop, and the next step is performed: determining the corner coordinates.

② If the tracking process has not completed a full cycle, further determine whether it has tracked to the yellow corner area. If the *R*, *G*, *B* values of the scan starting point (x-1, sy) satisfy Eq. (9.11), then this point is a blue point of the ruler, indicating that it has tracked near the yellow corner. At this time, during the scanning process, if the colour components and *Y* coordinate of the current scan point satisfy Eq. (9.12), stop tracking to the right, and use this scan point as the starting point for tracking upwards.

$$B > R \text{ and } B > G \tag{9.11}$$

$$\begin{cases} R > B \\ G > B \quad \text{or } y_1 \geq y \ (y_1 is\ Y \text{ coordinate of the current point}) \\ R < 250 \end{cases} \tag{9.12}$$

③ If the tracking process has not completed a full cycle, and the scan starting point is not a blue point of the ruler, continue scanning to detect boundary points. When the *R*, *G*, *B* values of the scan point satisfy Eq. (9.11), it indicates that the blue point at the blue-yellow boundary of the ruler has been scanned, stop scanning, record the current scan point, store it in the linked list, and this point is the target point to be tracked, marking this point with the marker colour F_c, and taking this point as the tracked point *P*., continue tracking to the right.

(2) Track upwards

After completing the rightward tracking, begin tracking upwards. The process in step ② of the rightward tracking provides the starting point for the upward tracking, taking this point as the tracked point $P\ (x, y)$, with $Y = y - 1$ as the scan line, perform a left-to-right scanning operation over the interval from $X = sx = x - step$ to $X = ex = x + step$, as shown in Fig. 9.9b. The scanning process for upward tracking is similar to that of rightward tracking.

(3) Track to the left

After tracking upwards to the yellow corner point, begin tracking to the left, using the starting point provided by the upward tracking as the tracked point $P\ (x, y)$, with $X = x - 1$ as the scan line, perform a bottom-to-top scanning operation over the interval from $Y = sy = y + step$ to $Y = ey = y - step$, as shown in Fig. 9.9d. The scanning process is similar to that of rightward tracking.

(4) Track downwards

After completing the leftward tracking, begin tracking downwards, using the starting point provided by the upward tracking as the tracked point $P\ (x, y)$, with $Y = y + 1$ as the scan line, perform a right-to-left scanning operation over the interval from $X = sx = x + step$ to $X = ex = x - step$, as shown in Fig. 9.9c. The scanning process is similar to that of rightward tracking. After completing the downward tracking, continue tracking to the right, and when the starting point of the entire process P_S (x_0, y_0) is reached, the tracking ends.

Here, the point $P(x, y)$ is the tracked point, the dashed line represents the scan line, and the arrow on the dashed line represents the scanning direction.

9.4.3 Determining the Coordinates of the Corner Points

After completing the blue-yellow boundary tracking, it is necessary to extract the coordinates of 4 corner points from all the boundary points. In this system, the ruler can be placed in either a diamond or rectangular orientation. Depending on the placement method, the 4 corner points have different coordinate characteristics on the image, which can be used to initially determine the coordinates of the 4 corner points.

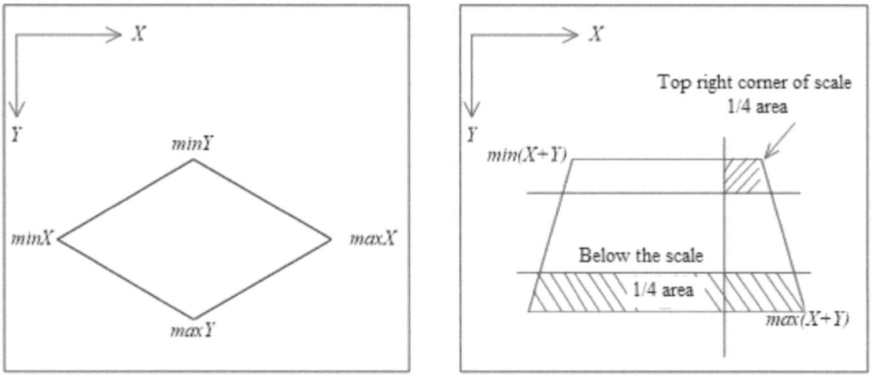

(a) Diamond placement (b) Rectangular placement

Fig. 9.10 The 4 corners of the ruler are determined according to the coordinate characteristics

In the diamond placement, the points with the maximum and minimum X coordinates and the maximum and minimum Y coordinates in the linked list can be considered as the 4 corner points of the ruler, as shown in Fig. 9.10a. In the rectangular placement, first determine the maximum and minimum values of $(X + Y)$, and use them as the corner points for the top-left and bottom-right corners, respectively. Then, extract the corner points for the top-right and bottom-left corners in the top-right 1/4 area and the bottom 1/4 area of the ruler, respectively, as shown in Fig. 9.10b. Then, $(X + \min Y - Y)$ is considered the point where the maximum value is obtained, regarded as the top-right corner point. The point where $(\mathrm{X} + 2\max \mathrm{Y} - \mathrm{Y})$ reaches its maximum value is considered the bottom-left corner point.

After initially determining the coordinates of the four corner points based on coordinate features, further use of the Hough transform is employed to locate the four corner points. The specific method is: targeting the pixel points on the two edges where the corner point is located, perform a known point Hough transform (see Chap. 7) separately. After the transformation is completed, fit the two lines where the two edges are located, and the intersection of the two lines is the corner point.

Since the calibration accuracy directly affects the subsequent measurement accuracy, and the calibration accuracy largely depends on the positioning accuracy of the calibration points, in order to improve the positioning accuracy of the calibration points, i.e., the four corner points, final precise positioning of the corner points is performed through pixel values.

Once the image coordinates of the four calibration points are determined, the ruler detection ends. The next step is to calculate the homography matrix between the image plane and the world plane through the image coordinates and world coordinates of the calibration points.

9.4.4 Homography Matrix Calculation

The four known calibration points provide eight linear equations in the form of Eq. (9.5) regarding the homography matrix $\boldsymbol{H}^{-1}$, expressed in matrix form (h is the vector form of the homography matrix $\boldsymbol{H}^{-1}$):

$$\boldsymbol{A}\boldsymbol{h} = \begin{bmatrix} A_1 \\ A_2 \\ A_3 \\ A_4 \end{bmatrix} \boldsymbol{h} = 0 \tag{9.13}$$

where, assuming the image coordinates and world coordinates of the four calibration points are (x_i, y_i) and (X_i, Y_i), $i = 1, 2, 3, 4$, then:

$$A_i = \begin{bmatrix} x_i \ y_i \ 1 \ 0 \ 0 \ 0 \ -x_i X_i \ -y_i X_i \ -X_i \\ 0 \ 0 \ 0 \ x_i \ y_i \ 1 \ -x_i Y_i \ -y_i Y_i \ -Y_i \end{bmatrix} \tag{9.14}$$

The vector $\boldsymbol{h}$ is the eigenvector corresponding to the smallest eigenvalue of $\boldsymbol{A}^T \boldsymbol{A}$. This system uses the cvFindHomography function provided by the open-source library OpenCV to obtain this eigenvector.

9.5 Calibration Result Analysis

By integrating the ruler detection algorithm and the homography matrix calculation algorithm into the system, using the hardware carrier of this system, an Android tablet, 30 ruler image samples were collected under normal lighting, strong light, dim light, and shadow (uneven lighting) conditions, and actual detection experiments were conducted. The experimental results are shown in Table 9.1.

From Table 9.1, it can be seen that the ruler detection algorithm of this system can be well adapted to ruler images taken under different lighting conditions, and the calibration error is controlled within 0.2%, with the calibration time controlled

Table 9.1 Camera calibration experimental results

Shooting state	Total number of experiments	Number of successful tests	Average calibration error (%)	Average calibration time (ms)
Normal illumination	30	30	0.170	244
Glare	30	30	0.166	587
Dark light	30	30	0.198	433
Shadow	30	30	0.183	365

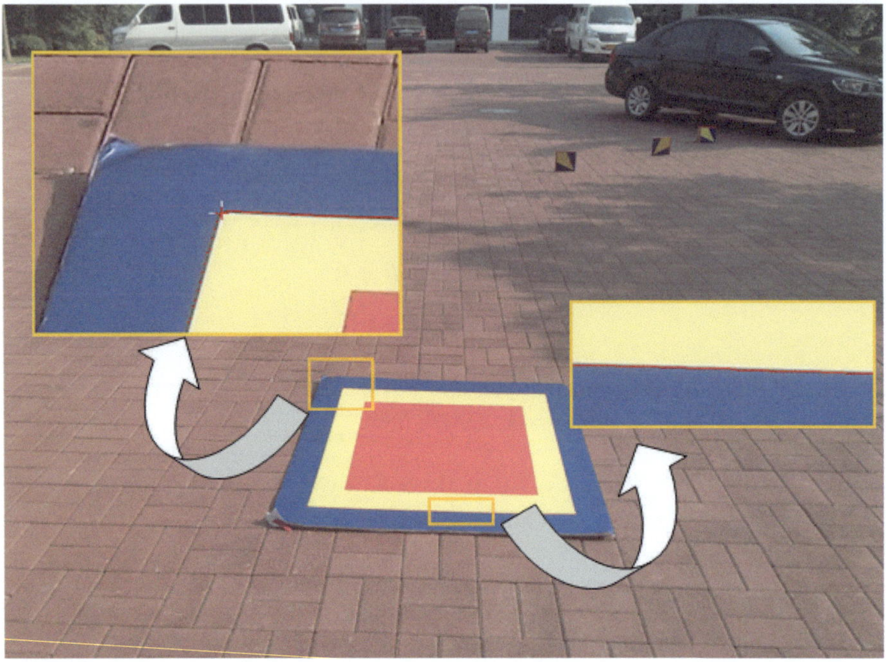

Fig. 9.11 Example of measuring scale under strong light irradiation (shot at 15:00)

within 1 s, basically meeting the general, accurate, and fast requirements proposed earlier.

Figure 9.11 is an example image of ruler detection under strong light conditions. Through the above ruler detection algorithm, the value of the homography matrix H^{-1} between the measurement plane and the image plane is obtained as shown in Eq. (9.15), with a calibration error of 0.142% and a calibration time of 268 ms.

$$\boldsymbol{H}^{-1} = \begin{bmatrix} -4.13 & -1.24 & 8904.71 \\ 0.27 & -9.25 & 16527.10 \\ 7.62 \times 10^{-4} & -0.028 & 1 \end{bmatrix} \tag{9.15}$$

9.6 Automatic Detection of Mark Points

This system employs specifically designed markers (as shown in Fig. 9.4) to transform uncertain targets into definite markers, thereby facilitating automatic detection. Once the markers are detected, the distance and area between the markers can be calculated.

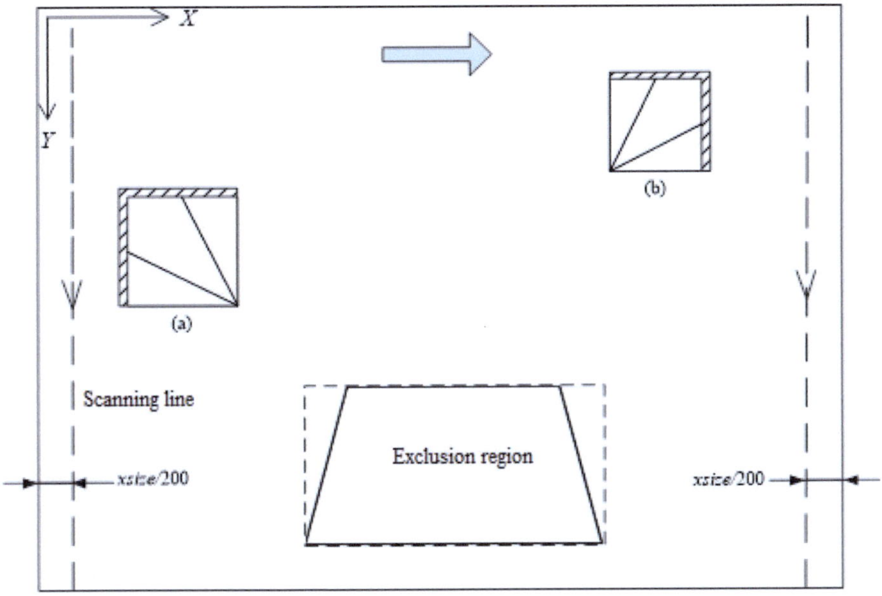

Fig. 9.12 Schematic diagram of identification point detection

To eliminate the interference of the calibration scale on marker detection, after the calibration scale detection is completed, the upper, lower, left, and right area ranges of the calibration scale are obtained to exclude the detection area. The dashed part in Fig. 9.12 represents the exclusion area.

(1) Locating the tracking starting point

By scanning the entire image, the pixels on the slanted edge at the bottom of the marker are detected. Since the marker's area is relatively small and is placed arbitrarily in the shooting scene, it may appear anywhere in the image. Therefore, the method of locating the tracking starting point adopts a fixed step size of *xsize/*200 (*xsize* is the image width), scanning the entire image from left to right, as shown in Fig. 9.12.

The specific steps for line scanning operation on the current column ($x = i$) are as follows.

① Read the red (R), green (G), and blue (B) components of the pixel points on the current scan line from top to bottom of the image, and store them in the arrays *buff_r* [*ysize*], *buff_g* [*ysize*], *buff_b* [*ysize*], where *ysize* is the image height.

② Define an integer j, representing the vertical coordinate of the current scan point, starting from $j = 0$, and scan the arrays *buff_r*, *buff_g*, *buff_b* element by element, i.e., read the red (R), green (G), and blue (B) values of the scan point from top to bottom of the image pixel by pixel. If the *RGB* component values of the current point satisfy Eq. (9.16), it indicates that the point carries the colour information of the red area of the marker and may be a red area point. Pause the scan and record the current j value as *ystart*.

$$\begin{cases} R - G > 50 \\ R - B > 50 \end{cases} \text{or} \begin{cases} \frac{R}{G} > 1.5 \\ \frac{R}{B} > 1.5 \end{cases} \tag{9.16}$$

③ Define an integer *num* to record the number of consecutive pixels that meet the set conditions, with an initial value of 0. Starting from $j = ysize - 1$, perform a local scan on the pixel points within the interval $[ystart, ysize - 1]$ on the current scan line, and determine whether the *RGB* component values of the current scan point satisfy Eq. (9.17). If satisfied, it indicates that the current point $P_1(i, j)$ may be a blue area point at the bottom of the marker. Further, determine whether the target point $P_2(i, j - step)$ at *step* pixels above this point satisfies Eq. (9.18). If satisfied, it indicates that point P_2 may be a yellow area point of the marker. Therefore, point P_1 can be temporarily considered as a blue area point, and the *num* value is incremented by 1; otherwise, the *num* value is cleared. When *num* > 3, it is considered that the identification point information is detected, and the current scanning point is a blue area point, record the current point $P_1(i, j)$.

$$B > R \text{ and } B > G \tag{9.17}$$

$$G > B \text{ and } R > B \tag{9.18}$$

④ Use the target point *step* pixels above point P_1 (i, j) as the starting point for scanning, perform a local scan on the pixels within the interval $[j\text{-}step, j]$ on the scanning line's vertical coordinate, and accurately locate the tracking starting point, as shown in Fig. 9.13. When the scanning point $P(i, y)$ satisfies Eq. (9.17), it indicates that this point is a blue point at the bottom slant edge, record this point, and use it as the tracking starting point for the next step.

(2) Determine the placement method of the identification point

As shown in Figs. 9.12 and 9.13, during actual measurement, the identification point may have two placement methods: right corner and left corner, corresponding to different tracking methods. Therefore, before tracking, it is necessary to first determine the current placement method of the identification point and then choose the corresponding tracking direction and method based on the judgment result.

In Fig. 9.13, $P(i, y)$ is the tracking starting point, obtain the pixel point *step* pixels to the right of point P, and perform a line scan operation on the target points within the range of 2 *step* above and below this point. Define a flag variable *flag*, initially set to 0. When the scanning point is a yellow point, i.e., satisfies Eq. (9.18), set *flag* to 1 and continue scanning. If, under the condition of $flag = 1$, a target point satisfying Eq. (9.17) is obtained, it indicates that this point is a blue point at the bottom slant edge of the identification point, stop scanning, and record this point as P_2 $(i + step, yy)$.

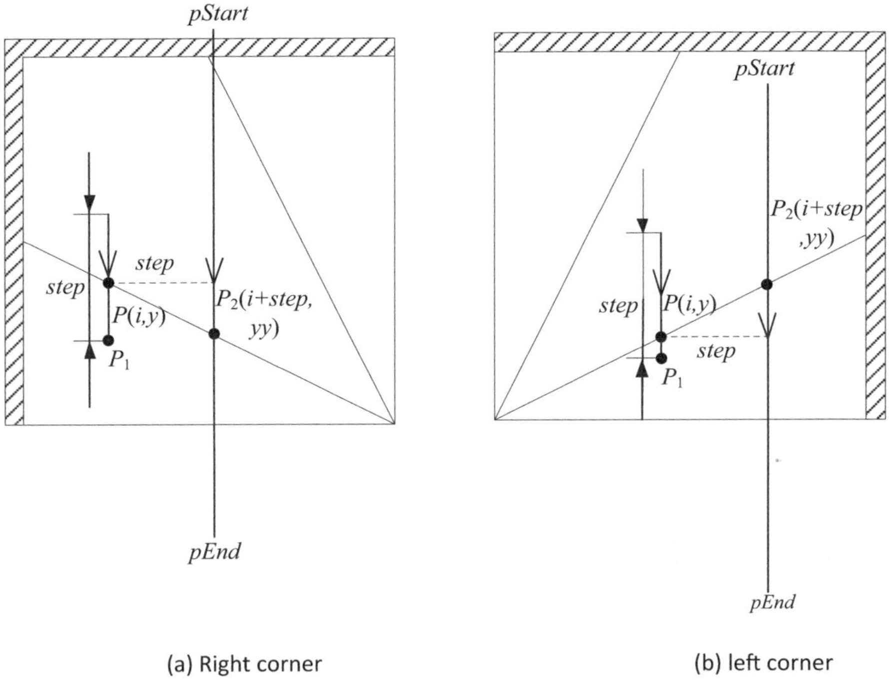

Fig. 9.13 Different placement of markers

As shown in Fig. 9.13a and b, for different placement methods of the identification point, the vertical coordinates *y* and *yy* of boundary points *P* and P_2 satisfy different relational expressions:

In Fig. 9.13a, when the yellow corner point is located at the bottom right corner of the identification point, *y* and *yy* satisfy: $y < yy$;

In Fig. 9.13b, when the yellow corner point is located at the bottom left corner of the identification point, *y* and *yy* satisfy: $y > yy$.

Therefore, this process will use the result of comparing the sizes of *y* and *yy* as the basis for determining the placement method of the identification point.

(3) Counterclockwise tracking of the yellow corner point

For the two different placement methods of the identification point in Fig. 9.13, different tracking algorithms are used.

① The corner point is located at the bottom right corner of the identification point. When the yellow corner point is located at the bottom right corner of the identification point, after obtaining the tracking starting point *P* (*x*, *y*), first track to the right, set $X = x + 1$ as the fixed scanning line, for *Y* belonging to [*y*-*step*, *y* + *step* … Scan the pixels within the interval to detect the boundary points located on the slanted edge at the bottom of the marker point, as shown in Fig. 9.14. The scanning direction is from *pStart* to *pEnd*. Before scanning, first determine the

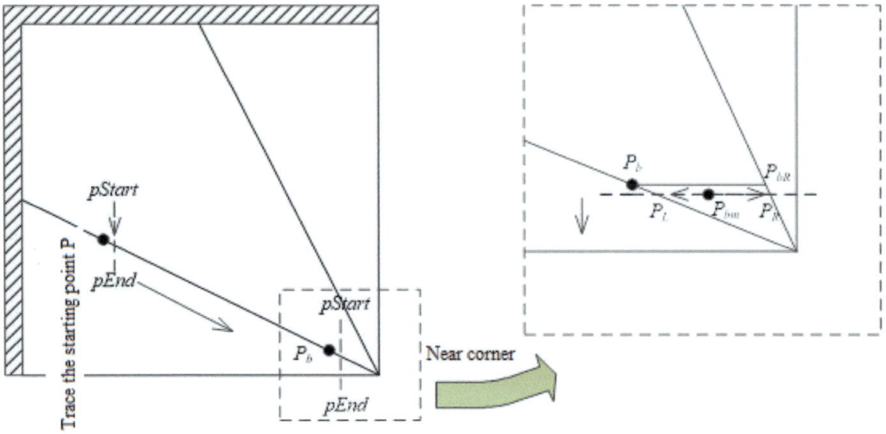

Fig. 9.14 Yellow corner tracking (corner is located at the lower right corner of the marker)

colour component values of the starting point *pStart* and the ending point *pEnd* to ascertain the current position of the scan line.

When the scan line is relatively far from the yellow corner point, *pStart* and *pEnd* are respectively the yellow point and the blue point. Therefore, when the scan point is a blue point, that is, when the *B* component in its *RGB* components is the largest, it is considered the boundary point to be detected. Record this point and mark it with the marking colour F_c of $R = 254$, $G = 0$, $B = 0$, and take this point as the starting point *P*, continuing to track the boundary point to the right of this point.

When the scan line is near the yellow corner point, *pStart* and *pEnd* may not be the yellow point and the blue point. When the *RGB* components of the starting point satisfy Eq. (9.19) or the *RGB* components of the ending point satisfy Eq. (9.20), it indicates that the tracking has reached near the corner point, and no further up and down scanning is performed. Stop tracking to the right, and use the current reference point P_b as the starting point to precisely locate the corner point.

$$\min(R, G, B) < 200 \text{ and } \min(R, G, B) \neq B \tag{9.19}$$

$$\max(R, G, B) > 50 \text{ and } \max(R, G, B) \neq B \tag{9.20}$$

Assume the coordinates of P_b are (bx, by), with $y = by$ as the fixed scan line, scan to the right. When the boundary point P_{bR} on the slanted edge at the top of the marker point is detected, stop scanning and obtain the midpoint P_m (mx, my) between P_b and P_{bR}, and start tracking downwards from point P_m. First, determine whether the point below P_{bm} $(mx, my + 1)$ is a yellow point, i.e., whether it satisfies Eq. (9.18). If it does, then with $y = my + 1$ as the fixed scan line, scan within the range of 2 *step* to the left and right for the boundary points of the yellow slanted edge P_L $(lx, my + 1)$ and P_R $(rx, my + 1)$, and P_L, P_R as the midpoint to continue tracking downwards from the new starting point. Among them, P_L and P_R are determined based on: these two points being blue points, meaning the colour component values of these two points satisfy Eq. (9.19). When P_{bm} $(mx, my + 1)$ is not a yellow point, it indicates that the tracking has reached near a corner point, and no further left or right scanning is performed, stopping the downward tracking, and obtaining the boundary points P_L and P_R from the previous detection, with X values lx and rx, searching on the column where P_{bm} is located, for the pixel point where X belongs to the interval $[lx, rx]$ and the R component reaches its maximum value, considering this point as the yellow corner point to be extracted.

② The corner point is located at the lower left of the marker point. To avoid mistakenly detecting other targets as marker points, the marker point features should be searched as widely as possible to improve the accuracy of marker point detection. Therefore, when the yellow corner point is located at the lower left of the marker point, after obtaining the starting point for tracking, do not directly track to the lower left, but instead track the vertex to be measured counterclockwise as shown in Fig. 9.15. The specific tracking method is as follows:

a. First, track to the right, the tracking method is similar to the right tracking process in ①, the only difference being: in the later stage of tracking, when the starting point of the scan is not a yellow point or the end point is not a blue point, it indicates that the right tracking has reached near the red edge (rather than near the yellow corner point), considering the current reference point as the boundary point P_{ls} at the rightmost side of the bottom slant edge, record this point, and stop tracking to the right.
b. Using P_{ls} as the starting point, search to the left for the pixel point P_{ds} located on the top slant edge of the marker point in the column where P_{ls} is located, i.e., the yellow-blue boundary point, the positioning method of this point is similar to the method of determining P_2 mentioned above.
c. After searching and determining P_{ds} (dsx, dsy), use this point as the starting point to begin tracking downwards, setting $Y = dsy + 1$ as the fixed scan line, scanning the pixel points where X belongs to the interval $[dsx\text{-}step, dsx + step]$, detecting the boundary points located on the top slant edge of the marker point, as shown in Fig. 9.15, with the scanning direction from *pStart* to *pEnd*. This process is similar to the right tracking process in ①, except that the method of determining the starting point *pStart* and the end point *pEnd* before scanning is different. In this process, it is determined by checking whether *pStart* is a blue point and *pEnd* is a yellow point, which is exactly the opposite of the right tracking process in ①. When tracking near the corner point, the method for precise positioning of the corner point is also consistent with the precise positioning process in ①, and will not be repeated here.

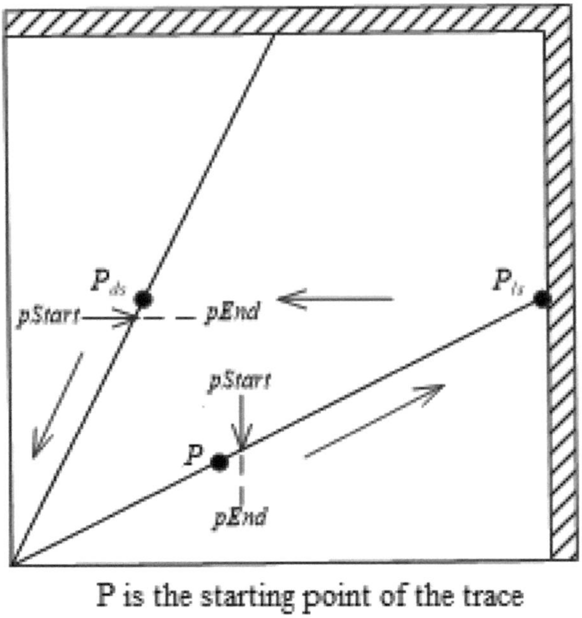

Fig. 9.15 Yellow corner tracking (corner is located in the lower left corner of the marker)

At this point, the detection of a marker point is complete. To avoid interference with the detection of other marker points, obtain the upper, lower, left, and right range of this marker point, and use the rectangle formed by this range as an exclusion area in the subsequent detection process, continuing to scan the image to detect the next marker point.

To implement the aforementioned feature point detection algorithm in programming and conduct practical tests on it, experimental results indicate that the algorithm can accurately extract feature point corners under varying lighting conditions or significant external environmental interference, demonstrating strong versatility. Figure 9.16 shows an example of feature point detection under strong light conditions, where the red car might cause significant interference in extracting the feature points. The algorithm successfully eliminates external environmental interference and accurately detects two feature points in different placements within the image.

9.7 Manual Target Selection

In addition to automatic detection of feature points, the system also includes a method for manually clicking on target locations. On a PC platform, this can be done by mouse click. On mobile devices, users can select the target point to be tested by finger touch and obtain its position information in the image. Unlike automatic extraction, manual

Fig. 9.16 Example of marker detection (taken at 14:00)

extraction does not require placing feature points at the target location beforehand, which reduces operation time to some extent. Users can select any point on the test plane as the measurement target with a simple finger tap, offering great flexibility. The following describes the main process of implementing the manual selection function on the Android platform.

To implement the manual point selection function on an image, use the setOnTouchListener (View.OnTouchListener listener) method to add a touch listener to the image display control, and override the onTouch (View view, MotionEvent event) method in the OnTouchListener class to add custom response events for actions such as ACTION_DOWN (finger press), ACTION_MOVE (finger move), and ACTION_UP (finger lift). The MotionEvent object event carries the position information of the current touch event, which can be obtained through the get*X*() and get*Y*() methods. The *X* and *Y* values obtained through these methods are the coordinates of the touch event relative to the top-left corner of the control, which can be converted into position coordinates in the image pixel coordinate system using the size ratio between the image and the control.

9.8 Distance Measurement Analysis

Using the method from the previous section, after detecting two target points on the image P_1 (x_1, y_1) and P_2 (x_2, y_2), and substituting them along with the homography matrix $\boldsymbol{H}^{-1}$ obtained from scale detection into Eq. (9.4), the image coordinates are restored to world coordinates, assuming the restored results are P_1 (X_1, Y_1) and P_2 (X_2, Y_2). The distance between the two points to be measured can be expressed by their Euclidean distance, as shown in Eq. (9.21).

$$d = \sqrt{(X_1 - X_2)^2 + (Y_1 - Y_2)^2} \tag{9.21}$$

After developing the above distance measurement algorithm, experiments are conducted to verify the algorithm and analyse the factors affecting the accuracy of distance measurement.

9.8.1 The Impact of Perspective Distortion on Measurement Accuracy

Figures 9.17 and 9.18 respectively show the measurement of the same scene taken by a camera at different angles from the same position. In Fig. 9.17, the degree of perspective distortion of the test plane is greater than in Fig. 9.18.

The world coordinates of the four feature points in the two images are calculated, with results shown in Table 9.2. In the Tables 9.1, 9.2, 9.3, and 9.4 respectively represent the feature points located at the bottom left, bottom right, top left, and top right positions in the image.

From Table 9.2, it can be seen that the positioning accuracy of the identification points in Fig. 9.18 is higher than that in Fig. 9.17. Further distance measurements of the two parallel lines formed by the identification points are conducted, with results shown in Table 9.3.

From Table 9.3, it is evident that the distance measurement accuracy in Fig. 9.18 is higher than that in Fig. 9.17, and the values are relatively stable. Therefore, when the degree of perspective distortion of the plane to be measured is small, the accuracy of point positioning and distance measurement is higher. When the imaging plane is parallel to the plane to be measured, i.e., when the camera's axis is perpendicular to the plane to be measured, the degree of perspective distortion is minimal, and at this time, the accuracy of point positioning and distance measurement can reach its best.

Fig. 9.17 Measurement image with large perspective distortion (shooting angle: 70.97°)

9.8.2 The Impact of the Distance Between Target Points and Calibration Points on Distance Measurement Accuracy

Figures 9.19 and 9.17 depict the same measurement scenario, but in Fig. 9.19, the four calibration points used are the four vertices of the rectangular area in the image, reducing the distance between the target point to be measured and the calibration points compared to Fig. 9.17.

In Fig. 9.19, the world coordinates of the four identification points are shown in Table 9.4, and the distances between the identification points are shown in Table 9.5. Comparing the data obtained from Fig. 9.17 in Tables 9.2 and 9.3, it can be seen that when the distance between the target to be measured and the calibration points is small, the positioning accuracy of the target point and the distance measurement accuracy are higher.

Through the analysis of factors affecting distance measurement accuracy, it is known that when the target to be measured is close to the calibration points, the measurement error is smaller, and there are no specific requirements for the image acquisition method. When the target to be measured is far from the calibration points, in order to obtain more accurate measurement results, the perspective distortion of

Fig. 9.18 Measurement image with low perspective distortion (shooting angle: 85.84°)

Table 9.2 Comparison of world coordinates and positioning errors of feature points

	Marker	Actual coordinate		Calculated coordinate		Average of X and Y error (%)
		X	Y	X	Y	
Figure 9.17	1	−110	−260	−101.19	−268.02	5.55
	2	190	−260	209.11	−269.37	6.83
	3	−110	−510	−95.75	−532.97	8.73
	4	190	−510	224.15	−551.69	13.07
Figure 9.18	1	−110	−260	−106.68	−258.41	2.17
	2	190	−260	202.18	−256.58	3.86
	3	−110	−510	−101.00	−501.00	4.97
	4	190	−510	211.66	−512.99	5.99

Table 9.3 Results of distance measurements under different degrees of perspective distortion

	Measuring object	Actual distance (cm)	Measuring distance (cm)	Relative error (%)
Figure 9.17	1	300	310.31	3.44
	2	300	320.45	6.82
Figure 9.18	1	300	298.88	0.37
	2	300	299.90	0.03

Table 9.4 World coordinates and positioning errors of identification points after moving calibration points closer

	Marker	Actual coordinate		Calculated coordinate		Average of X and Y error (%)
		X	Y	X	Y	
Figure 9.19	1	−100	−40	−102.18	−39.22	2.06
	2	200	−40	205.98	−39.34	2.32
	3	−100	−290	−99.69	−289.69	0.21
	4	200	−290	205.80	−300.38	3.24

Fig. 9.19 Measurement image after the marker point is moved closer

Table 9.5 Distance measurement results between identification points after moving calibration points closer

	Measuring object	Actual distance (cm)	Measuring distance (cm)	Relative error (%)
Figure 9.19	1	300	307.64	2.54
	2	300	305.57	1.86

the plane to be measured should be minimized by increasing the shooting angle when acquiring measurement images.

Additionally, in the above experiments, the distance between the camera and the target object is within 10m. If the distance between the camera and the target object is relatively far, such as 50 m, the actual distance represented by one pixel is larger, and the measurement error will also increase.

9.9 Area Measurement Algorithm

On the detection platform, the image area measurement method calculates the number of pixels within a region and combines it with a calibration value representing the area size of one pixel. This measurement method requires the camera's optical axis to be perpendicular to the plane to be measured. In portable measurement situations, since the camera's position is not fixed and the measurement field of view is large, only several contour points of the region to be measured can be selected. The area of the minimum enclosing convex polygon of the contour points is calculated to estimate the area of the region. Specific operations include: ① obtaining the contour point set of the region to be measured; ② performing minimum convex polygon fitting on the obtained contour point set; ③ calculating the area of the fitted convex polygon.

9.9.1 Obtaining the Contour Point Set of the Region to Be Measured

First, place a ruler used for camera calibration on the plane to be measured, then place several identification points around the region to be measured. Use the identification point detection algorithm described in Sect. 9.6 to extract the positions of the identification points, or manually select several region contour points as described in Sect. 9.7. After obtaining the image coordinates of the contour points, substitute the homography matrix $\boldsymbol{H}^{-1}$obtained from ruler detection and the image coordinates of each contour point into Eq. (9.4) to complete the conversion from image coordinates to world coordinates. Then, perform minimum convex polygon fitting on the obtained contour point set.

9.9.2 Minimum Convex Polygon Fitting

A convex hull is a concept in computational geometry. For a set of points on a two-dimensional plane, the convex hull is the smallest convex polygon formed by connecting the outermost points, such that it can contain all the points in the set. The process of performing minimum convex polygon fitting on a plane point set can be regarded as the process of finding the convex hull.

Research on the convex hull of plane point sets began early, and currently, experts and scholars have proposed numerous algorithms for finding convex hulls, such as the divide-and-conquer algorithm, Jarvis march, and Graham scan. Among them, the Graham scan is a commonly used convex hull detection algorithm and is considered the best algorithm for constructing convex hulls. Therefore, this system uses this algorithm to find the convex hull of the contour point set.

The Graham scan method was invented by mathematician Graham in 1972. This method constructs the convex hull of a set of points on a plane by determining whether the circuit formed by any three points on the plane is a left turn or a right turn. It mainly includes two steps: angle sorting and angle scanning.

(1) Angle Sorting

First, select the point with the smallest y-coordinate from the plane point set P. If there are multiple such points, select the point with the smallest x-coordinate among them and denote this point as p_0. Then, use the point p_0 as the origin to perform a coordinate transformation on the point set P. For the transformed points, calculate their angles in polar coordinates with p_0 as the origin. Then sort the points in $P - p_0$ in ascending order. If $P - p_0$ contains two or more points with the same angle, select the point closest to p_0 first. Denote the sorted set of points as $P' = \{p_1, p_2, \ldots, p_{n-1}\}$, where p_1 and p_{n-1} represent the points forming the minimum and maximum angles with p_0, respectively, as shown in Fig. 9.20.

(2) Angle Scanning

Initialise the stack as $H(P) = \{p_{n-1}, p_0\}$, with p_{n-1} as the top element of the stack. Then, start scanning in order of increasing polar angle, beginning from p_0 and ending

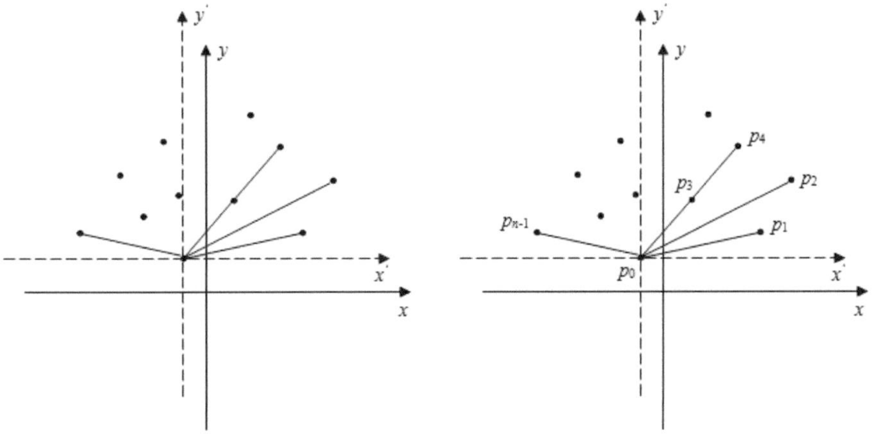

(a) Re-determine the coordinate system

(b) angular order

Fig. 9.20 Amplitude ordering of point set P

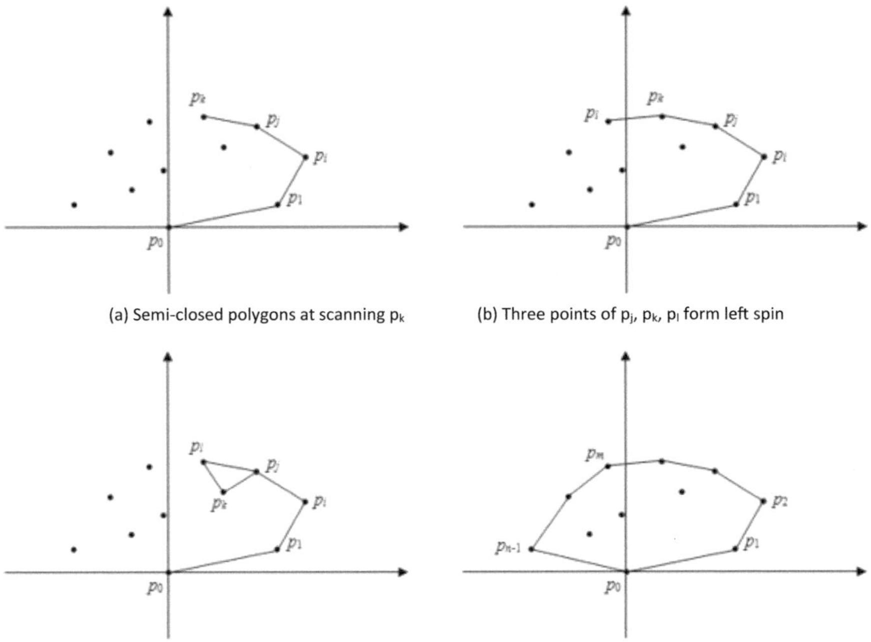

(a) Semi-closed polygons at scanning p_k (b) Three points of p_j, p_k, p_l form left spin

(c) Three points of p_j, p_k, p_l form right spin (d) Complete convex hull

Fig. 9.21 Schematic diagram of convex hull calculation by Graham scanning method

at p_{n-1}. If at a certain moment, the elements in the stack are $H(p) = \{p_0, p_1, \ldots, p_i, p_j, p_k\}$, with p_k as the top element, then the elements in the stack form a closed convex polygon. Suppose the point being scanned at a certain moment is p_l. If p_j, p_k, p_l form a left turn path, then the edge formed by the path p_j, p_k, p_l is a convex edge, at which point p_k p_l Push p_l into the stack, which forms an edge in the convex polygon, and then scan the next point. If the path formed by the three points p_j, p_k, p_l is a right turn, then p_k is a point inside the convex hull, and p_k should be popped from the stack. At this point, the scan line is still at p_l, and then process and judge the three points p_i, p_j, p_k until it is determined that the current points in the stack are vertices of a convex polygon, as shown in Fig. 9.21.

9.9.3 Calculation of Polygon Area

After constructing the convex hull of the contour point set of the region to be tested, calculate the area of the resulting convex polygon. The polygon to be calculated may have any complex shape and is irregular in most cases. Therefore, the main task of this process is to solve the area of an irregular polygon. This system uses the vertex coordinates to calculate the area of the polygon.

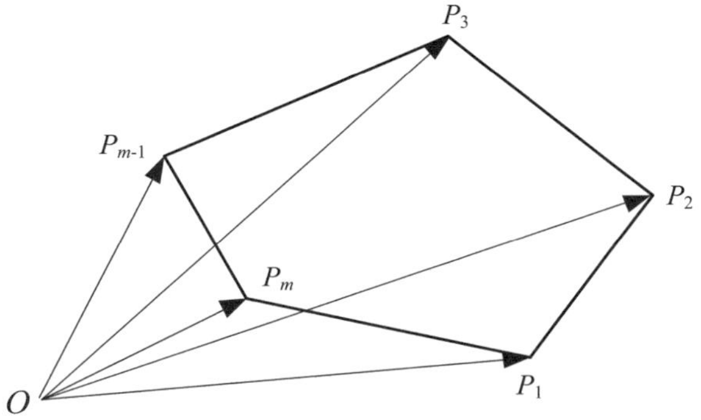

Fig. 9.22 Polygon vector diagram

Let Ω be an m-sided polygon (as shown in Fig. 9.22), with vertices P_k ($k = 1, 2, \ldots, m$) arranged in the positive direction along the boundary, $P_{m+1} = P_1$, with coordinates as follows:

$$(x_1, y_1), (x_2, y_2), \ldots, (x_m, y_m) \tag{9.22}$$

As shown in Fig. 9.22, establish the vector diagram of the polygonal region Ω. In this diagram, the origin of coordinates and any two adjacent vertices of the polygon Ω form a triangle. The formed triangle $\triangle OP_k P_{k+1}$ can be divided into two categories: one type of triangle contains components of Ω, such as $\triangle OP_1P_2$, and the other type does not contain components of Ω, such as $\triangle OP_{m\text{-}1}P_m$. By summing the areas of all triangles of the first type and subtracting the sum of the areas of all triangles of the second type, the area of the polygon Ω can be obtained.

The area of a triangle can be obtained by the cross product of two plane vectors formed by its three vertices. In the triangle $\triangle OP_kP_{k+1}$, when $\overrightarrow{P_kP_{k+1}}$ in the positive direction, the triangle belongs to the first type of triangle, and the cross product value is positive; otherwise, the triangle belongs to the second type of triangle, and the cross product value is negative. Therefore, by calculating the cross product of the two plane vectors formed by the vertices, the area of all triangles can be determined. The area calculation result for the first type of triangle will be positive, while for the second type, it will be negative. Summing the area calculation results of all triangles gives the area of the polygon to be measured. Based on this principle, the area formula for any polygon can be derived through the following process.

Let vector

$$\overrightarrow{OP_k} = \{x_k, y_k, 0\}\overrightarrow{OP_{k+1}} = \{x_{k+1}, y_{k+1}, 0\} \tag{9.23}$$

The cross product of vectors is calculated as:

$$\overrightarrow{OP_k} \times \overrightarrow{OP_{k+1}} = \{0, 0, x_k y_{k+1}, -x_{k+1} y_k\} \tag{9.24}$$

Therefore, the area formula for any polygon is:

$$S_n = \sum_{k=1}^{m} S_{\Delta OP_k P_{k+1}} = \frac{1}{2}\left|\sum_{k=1}^{m} \overrightarrow{OP_k} \times \overrightarrow{OP_{k+1}}\right| = \frac{1}{2}\sum_{k=1}^{m}(x_k y_{k+1} - x_{k+1} y_k) \quad (9.25)$$

9.9.4 Measurement Example

Figure 9.23 shows an example of measuring the area of a quadrilateral, where four marker points form a quadrilateral with an actual area of 30,000.00 cm^2. In this figure, both the scale and the four marker points are successfully detected, and the measured area is 30,013.89 cm^2, with a relative error of 0.05%. The measurement result is quite accurate. Similarly, the main source of area measurement error is the positioning error of the target points to be measured. Therefore, the discussion of area measurement error is consistent with that of distance measurement and will not be repeated here.

Fig. 9.23 Example of area measurement

References

1. Ouyang Di. Development of a Geometric Parameter Measurement System Based on Machine Vision [D]. Beijing: China Agricultural University, 2013.
2. Liu Yang. Research on High-Speed Image Detection Algorithm for Target Objects in Natural Environments [D]. Beijing: China Agricultural University, 2014.

Chapter 10
Binocular Vision Measurement

This chapter introduces the hardware components, fundamental principles, calibration techniques, and 3D reconstruction of binocular vision measurement. The calibration methods will cover the commonly used direct linear calibration method and Zhang Zhengyou's calibration method, as well as the parameter conversion between the two. Finally, the calibration methods and the measurement accuracy of binocular vision will be discussed through actual measurements [1].

As shown in Fig. 10.1, the functional modules of the binocular vision measurement system include: left and right vision cameras, computer, tripod, calibration device, light source, etc. The functions of each module are shown in Table 10.1.

The processing of the binocular vision system can be summarised as binocular image acquisition, camera calibration, target point acquisition, and 3D reconstruction of target points.

10.1 Structure of the Binocular Vision System

Generally, the structure of the binocular vision system can be classified into parallel stereo vision models and convergent stereo vision models, depending on whether the optical axes of the cameras are parallel. The choice can be made according to the measurement scene and the requirements for measurement accuracy.

10.1.1 Parallel Stereo Vision Model

A parallel stereo vision model refers to a binocular vision system where the optical axes of two cameras are placed parallel, making the convergence distance at infinity. The simplest stereo imaging system model is the parallel stereo vision model, which

© Chemical Industry Press 2026

B. Chen and S. Chen, *Machine Vision Technology*, Advanced and Intelligent Manufacturing in China, https://doi.org/10.1007/978-981-95-1184-6_10

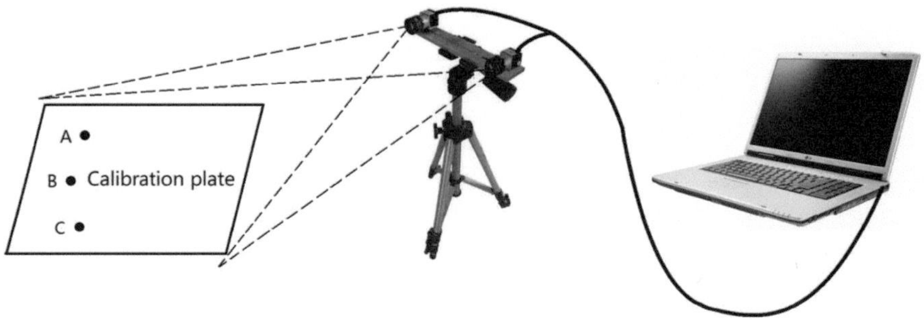

Fig. 10.1 Composition of binocular vision measurement system

Table 10.1 Functions of each part of the binocular vision measurement system

Name	Function
Left and right vision camera	Used to capture left and right visual images
Computer	Camera calibration, synchronous image acquisition, image data processing, 3D reconstruction, data preservation
Calibration device (Calibration stand or black and white checkerboard)	Carry out camera calibration and obtain camera internal and external parameters
Tripod	Fix the camera, adjust its height and angle
Illuminant	Ensure clear image capture (can be omitted if necessary)

is referred to as such when two identical cameras are placed parallel, as shown in Fig. 10.2.

The schematic diagram is shown in Fig. 10.3. Assume that cameras C_1 and C_2 are identical, meaning their internal parameters are completely the same. The x-axes of the two cameras coincide, and the y-axes are parallel. Therefore, by translating one of the cameras along its x-axis by a certain distance, it can completely coincide with the other camera. As shown in Fig. 10.3, $P(x_1, y_1, z_1)$ is any point in space, and through the optical imaging process of the left and right cameras, the imaging points on the left and right projection planes are p_1 and p_2, respectively. According to the imaging principle, the vertical coordinates of points p_1 and p_2 are equal, and the difference in horizontal coordinates is the distance between the two imaging coordinate systems.

In the parallel stereo vision model, assume the distance between the two imaging coordinate systems, i.e., the difference in the horizontal coordinate of a certain point, is *b*. The coordinate system of C_1 is $O_1\, x_1\, y_1\, z_1$, and the coordinate system of C_2 is $O_2\, x_2\, y_2\, z_2$. Thus, the coordinates of any point *P* in space in the C_1 coordinate system are (x_1, y_1, z_1), and in the C_2 coordinate system are $(x_1–\mathrm{b}, y_1, z_1)$. Therefore, if the internal parameters of the camera are known, the three-dimensional coordinates of point *P* can be derived as shown in Eq. (10.1).

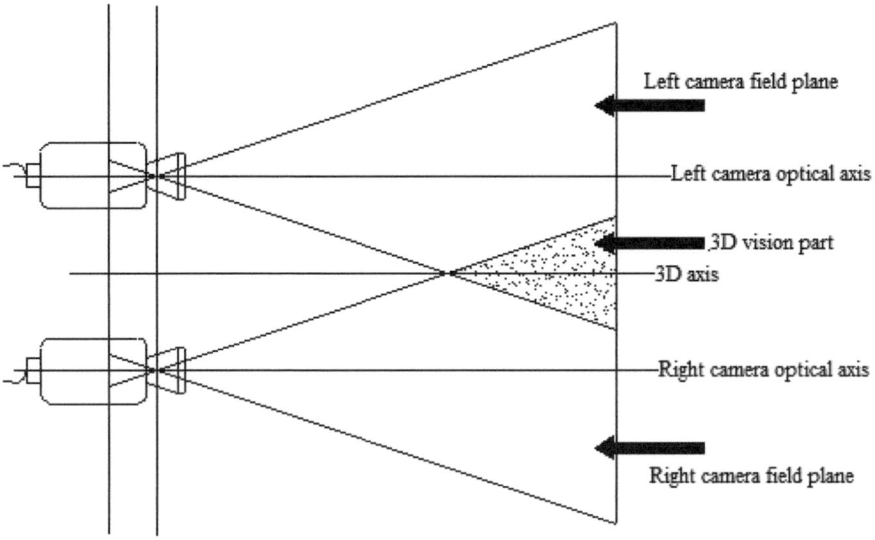

Fig. 10.2 Parallel stereo vision model

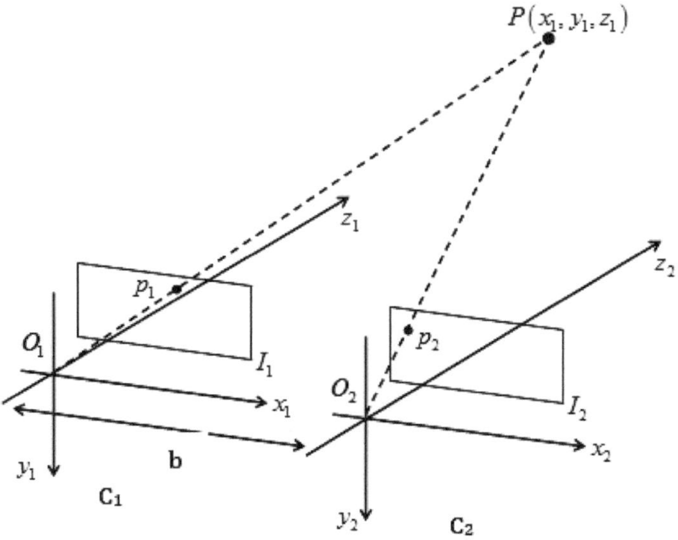

Fig. 10.3 Schematic diagram of parallel stereo vision model

$$\begin{cases} x_1 = \frac{b(u_1 - u_0)}{u_1 - u_2} \\ y_1 = \frac{ba_x(v_1 - v_0)}{a_y(u_1 - u_2)} \\ z_1 = \frac{ba_x}{u_1 - u_2} \end{cases} \quad (10.1)$$

Here, u_0, v_0, a_x, a_y are the internal parameters of the camera. (u_1, v_1), (u_2, v_2) are the image coordinates of p_1 and p_2, respectively. It can be seen that from p_1 and p_2 image coordinates (u_1, v_1) and (u_2, v_2), the three-dimensional coordinates of the spatial point $P(x_1, y_1, z_1)$ can be determined.

In Eq. (10.1), b is the baseline length, u_1–u_2 is called the disparity. Disparity refers to the difference in the position of the projection points of point P in the left and right images due to the different positions of the two cameras in a binocular vision system. As seen from Eq. (10.1), the further the distance of point P (i.e., z_1 is larger), the smaller the disparity. Therefore, when point P approaches infinity, O_1 P and O_2 P tend to be parallel, and the disparity tends to zero.

10.1.2 Convergent Stereo Vision Model

In the parallel stereo vision model, the optical axes of the cameras are parallel, thus the geometric relationship of imaging is also the simplest. However, in reality, it is difficult to achieve an absolutely parallel stereo camera system because, during the actual installation of cameras, we cannot see the optical axes of the cameras, and therefore cannot adjust the relative positions of the cameras to the ideal situation shown in Fig. 10.3. Generally, two arbitrarily placed cameras as shown in Fig. 10.4 are used to form a binocular stereo vision system.

The principle of the convergent stereo vision model is shown in Fig. 10.5.

In the convergent stereo vision model, it is assumed that p_1 and p_2 are corresponding points of the same spatial point P on the left and right images, respectively.

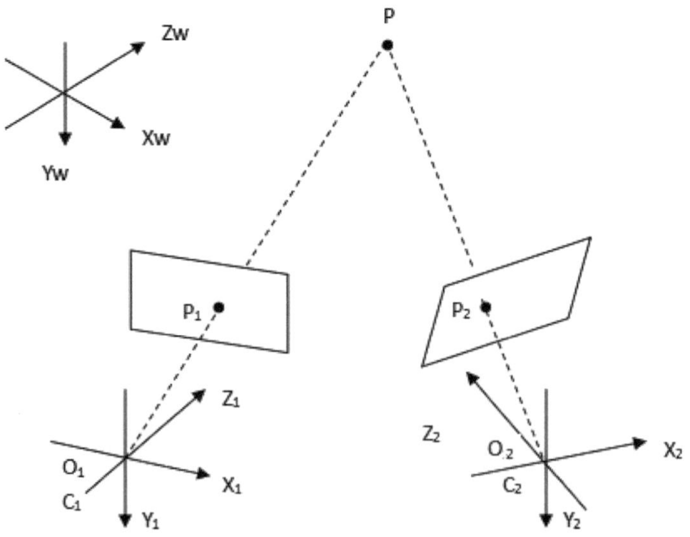

Fig. 10.4 Schematic diagram of convergent stereo vision mode

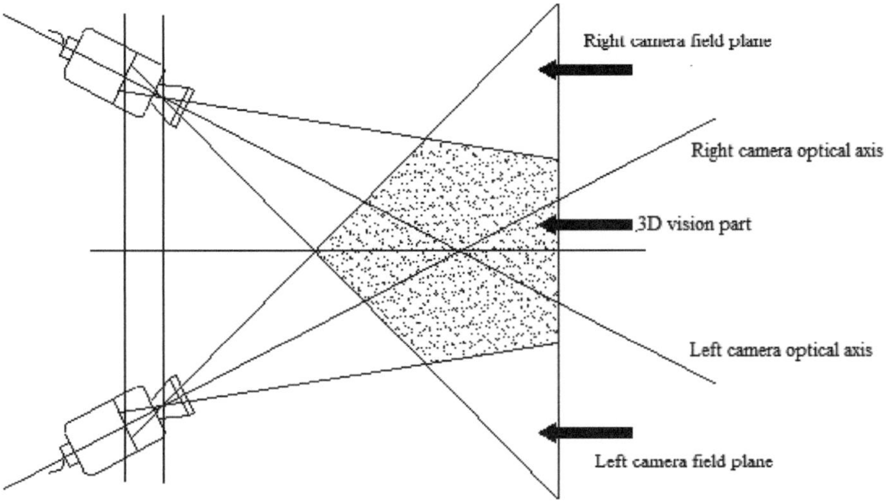

Fig. 10.5 Convergent stereo vision model

Moreover, it is assumed that the calibration results of cameras C_1 and C_2 are known, i.e., their projection matrices are known as M_1 and M_2, respectively. Thus, the relationship between the spatial point and the image points in the left and right images is given by Eqs. (10.2) and (10.3).

$$Z_{c1}\begin{bmatrix} u_1 \\ v_1 \\ 1 \end{bmatrix} = M_1\begin{bmatrix} X \\ Y \\ Z \\ 1 \end{bmatrix} = \begin{bmatrix} m_{11}^1 & m_{12}^1 & m_{13}^1 & m_{14}^1 \\ m_{21}^1 & m_{22}^1 & m_{23}^1 & m_{24}^1 \\ m_{31}^1 & m_{32}^1 & m_{33}^1 & m_{34}^1 \end{bmatrix}\begin{bmatrix} X \\ Y \\ Z \\ 1 \end{bmatrix} \tag{10.2}$$

$$Z_{c2}\begin{bmatrix} u_2 \\ v_2 \\ 1 \end{bmatrix} = M_2\begin{bmatrix} X \\ Y \\ Z \\ 1 \end{bmatrix} = \begin{bmatrix} m_{11}^2 & m_{12}^2 & m_{13}^2 & m_{14}^2 \\ m_{21}^2 & m_{22}^2 & m_{23}^2 & m_{24}^2 \\ m_{31}^2 & m_{32}^2 & m_{33}^2 & m_{34}^2 \end{bmatrix}\begin{bmatrix} X \\ Y \\ Z \\ 1 \end{bmatrix} \tag{10.3}$$

where, $(u_1, v_1, 1)$ and $(u_2, v_2, 1)$ are the homogeneous coordinates of points p_1 and p_2 in the image coordinate system respectively; $(X, Y, Z, 1)$ is the homogeneous coordinate of point P in the world coordinate system; m_{ij}^k ($k = 1, 2$; $i = 1,\ldots, 3$; $j = 1,\ldots, 4$.) are the elements in the i-th row and j-th column of M_k respectively. According to the linear model formula (9.1) introduced in Chap. 9 on monocular vision measurement, Z_{c1} and Z_{c2} can be eliminated in the above formula, resulting in four linear equations about X, Y, Z as shown in formulas (10.4) and (10.5).

$$\begin{cases} \left(u_1 m_{31}^1 - m_{11}^1\right)X + \left(u_1 m_{32}^1 - m_{12}^1\right)Y + \left(u_1 m_{33}^1 - m_{13}^1\right)Z = m_{14}^1 - u_1 m_{34}^1 \\ \left(v_1 m_{31}^1 - m_{21}^1\right)X + \left(v_1 m_{32}^1 - m_{22}^1\right)Y + \left(v_1 m_{33}^1 - m_{23}^1\right)Z = m_{24}^1 - v_1 m_{34}^1 \end{cases} \tag{10.4}$$

$$\begin{cases} (u_2m_{31}^2 - m_{11}^2)X + (u_2m_{32}^2 - m_{12}^2)Y + (u_2m_{33}^2 - m_{13}^2)Z = m_{14}^2 - u_{21}m_{34}^2 \\ (v_2m_{31}^2 - m_{21}^2)X + (v_2m_{32}^2 - m_{22}^2)Y + (v_2m_{33}^2 - m_{23}^2)Z = m_{24}^2 - v_2m_{34}^2 \end{cases} \tag{10.5}$$

The geometric meaning of formulas (10.4) and (10.5) is the line passing through $o_1 p_1$ and $o_2 p_2$. Since the spatial point $p(X, Y, Z)$ is the intersection of $o_1 p_1$ and $o_2 p_2$, it must satisfy both equations simultaneously. Therefore, the coordinates (X, Y, Z) of the spatial point P can be obtained by solving these two equations together. However, in practical applications, to reduce errors, the least squares method is usually used to determine the three-dimensional coordinates of the spatial point.

The convergent stereo vision model can adjust the angle of the camera's optical axis to allow the binocular vision system to obtain the maximum field of view without affecting the accuracy of the results. Therefore, the convergent stereo vision model is generally adopted.

10.2 Camera Calibration

Camera calibration refers to the process of establishing the relationship between the pixel positions in a camera image and the positions of target points. Based on the camera model, the camera's parameters are solved using the image coordinates and world coordinates of known feature points. This is the primary issue that must be tackled in computer stereo vision research and a critical step in binocular vision 3D reconstruction. The accuracy of this process directly affects the measurement precision of the stereo vision system, making the calibration of stereo cameras indispensable. This section introduces the direct linear calibration method and the Zhang Zhengyou calibration method.

Camera parameters are determined by the camera's position, attribute parameters, and imaging model, including intrinsic and extrinsic parameters. The intrinsic parameters describe the relationship between the camera coordinate system and the ideal coordinate system, including attributes such as focal length, optical centre, and distortion factors. The extrinsic parameters represent the camera's position and orientation in the world coordinate system. The extrinsic parameters include the rotation matrix R and the translation matrix T, which describe the transformation relationship between the camera and the world coordinate system. The process of obtaining the intrinsic and extrinsic parameters through experiments and calculations is known as camera calibration.

10.2.1 Direct Linear Calibration Method

In the early 1970s, Abdel-Aziz and Karara proposed the direct linear transformation (DLT) method for camera calibration. This method disregards errors arising from camera distortion and directly employs a linear imaging model to solve a system of linear equations to derive the camera's parameters.

The advantage of the DLT method is its fast computation speed, simplicity, and ease of implementation. However, its disadvantage is that it does not account for lens distortion, making it unsuitable for lenses with significant distortion coefficients, as this would result in substantial errors.

The DLT calibration method requires placing a specially designed cubic calibration template in front of the camera to be calibrated, where the positions of the calibration points on the template relative to the world coordinate system are known. Thus, the camera's parameters can be obtained using the linear model described in Sect. 9.2.2.

First, the algorithm for obtaining the projection matrix M from the stereo calibration reference image is introduced, where Eq. (9.1) can be rewritten as Eq. (10.6).

$$Z_c\begin{bmatrix} u_i \\ v_i \\ 1 \end{bmatrix} = \begin{bmatrix} m_{11} & m_{12} & m_{13} & m_{14} \\ m_{21} & m_{22} & m_{23} & m_{24} \\ m_{31} & m_{32} & m_{33} & m_{34} \end{bmatrix}\begin{bmatrix} X_{wi} \\ Y_{wi} \\ Z_{wi} \\ 1 \end{bmatrix} \tag{10.6}$$

In this equation, (X_{wi}, Y_{wi}, Z_{wi}) are the coordinates of the i-th point in space; (u_i, v_i) are the image coordinates of the i-th point; m_{ij} are the elements of the i-th row and j-th column of the projection matrix M for any point in space. From Eq. (10.6), three sets of linear equations can be derived, as shown in Eq. (10.7).

$$\begin{cases} Z_c u_i = m_{11}X_{wi} + m_{12}Y_{wi} + m_{13}Z_{wi} + m_{14} \\ Z_c v_i = m_{21}X_{wi} + m_{22}Y_{wi} + m_{23}Z_{wi} + m_{24} \\ Z_c = m_{31}X_{wi} + m_{32}Y_{wi} + m_{33}Z_{wi} + m_{34} \end{cases} \tag{10.7}$$

By eliminating Z_c from the above equations, two linear equations regarding m_{ij} are obtained.

This equation indicates that if there are n calibration points in three-dimensional space, where the spatial coordinates of each calibration point are (X_{wi}, Y_{wi}, Z_{wi}), the image coordinates are (u_i, v_i) $(i = 1, \ldots, n)$, then $2n$ linear equations regarding the elements of the M matrix can be obtained, and these $2n$ linear equations can be expressed in the matrix form as shown in Eqs. (10.8) and (10.9).

$$\begin{cases} X_{wi}m_{11} + Y_{wi}m_{12} + Z_{wi}m_{13} + m_{14} - u_iX_{wi}m_{31} - u_iY_{wi}m_{32} - u_iZ_{wi}m_{33} = u_im_{34} \\ X_{wi}m_{21} + Y_{wi}m_{22} + Z_{wi}m_{23} + m_{24} - v_iX_{wi}m_{31} - v_iY_{wi}m_{32} - v_iZ_{wi}m_{33} = v_im_{34} \end{cases} \tag{10.8}$$

$$
\begin{bmatrix}
X_{w1} & Y_{w1} & Z_{w1} & 1 & 0 & 0 & 0 & 0 & -u_1X_{w1} & -u_1Y_{w1} & -u_1Z_{w1} \\
0 & 0 & 0 & 0 & X_{w1} & Y_{w1} & Z_{w1} & 1 & -v_1X_{w1} & -v_1Y_{w1} & -v_1Z_{w1} \\
 & & & & \cdots & \cdots & \cdots & \cdots & & & \\
 & & & & \cdots & \cdots & \cdots & \cdots & & & \\
X_{wn} & Y_{wn} & Z_{wn} & 1 & 0 & 0 & 0 & 0 & -u_nX_{wn} & -u_nY_{wn} & -u_nZ_{wn} \\
0 & 0 & 0 & 0 & X_{wn} & Y_{wn} & Z_{wn} & 1 & -v_nX_{wn} & -v_nY_{wn} & -v_nZ_{wn}
\end{bmatrix}
\begin{bmatrix} m_{11} \\ m_{12} \\ m_{13} \\ m_{14} \\ m_{21} \\ m_{22} \\ m_{23} \\ m_{24} \\ m_{31} \\ m_{32} \\ m_{33} \end{bmatrix}
= \begin{bmatrix} u_1m_{34} \\ u_1m_{34} \\ \cdots \\ \cdots \\ u_nm_{34} \\ v_nm_{34} \end{bmatrix} \tag{10.9}
$$

From Eq. (10.8), it can be seen that multiplying the M matrix by any non-zero constant does not affect the relationship between (X_{wi}, Y_{wi}, Z_{wi}) and (u_i, v_i), therefore, assume $m_{34} = 1$, thus obtaining $2n$ linear equations regarding the other elements of the M matrix, where the linear equations contain 11 unknowns, and the unknowns are represented as a vector, i.e., an 11-dimensional vector m, simplifying Eq. (10.9) to Eq. (10.10).

$$
Km = U \tag{10.10}
$$

where, K is the $2n \times 11$ matrix on the left side of Eq. (10.9); U is the $2n$-dimensional vector on the right side of Eq. (10.9); K, U are known vectors. When $2n > 11$, the above linear equations can be solved using the least squares method as:

$$
m = (K^TK)^{-1}K^TU \tag{10.11}
$$

The m vector and $m_{34} = 1$ constitute the M matrix to be solved. From Eqs. (10.6) to (10.11), it can be seen that if at least 6 feature points in space and their corresponding image point coordinates are known, the projection matrix M can be determined. Generally, more than 8 known points are selected on the calibration reference object, so that the number of equations far exceeds the number of unknowns, thereby reducing the error caused by solving with the least squares method.

10.2.2 Zhang's Calibration Method

Zhang's Calibration Method, also known as the Zhang method, is a planar calibration technique proposed by Dr. Zhengyou Zhang of Microsoft Research in 1998, which is between traditional calibration methods and self-calibration methods. It avoids the high equipment requirements and cumbersome operation of traditional calibration methods, while offering higher accuracy and better robustness than self-calibration. The main steps of this method are as follows:

① Print a black and white chessboard pattern and attach it to a rigid plane as a calibration board;
② Move the calibration board or the camera, and take several photos from different angles (theoretically, the more photos, the smaller the error);
③ Detect the corner points in each photo to determine the image coordinates and actual coordinates of the corner points.
④ Without considering radial distortion, i.e., using the linear model of the camera, obtain the internal parameters of the camera and the external parameters of the first image by solving linear equations based on the orthogonality of the rotation matrix.
⑤ Estimate the radial distortion coefficients of the camera using the least squares method.
⑥ Optimise the internal and external parameters according to the criterion of minimising the reprojection error.

The basic principles of the above steps are introduced below.

(1) Calculate the initial values of internal and external parameters Unlike the direct linear calibration method, which obtains the projection matrix M as the calibration result by solving a set of linear equations, the Zhang Zhengyou calibration method results in the internal and external parameters of the camera, as shown in Eq. (10.12).

$$\boldsymbol{A} = \begin{bmatrix} \alpha & \gamma & u_0 \\ 0 & \beta & v_0 \\ 0 & 0 & 1 \end{bmatrix}, \boldsymbol{R} = \begin{bmatrix} r_{11} & r_{12} & r_{13} \\ r_{21} & r_{22} & r_{23} \\ r_{31} & r_{32} & r_{33} \end{bmatrix}, \boldsymbol{T} = \begin{bmatrix} t_1 & t_2 & t_3 \end{bmatrix}^T \tag{10.12}$$

where $\boldsymbol{A}$ is the internal parameter matrix of the camera; $\alpha = f/dx$, $\beta = f/dy$, f is the focal length, dx and dy are the width and height of the pixels, respectively; γ represents the deviation of the pixel scale in the x and y directions, and if this parameter is not considered, it can be set to $\gamma = 0$; (u_0, v_0) is the principal point; R is the external rotation matrix, and T is the translation vector.

The basic principle of the Zhang Zhengyou calibration method is explained below. According to the pinhole imaging principle, the homogeneous transformation from world coordinate points to ideal pixel points is shown in Eq. (10.13).

$$s\begin{bmatrix} u \\ v \\ 1 \end{bmatrix} = \boldsymbol{A}\begin{bmatrix} \boldsymbol{R} & t \end{bmatrix}\begin{bmatrix} X_W \\ Y_W \\ Z_W \\ 1 \end{bmatrix} = \boldsymbol{A}\begin{bmatrix} r_1 & r_2 & r_3 & t \end{bmatrix}\begin{bmatrix} X_W \\ Y_W \\ Z_W \\ 1 \end{bmatrix} \quad (10.13)$$

Assuming the plane where the calibration template is located is the $Z_W = 0$ plane of the world coordinate system, then Eq. (10.14) can be obtained.

$$s\begin{bmatrix} u \\ v \\ 1 \end{bmatrix} = \boldsymbol{A}\begin{bmatrix} r_1 & r_2 & r_3 & t \end{bmatrix}\begin{bmatrix} X_W \\ Y_W \\ 0 \\ 1 \end{bmatrix} = \boldsymbol{A}\begin{bmatrix} r_1 & r_2 & t \end{bmatrix}\begin{bmatrix} X \\ Y \\ 1 \end{bmatrix} \quad (10.14)$$

Let $\overline{M} = \begin{bmatrix} X & Y & 1 \end{bmatrix}^T, \overline{m} = \begin{bmatrix} u & v & 1 \end{bmatrix}^T$, then we have $s\overline{m} = \boldsymbol{H}\overline{M}$, where:

$$\boldsymbol{H} = \boldsymbol{A}\begin{bmatrix} r_1 & r_2 & t \end{bmatrix} = \begin{bmatrix} h_1 & h_2 & h_3 \end{bmatrix} = \begin{bmatrix} h_{11} & h_{12} & h_{13} \\ h_{21} & h_{22} & h_{23} \\ h_{31} & h_{32} & h_{33} \end{bmatrix} \quad (10.15)$$

$\boldsymbol{H}$ is the homography matrix, representing the mapping relationship between points on the template and their image points. If the coordinates of the template points in space and on the image are known, $\overline{m}$ and $\overline{M}$ can be determined, thus solving the homography matrix, with each template corresponding to a homography matrix. In Chap. 9 on monocular vision, the homography matrix and its solution method were introduced. s is the scale factor, which does not change the value of homogeneous coordinates.

The following introduces the principle of solving the camera's internal and external parameters through the homography matrix. Equation (10.15) can be rewritten as Eq. (10.16).

$$\begin{bmatrix} h_1 & h_2 & h_3 \end{bmatrix} = \lambda\boldsymbol{A}\begin{bmatrix} r_1 & r_2 & t \end{bmatrix} \quad (10.16)$$

where λ is the scale factor. Since r_1 and r_2 are unit orthogonal vectors, we have:

$$\begin{aligned} h_1^T A^{-T} A^{-1} h_2 &= 0 \\ h_1^T A^{-T} A^{-1} h_1 &= h_2^T A^{-T} A^{-1} h_2 \end{aligned} \quad (10.17)$$

Since h_1 and h_2 in Eq. (10.17) are solved through homography, the only unknown left is the internal parameter matrix $\boldsymbol{A}$. The internal parameter matrix $\boldsymbol{A}$ contains 5 parameters: *fx*, *fy*, *cx*, *cy*, γ. To fully solve these five unknowns, three homography matrices are needed. Three homography matrices under two constraints can generate six equations, thus solving all five internal parameters. How can three different homography matrices be obtained? The answer is to use three photos of the calibration object plane. Three different photos can be obtained by changing the relative position between the camera and the calibration board; alternatively, set $\gamma = 0$ and use two photos to calculate the internal parameters.

Next, some mathematical transformations are made to the obtained equations, let:

$$B = A^{-T}A^{-1} = \begin{bmatrix} B_{11}B_{12}B_{13} \\ B_{12}B_{22}B_{23} \\ B_{13}B_{23}B_{33} \end{bmatrix}$$

$$= \begin{bmatrix} \frac{1}{\alpha^2} & -\frac{\gamma}{\alpha^2\beta} & \frac{v_0\gamma - u_0\beta}{\alpha^2\beta} \\ -\frac{\gamma}{\alpha^2\beta} & \frac{\gamma^2}{\alpha^2\beta^2} + \frac{1}{\beta^2} & -\frac{\gamma(v_0\gamma - u_0\beta)}{\alpha^2\beta^2} - \frac{v_0}{\beta^2} \\ \frac{v_0\gamma - u_0\beta}{\alpha^2\beta} & -\frac{\gamma(v_0\gamma - u_0\beta)}{\alpha^2\beta^2} - \frac{v_0}{\beta^2} & \frac{(v_0\gamma - u_0\beta)^2}{\alpha^2\beta^2} + \frac{v_0^2}{\beta^2} + 1 \end{bmatrix} \tag{10.18}$$

It can be seen that $\boldsymbol{B}$ is a symmetric matrix, so B effective elements are reduced to 6 (since there are three pairs of symmetrical elements that are equal, solving for the following 6 elements will yield the complete B), let these six elements form the vector b:

$$b = \begin{bmatrix} B_{11} & B_{12} & B_{22} & B_{13} & B_{23} & B_{33} \end{bmatrix}^T \tag{10.19}$$

Let the i-th column vector of $\boldsymbol{H}$ be $h_i = \begin{bmatrix} h_{i1} & h_{i2} & h_{i3} \end{bmatrix}$, then

$$h_i^T B h_i = V_{ij}^T b \tag{10.20}$$

where

$$\begin{aligned} V_{ij} = \big[& h_{i1}h_{j1} \;\; h_{i1}h_{j2} + h_{i2}h_{j1} \;\; h_{i2}h_{j2} \\ & h_{31}h_{j1} + h_{i1}h_{j3} \;\; h_{31}h_{j1} + h_{i3}h_{j3} \;\; h_{i3}h_{j3} \big]^T \end{aligned} \tag{10.21}$$

The above intrinsic parameter constraints are written as two equations in terms of b, as shown in Eq. (10.22).

$$\begin{bmatrix} V_{12}^T \\ V_{11}^T - V_{22}^T \end{bmatrix} b = 0 \tag{10.22}$$

Assuming there are n images, the simultaneous equations can yield a linear equation: $Vb = 0$.

Where V is a $2n \times 6$ matrix. If $n \geq 3$, more than 6 equations can be listed, thus solving for the camera's internal parameters. Then, using the intrinsic parameters and the homography matrix $\boldsymbol{H}$, the external parameters of each image can be calculated, as shown in Eq. (10.23). In this way, both the internal and external parameters of the camera are solved.

$$\begin{cases} r_1 = \lambda A^{-1} h_1 \\ r_2 = \lambda A^{-1} h_2 \\ r_3 = r_1 \times r_2 \\ t = \lambda A^{-1} h_3 \\ Where\ \lambda = \frac{1}{A^{-1} h_1} = \frac{1}{A^{-1} h_2} \end{cases} \tag{10.23}$$

(2) Maximum Likelihood Estimation

The above derivation results are based on ideal conditions, but due to the possible presence of Gaussian noise, maximum likelihood estimation is used for optimization. Suppose n images containing a chessboard pattern are collected for calibration, with m chessboard corner points in each image. Let the corner point M_j on the i-th image be projected onto the image under the camera matrix obtained from the above calculation:

$$\overline{m} = \left(A, R_i, t_i, M_{ij}\right) = A[R|t]M_{ij} \tag{10.24}$$

where R_i and t_i are the rotation matrix and translation vector corresponding to the i-th image, and A is the intrinsic parameter matrix. The probability density function of the corner point m_{ij} is:

$$f\left(m_{ij}\right) = \frac{1}{\sqrt{2\pi}} e^{\frac{-(\overline{m}\left(A, R_i, t_i, M_{ij}\right) - m_{ij})^2}{\sigma^2}} \tag{10.25}$$

Construct the likelihood function:

$$\begin{aligned} L\left(A, R_i, t_i, M_{ij}\right) &= \prod_{i=1,j=1}^{n,m} f\left(m_{ij}\right) \\ &= \frac{1}{\sqrt{2\pi}} e^{\frac{-\sum_{i=1}^{n} \sum_{j=1}^{m} (\overline{m}\left(A, R_i, t_i, M_{ij}\right) - m_{ij})^2}{\sigma^2}} \end{aligned} \tag{10.26}$$

Let L reach its maximum, which means making expression (10.27) minimal. Here, the LM (Levenberg–Marquardt) algorithm for multi-parameter nonlinear system optimisation problems is used to iteratively find the optimal solution.

$$\sum_{i=1}^{n} \sum_{j=1}^{m} \| \overline{m} = \left(A, R_i, t_i, M_{ij}\right) - m_{ij} \|^2 \tag{10.27}$$

(3) Radial Distortion Estimation

The Zhang calibration method only focuses on the most significant radial distortion, with the mathematical expression as:

$$\begin{cases} u' = u + (u - u_0)\left[k_1\left(x^2 + y^2\right) + k_2(x^2 + y^2)^2\right] \\ v' = v + (v - v_0)\left[k_1\left(x^2 + y^2\right) + k_2(x^2 + y^2)^2\right] \end{cases} \tag{10.28}$$

$$\begin{cases} u' = u_0 + \alpha x' + \gamma y' \\ v' = v_0 + \beta y' \end{cases} \tag{10.29}$$

where (u, v) are the ideal undistorted pixel coordinates, (u', v') are the actual distorted pixel coordinates. (u_0, v_0) represent the principal point, (x, y) are the ideal undistorted continuous image coordinates, (x', y') are the actual distorted continuous image coordinates. k_1 and k_2 are the first two orders of distortion parameters. Converted into matrix form:

$$\begin{bmatrix} (u - u_0)\left(x^2 + y^2\right)(u - u_0)(x^2 + y^2)^2 \\ (v - v_0)\left(x^2 + y^2\right)(v - v_0)(x^2 + y^2)^2 \end{bmatrix} \begin{bmatrix} k_1 \\ k_2 \end{bmatrix} = \begin{bmatrix} u' - u \\ v' - v \end{bmatrix} \tag{10.30}$$

Denoted as:

$$Dk = d \tag{10.31}$$

Then it can be obtained:

$$k = [k_1 k_2]^T = (D^T D)^{-1} D^T d \tag{10.32}$$

The distortion coefficient k is calculated.

Using the maximum likelihood approach to optimise the obtained results, that is, like the previous step, the LM method calculates the parameter values that minimise the following function:

$$\sum_{i=1}^{n} \sum_{j=1}^{m} \left\| \overline{m} = \left(A, k_1, k_2, R_i, t_i, M_{ij}\right) - m_{ij} \right\|^2 \tag{10.33}$$

The above is the complete process of obtaining the camera's internal parameters, external parameters, and distortion coefficients using the Zhang Zhengyou calibration method.

10.2.2.1 Conversion of Camera Parameters and Projection Matrix

The result obtained by the direct linear calibration method is the projection matrix M, while the result obtained by the Zhang Zhengyou calibration method is the camera's internal and external parameters. In fact, the 11 parameters in the projection matrix M do not have specific physical significance, and thus are also referred to as implicit parameters. The camera's internal and external parameters obtained by the Zhang Zhengyou calibration method can be converted into the projection matrix M.

Let m_i^T $(i = 1 \sim 3)$ are the row vectors composed of the first three elements of the i-th row of the projection matrix $\boldsymbol{M}$; m_{i4} $(i = 1 \sim 3)$ are the elements in the fourth column of the i-th row of the matrix $\boldsymbol{M}$; r_i^T $(i = 1 \sim 3)$ are the i-th rows of the rotation matrix R, and t_x, t_y, t_z are the three components of the translation vector $\boldsymbol{t}$. If we set $\gamma = 0$, then the relationship between the matrix $\boldsymbol{M}$ and the internal and external parameters of the camera is as shown in Eq. (10.34).

$$m_{34}\begin{bmatrix} m_1^T & m_{14} \\ m_2^T & m_{24} \\ m_3^T & 1 \end{bmatrix} = \begin{bmatrix} a_x & 0 & u_0 & 0 \\ 0 & a_y & v_0 & 0 \\ 0 & 0 & 1 & 0 \end{bmatrix}\begin{bmatrix} r_1^T & t_x \\ r_2^T & t_y \\ r_3^T & t_z \\ 0^T & 1 \end{bmatrix} \tag{10.34}$$

where $m_{34} = t_z$, thus the relationship between the projection matrix M and the internal and external parameters can be derived as follows:

$$\begin{aligned}
\boldsymbol{m}_{11} &= (\boldsymbol{a}_x\boldsymbol{r}_{11} + \boldsymbol{u}_0\boldsymbol{r}_{31})/\boldsymbol{t}_z \\
\boldsymbol{m}_{12} &= (\boldsymbol{a}_x\boldsymbol{r}_{12} + \boldsymbol{u}_0\boldsymbol{r}_{32})/\boldsymbol{t}_z \\
\boldsymbol{m}_{13} &= (\boldsymbol{a}_x\boldsymbol{r}_{13} + \boldsymbol{u}_0\boldsymbol{r}_{33})/\boldsymbol{t}_z \\
\boldsymbol{m}_{14} &= (\boldsymbol{a}_x\boldsymbol{t}_x + \boldsymbol{u}_0\boldsymbol{t}_z)/\boldsymbol{t}_z \\
\boldsymbol{m}_{21} &= \left(\boldsymbol{a}_y\boldsymbol{r}_{21} + \boldsymbol{v}_0\boldsymbol{r}_{31}\right)/\boldsymbol{t}_z \\
\boldsymbol{m}_{22} &= \left(\boldsymbol{a}_y\boldsymbol{r}_{22} + \boldsymbol{v}_0\boldsymbol{r}_{32}\right)/\boldsymbol{t}_z \\
\boldsymbol{m}_{23} &= \left(\boldsymbol{a}_y\boldsymbol{r}_{23} + \boldsymbol{v}_0\boldsymbol{r}_{33}\right)/\boldsymbol{t}_z \\
\boldsymbol{m}_{24} &= \left(\boldsymbol{a}_y\boldsymbol{t}_y + \boldsymbol{v}_0\boldsymbol{t}_z\right)/\boldsymbol{t}_z \\
\boldsymbol{m}_{31} &= \boldsymbol{r}_{31}/\boldsymbol{t}_z \\
\boldsymbol{m}_{32} &= \boldsymbol{r}_{32}/\boldsymbol{t}_z \\
\boldsymbol{m}_{33} &= \boldsymbol{r}_{33}/\boldsymbol{t}_z
\end{aligned}$$

10.3 Calibration Measurement Experiment

Within the same scene, the camera is calibrated using both the direct linear calibration method and Zhang's calibration method. Then, the two sets of calibration results are used to perform three-dimensional measurements of target points, analysing the calibration accuracy and comparing the differences between the two calibration methods.

Figure 10.6 shows the binocular vision image acquisition system used for the experiment. To obtain the maximum field of view through the two cameras, a convergent stereo vision model is used, which allows for the adjustment of the camera bracket to change the distance and optical axis angle between the two cameras, thereby adjusting the field of view.

The camera selected for the experiment is the Canon 550D DSLR camera. The specific parameters of this camera are shown in Table 10.2.

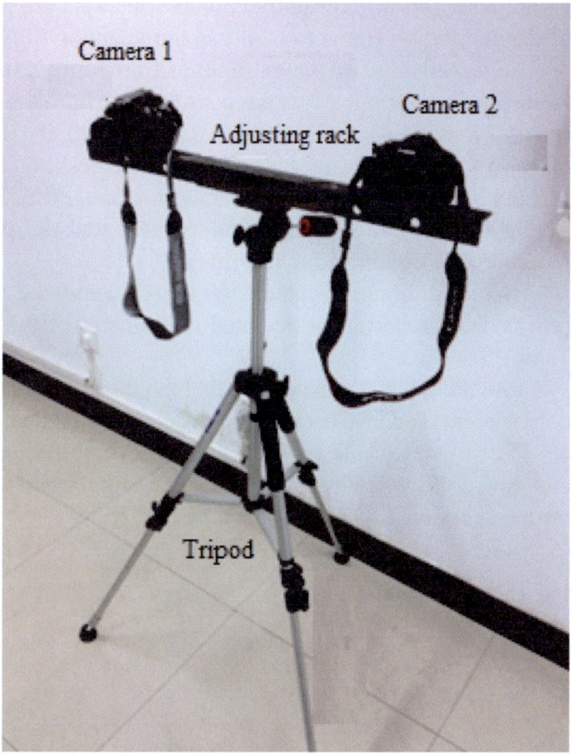

Fig. 10.6 Binocular vision image acquisition system

Table 10.2 CANON 550D camera parameter table

Item	Parameter	Item	Parameter
Sensor type	CMOS	Image type	JPEG
Effective pixel	18 million	Interface type	USB2.0 I/O (including SD card)
Maximum resolution	5184×3456	Overall dimension	$128.8 \times 97.5 \times 75.3$ mm
Maximum frame rate	60 frames per second	Exposure compensation	Manual automatic enveloping exposure

10.3.1 Direct Linear Calibration Method Experiment

The calibration frame shown in Fig. 10.7 is used, where the X, Y, and Z axes are perpendicular to each other and not easily deformed, ensuring calibration accuracy. Since the image coordinates of each image are obtained by clicking the calibration points with a mouse, brightly coloured markers are attached to the 8 corner points of the calibration frame to facilitate the selection of these points.

The dimensions of the calibration frame are 520 mm × 520 mm × 520 mm. Assuming corner point 1 is the coordinate origin, the relative coordinates of corner points 1 to 8 are (0, 0, 0), (520, 0, 0), (520, 520, 0), (0, 520, 0), (0, 0, 520), (520, 0, 520), (520, 520, 520), (0, 520, 520). On the desktop, A, B, C, D

Four points are used to determine a plane, ensuring that the subsequent Zhang Zhengyou calibration method experiment and the placement of the object to be measured are conducted above this plane.

After the calibration calculation is completed, the comparison results of the theoretical coordinates and calculated coordinates of the 8 corner points are shown in Table 10.3.

From the table, it can be seen that the three-dimensional coordinate error of the 8 corner points reconstructed through calibration calculation can be controlled within 1%. The causes of the error include distortion during the image formation process and deviations when manually selecting target points. For feature points with larger errors, accuracy can be improved by re-clicking their pixel points.

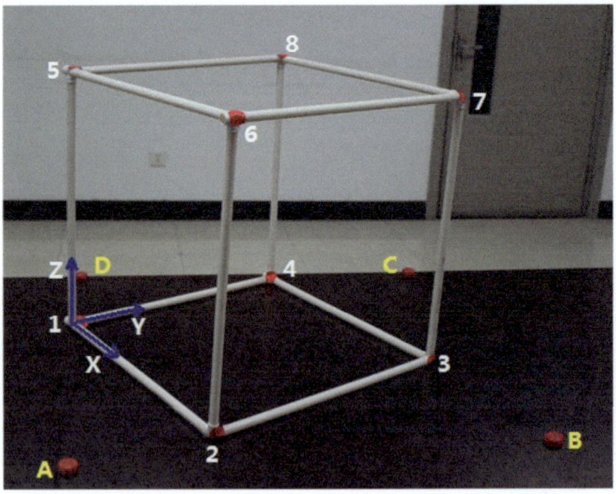

Fig. 10.7 Calibration frame of direct linear calibration method

Table 10.3 Comparison of calibration point reconstruction calculation results and actual results (mm)

Fixed point	Actual coordinates			Calculated coordinates			Average error in X, Y, Z direction (%)
	X	Y	Z	X	Y	Z	
1	0	0	0	−3.38	2.74	0.04	0.39
2	520	0	0	518.52	−3.53	−0.72	0.37
3	520	520	0	523.59	515.43	2.36	0.67
4	0	520	0	−0.77	524.51	−1.79	0.45
5	0	0	520	1.71	−3.62	518.67	0.43
6	520	0	520	522.97	4.40	521.04	0.54
7	520	520	520	514.71	524.27	519.05	0.67
8	0	520	520	2.70	515.92	521.27	0.52

10.3.2 Zhang Zhengyou Calibration Method Experiment

A chessboard is used for the calibration experiment of the Zhang Zhengyou calibration method, with the calibration steps as follows.

① Create a planar calibration template. The calibration template is a printed 8 × 7 black and white square chessboard, with each square measuring 51 mm × 51 mm. The chessboard template is affixed to a hard plastic board to ensure the template is flat.
② The left and right cameras capture images of the calibration template. In this experiment, 9 images were taken from different positions.
③ Chessboard corner detection. Corner detection is performed to obtain the two-dimensional image coordinate data of the chessboard corners, using the Harris corner detection algorithm. The detection results are shown in Fig. 10.8.

To verify the calibration accuracy, the reprojection error is used to calculate the error of the camera's internal and external parameters. The reprojection error refers to the sum of the squares of the differences between the corner coordinates extracted from the calibration template and the image coordinates calculated through projection. The calculation formula is shown in Eq. (10.35).

$$E = \frac{\sum_{i=1}^{n} \sqrt{(U_i - u_i)^2 + (V_i - v_i)^2}}{n} \tag{10.35}$$

where, n is the number of calibration points, (U_i, V_i) are the corner coordinates extracted from the image, and (u_i, v_i) are the image coordinates obtained by projecting the actual three-dimensional coordinates using the calibration results. This allows the error of an image to be obtained, and by averaging the errors of all images captured by each camera, the calibration accuracy of each camera is obtained. After calculation,

(a) Left camera

(b) Right camera

Fig. 10.8 Checkerboard calibration image and corner detection results

the projection error of the left camera is 0.2200, and the projection error of the right camera is 0.2224, with the error level being less than one pixel.

The causes of error include:

① The processing accuracy of the calibration template. The processing quality of the chessboard template is the main factor affecting the accuracy of corner extraction by the image processing algorithm.
② The placement of the calibration template. When photographing the chessboard template, it should fill the field of view as much as possible, so several positions in the field of view should be selected for shooting. Additionally, the chessboard template should be tilted in different directions, with a 45° tilt being optimal.
③ The number of calibration images taken. Generally, the more images, the higher the calibration accuracy. However, this increases the amount of calibration calculation. Moreover, when the number of images increases to a certain extent, the calibration accuracy will tend to stabilize. According to the experiment, selecting 9 images can achieve a good balance between calibration accuracy and calculation amount.
④ The synchronous shooting of the binocular vision system. Although the calibration images used in this experiment are taken statically, the synchronization of the camera shooting can still affect the calibration accuracy, so synchronous acquisition should be used.

The data results obtained from the above calibration calculation are:

① The internal parameter matrix is:

$$A_1 = \begin{bmatrix} 803.39 & 0 & 313.61 \\ 0 & 803.06 & 248.87 \\ 0 & 0 & 1.00 \end{bmatrix}$$

$$D_1 = \begin{bmatrix} -0.1719 & 0.3721 & -0.0006 & -0.0161 \end{bmatrix}$$

$$D_2 = \begin{bmatrix} -0.1574 & 0.7096 & -0.0042 & -0.0083 \end{bmatrix}$$

$$A_2 = \begin{bmatrix} 899.71 & 0 & 349.13 \\ 0 & 899.49 & 249.99 \\ 0 & 0 & 1.00 \end{bmatrix}$$

where, A_1, A_2 and D_1, D_2 are the intrinsic parameter matrices and radial distortion matrices of the left and right cameras, respectively.

② The extrinsic parameter matrices are:

$$R_1 = \begin{bmatrix} -0.0154 & 0.8065 & 0.5910 \\ 0.9598 & -0.1536 & 0.2347 \\ 0.28016 & 0.5709 & -0.7717 \end{bmatrix}$$

$$R_2 = \begin{bmatrix} -0.0154 & 0.8065 & 0.5910 \\ 0.9598 & -0.1536 & 0.2347 \\ 0.28016 & 0.5709 & -0.7717 \end{bmatrix}$$

$$T_1 = \begin{bmatrix} -146.64 & 0.2320 & 1417.01 \end{bmatrix}$$

$$T_2 = \begin{bmatrix} -356.99 & -0.61725 & 1492.22 \end{bmatrix}$$

where, R_1, R_2 and T_1, T_2 are the extrinsic parameter matrices and translation matrices of the left and right cameras, respectively.

10.3.3 Three-Dimensional Measurement Experiment

The objects to be measured in this experiment comprise square boxes. The accuracy of the three-dimensional measurement algorithm is tested by measuring the distances between any two of the six corner points of the box. Since the corner points of the objects to be measured are selected by mouse during the experiment, in order to reduce the error caused by manually selecting target points, the experimental results use the average value of ten measurements.

The first experiment uses a square box as the object to be measured, with clear markings on the six corner points that can be observed simultaneously by the left and right cameras, as shown in Fig. 10.9, measuring the distances between points 1 to 6.

The calibration results obtained by the direct linear calibration method and Zhang's calibration method are used for three-dimensional measurement. The average value of ten calculations is taken, and the comparison of the distances between each point and the actual distances is shown in Table 10.4.

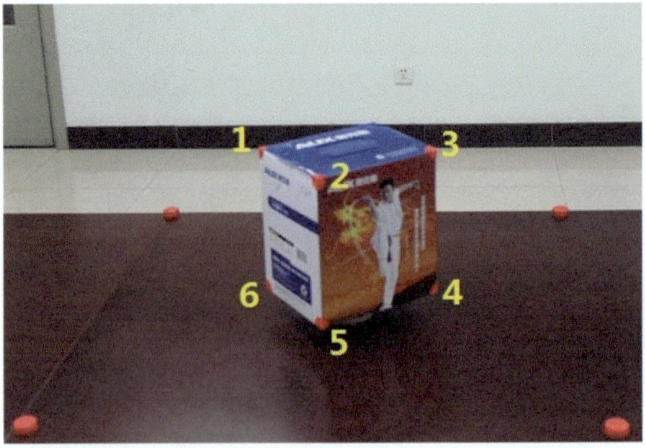

Fig. 10.9 Measurement of corner distance of object 1 to be measured

Table 10.4 Comparison of actual and measured distances between target points of object 1 (mm)

	1–2	5–6	2–3	4–5	1–6	2–5	3–4
Actual distance	175.00	175.00	215.00	215.00	245.00	245.00	245.00
Zhang	173.63	171.90	217.08	213.36	247.31	244.20	246.10
Error (%)	0.78	1.77	0.97	0.76	0.94	0.32	0.45
DLT	173.18	171.27	218.47	213.17	247.30	243.35	246.48
Error (%)	0.82	2.13	1.61	0.85	0.94	0.67	0.60

To reduce error and increase sample size, the same experimental method is used to measure the rectangular box shown in Fig. 10.10. Similarly, the result of this experiment is also the average value obtained after ten measurements and calculations.

The experimental results obtained from this experiment are shown in Table 10.5.

Figures 10.11 and 10.12 are the statistical comparison results of measurement errors between the distances of each point of objects 1 and 2 using the two calibration methods, respectively.

From the results of the two experiments, the following points can be analysed:

① The trend of the curves for both methods is the same, and the difference in error values is very small, indicating that the method of converting the camera's

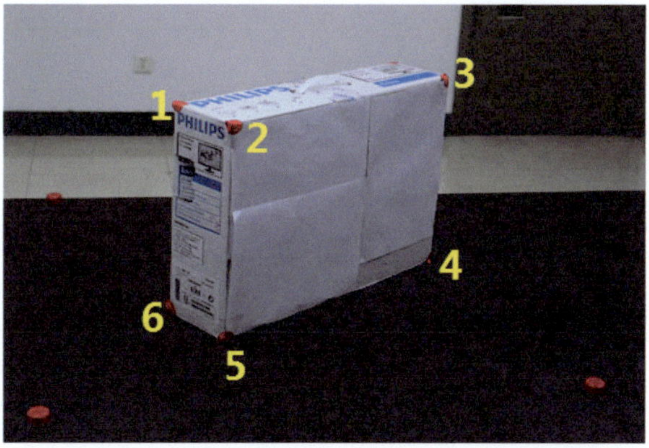

Fig. 10.10 Measurement of corner distance of object 2 to be measured

Table 10.5 Comparison of actual and measured distances between target points of object 2 (mm)

	1–2	5–6	2–3	4–5	1–6	2–5	3–4
Actual distance	133.00	133.00	513.00	513.00	352.00	352.00	352.00
Zhang	130.13	137.52	527.63	524.87	356.25	351.87	352.40
Error (%)	2.87	3.39	2.85	2.31	1.21	0.04	0.11
DLT	129.06	135.99	524.38	526.12	355.35	352.22	355.67
Error (%)	2.96	2.25	2.22	1.19	0.95	0.06	1.04

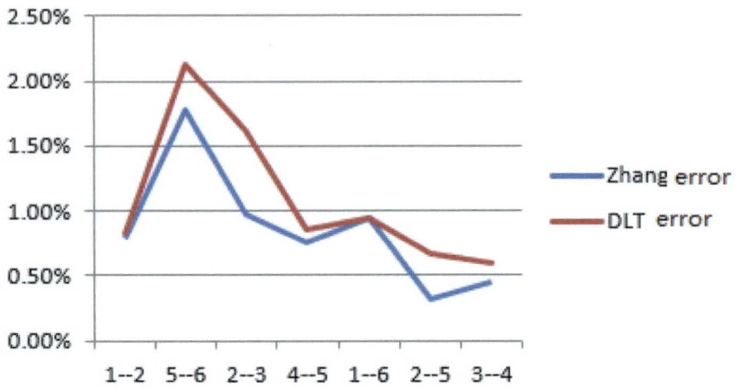

Fig. 10.11 Comparison of distance errors between points in object to be measured 1

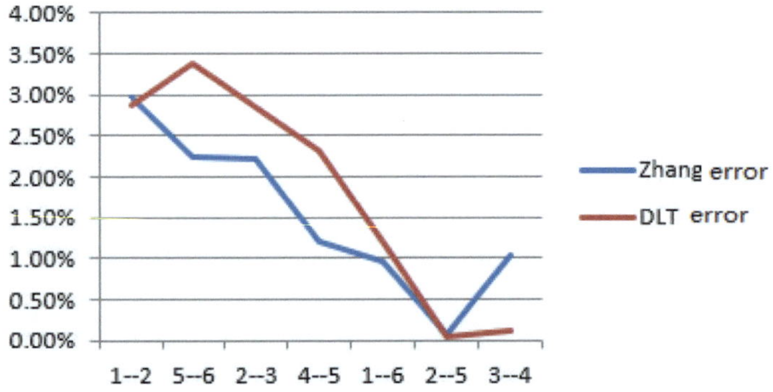

Fig. 10.12 Comparison of distance errors between points in object to be measured 2

intrinsic and extrinsic parameters obtained by Zhang's calibration method into a projection matrix M is feasible.

② The measurement error of target points parallel to the imaging plane is the smallest: the measurement errors of the distances between target points 1—6, 2—5, and 3—4 are all below 1%, which is much smaller than the errors between other points. The reason is that the direction of these three pairs of distances is parallel to the camera's imaging plane. Since the depth information of the image is lost during the process of converting spatial points from the world coordinate system to the image coordinate system, the information of points parallel to the camera's imaging plane is not significantly affected.

③ The distance error between each vertex is kept within 3%. The main causes of this error are the human error introduced when manually selecting feature points and target points, as well as the image distortion produced during the camera imaging process.

④ By comparing the error curves of the DLT calibration method and Zhang Zhengyou's calibration method, it can be seen that the mean error of Zhang

Zhengyou's calibration method is smaller than that of the DLT calibration method. The reason is that the DLT calibration method is based on a linear camera model and does not consider the impact of image distortion. In contrast, Zhang Zhengyou's calibration method completes the task of image correction during the calibration process. Therefore, Zhang Zhengyou's calibration accuracy is higher.

Reference

1. Tian Hao. Distance Measurement Based on Machine Vision [D]. Beijing: China Agricultural University, 2013.

Chapter 11
Motion Image Processing

The methods introduced in the previous chapters are all aimed at processing single static images, while this chapter introduces the processing of moving images. Moving images refer to a sequence of images composed of multiple static images with the addition of a time dimension.

11.1 Optical Flow Method

11.1.1 Basic Concept of Optical Flow

The concept of optical flow was first introduced by Gibson in 1950. It is the instantaneous velocity of pixel movement of a spatially moving object on the observation imaging plane. It utilizes the changes in pixel values over time within an image sequence and the correlation between adjacent frames to establish the correspondence between the previous frame and the current frame, thereby calculating the motion information of objects between adjacent frames. Generally speaking, optical flow is caused by the movement of the foreground object in the scene, the movement of the camera, or the combined movement of both. This includes the concepts of motion field and optical flow field. The motion field refers to the movement of objects in the three-dimensional real world, while the optical flow field is the projection of the motion field on the two-dimensional image plane.

In 1981, Horn and Schunck creatively linked the two-dimensional velocity field with the grayscale, introducing the optical flow constraint equation, and obtained the basic algorithm for optical flow calculation. Various optical flow calculation methods have been proposed based on different theoretical foundations, with varying algorithm performance. Barron and others summarized various optical flow calculation

© Chemical Industry Press 2026

B. Chen and S. Chen, *Machine Vision Technology*, Advanced and Intelligent Manufacturing in China, https://doi.org/10.1007/978-981-95-1184-6_11

techniques and divided them into four categories based on the differences in theoretical foundations and mathematical methods: gradient-based methods, matching-based methods, energy-based methods, and phase-based methods. In recent years, neural dynamic methods have also attracted considerable attention from scholars.

(1) Gradient-Based Methods Gradient-based methods, also known as differential methods, use the spatiotemporal derivatives (i.e., spatiotemporal gradient functions) of the grayscale of sequence images (or their filtered forms) to calculate the velocity vector of pixels. Due to their simplicity and good results, these methods have been widely researched and applied. Although many gradient-based optical flow estimation methods have achieved good optical flow estimation, there are still many issues in applying optical flow for real-time detection and automatic tracking of targets due to the manual selection of adjustable parameters, the difficulty in selecting reliability evaluation factors, and the impact of preprocessing on optical flow calculation results.

(2) Matching-Based Methods Matching-based optical flow calculation methods include feature-based and region-based methods. Feature-based methods continuously locate and track the main features of the target, showing robustness to large movements and brightness changes of the target. The problem is that optical flow is usually very sparse, and feature extraction and precise matching are also very difficult. Region-based methods first locate similar regions and then calculate optical flow through the displacement of similar regions. This method has been widely applied in video encoding. However, the optical flow it calculates is still not dense. Additionally, both methods have difficulties in estimating sub-pixel accuracy optical flow, and the computational load is large. When considering optical flow accuracy and density, matching-based methods are applicable.

(3) Energy-Based Methods Based on the energy method, the input image sequence must first undergo spatiotemporal filtering, which is an integration of time and space. For a uniform flow field, this spatiotemporal integration is essential to obtain accurate velocity estimates. However, doing so reduces the spatial and temporal resolution of optical flow estimation, especially when the spatiotemporal integration region contains several motion components (such as motion edges), the estimation accuracy will deteriorate. Additionally, energy-based optical flow techniques also face the issue of high computational load. This method involves a large number of filters, which are currently the main computational expense. However, it is expected that with the development of corresponding hardware, filtering will no longer be a significant limiting factor in the near future, and all these techniques can be implemented at frame rate.

(4) Phase-based Method Fleet and Jepson first conceptually proposed the use of phase information for optical flow computation. Since velocity is determined based on the phase characteristics of the band-pass filter output, it is called the phase method. They define component velocity according to the instantaneous motion perpendicular to the iso-phase contours in the output of the band-pass

velocity-tuned filter. The band-pass filter separates the input signal according to scale, velocity, and orientation.

(5) Neurodynamic Method The original intention of machine vision research is to mimic the functions of the human visual system; however, there is a stark contrast between human image understanding and recognition capabilities and those of computers. Visual scientists are eager to learn from human methods of processing images to overcome this predicament. For optical flow computation, if the previous energy or phase-based models have some biological plausibility, then the neurodynamic models of visual motion perception established using neural networks in recent years are a more direct simulation of the functions and structures of the biological visual system. Although using these neurodynamic models to measure optical flow is still immature, these methods and their conclusions lay a good foundation for further research and are a very meaningful attempt to introduce neural mechanisms into motion computation.

11.1.2 Principles of Optical Flow Method for Target Tracking

① Process a continuous sequence of video frames.
② For each video sequence, use a certain target detection method to detect potential foreground targets.
③ If a foreground target appears in a certain frame, identify its representative key feature points (which can be randomly selected or corner points used as feature points).
④ For any two adjacent video frames thereafter, find the best position of the key feature points from the previous frame in the current frame, thereby obtaining the position coordinates of the foreground target in the current frame.
⑤ By iterating in this manner, target tracking can be achieved.

In practical applications, due to occlusion, multiple light sources, transparency, and noise, the grayscale conservation assumption condition of the basic equation of the optical flow field cannot be satisfied, and the correct optical flow field cannot be solved. At the same time, most optical flow computation methods are quite complex and computationally intensive, which cannot meet real-time requirements. Therefore, they are generally not adopted by monitoring systems with high accuracy and real-time requirements.

11.2 Template Matching

A template is a predefined small image, and template matching involves searching for this small image within a larger image.

Taking a grayscale image as an example, the template $T(M \times N$ pixels) is overlaid on the search image $S(W \times H$ pixels) and translated. The area of the search image covered by the template is called the sub-image $S_{i,j}$, where i, j are the coordinates of the top-left corner of the sub-image on the search image S. The search range is: $1 \leq I \leq W\text{-}M$, $1 \leq j \leq H\text{-}N$.

The template matching process is completed by comparing the similarity between T and $S_{i,j}$.

The degree of matching between template T and sub-image $S_{i,j}$ can be measured using one of the following two metrics.

$$D(i,j) = \sum_{m=1}^{M} \sum_{n=1}^{N} \left[S_{i,j}(m,n) - T(m,n)\right]^2 \tag{11.1}$$

$$D(i,j) = \sum_{m=1}^{M} \sum_{n=1}^{N} \left|S_{i,j}(m,n) - T(m,n)\right| \tag{11.2}$$

Compared to formula (11.1), formula (11.2) requires less computation and has a faster matching speed. When the calculated D value is less than the set threshold, the match is considered successful.

The aforementioned matching method is limited to cases without rotation. If the template image has directional changes on the matched image, it is necessary to perform rotation calculations for each matching point at each angle. For example, if rotation matching calculations are performed at 5° intervals, a full circle of 360° would require 72 matching calculations for each point, which is very time-consuming.

11.3 Motion Image Processing Example

11.3.1 Real-Time Image Detection of Badminton Techniques and Tactics [1]

(1) Technical Objectives and Key Points

During badminton matches and training sessions, coaches need to promptly and accurately capture on-site statistics of techniques and tactics, enabling them to adjust strategies based on the opponent's performance. This system requires real-time image acquisition and analysis of the badminton match site to determine the technical type of each rally, and to statistically analyse and save the data, achieving intelligent on-the-spot tactical statistics for badminton matches.

To facilitate portability, this system uses a camera and a laptop to achieve the above purpose by analysing two-dimensional sequence images. The main technical points are as follows:

① Real-time acquisition and storage of motion images. It is necessary to control the frame rate (frames/second) of the acquisition on-site and to save the acquired images in the form of continuous image files or video files to the hard disk in real-time.

② Extraction of moving parts in each frame image. To analyse the shuttlecock and players, it is first necessary to accurately detect the moving areas in each frame image, providing a basis for subsequent matching connections.

③ Connection of motion trajectories in sequence images. By connecting the centroid points of the moving areas in the sequence images, identify the start and end points of the motion trajectory line and save the data of each trajectory.

④ Recognition and analysis of badminton motion trajectories. By analysing and judging each trajectory, identify the motion trajectory of the shuttlecock from numerous motion trajectories.

⑤ Calculation of the centre of gravity of players in each frame image. Obtain the centre of gravity positions of players on both sides of the court in the image to provide a basis for the shuttlecock trajectory's ball type judgement.

⑥ Recognition of the court area. Determine the range of the court in the collected images and provide the parameters of the court area as a basis for the shuttlecock trajectory's ball type judgement.

⑦ Ball type judgement of the shuttlecock trajectory. Based on the obtained shuttlecock trajectory, players' centre of gravity, court parameters, and other data, analyse the shuttlecock trajectory to determine its ball type, such as a high shot or a lob.

⑧ Real-time display of technical statistical data. Display the results of statistical analysis in a detailed and intuitive manner in real-time for coaches and players to reference.

⑨ Playback of motion trajectories. Be able to replay the analysis and processing process, clearly and intuitively displaying the motion trajectory of the shuttlecock.

(2) Video Image Acquisition

The image acquisition device selected is the Basler A601f CMOS camera. The main performance features of this camera are as follows: the data output interface is IEEE1394, the maximum resolution is 659 × 493 pixels, the maximum acquisition frame rate is 60 frames/s, and the image type is greyscale image. The acquisition and processing equipment uses a laptop with a Pentium 2.4 GHz CPU and 256MB of memory. The motion image processing development platform uses the two-dimensional motion image measurement system MIAS from Beijing Modern Fubo Technology Co., Ltd., and the software development tool is Microsoft's Visual C++.

The parameters of the detection object are as follows: shuttlecock length: 62 ~ 70 mm; shuttlecock top diameter: 58 ~ 68 mm; shuttlecock racket face diameter: 25 ~ 28 mm; court area: 13.4 m × 6.1 m; net height: 1.524 m.

In order to capture a panoramic image of the badminton court, the camera is set at a distance of 5 m from the court, a height of 4 m from the ground, and an angle of approximately 45° with the ground.

This system employs a multithreading approach to achieve parallel operation of collection, analysis, and storage, thereby fully utilising CPU resources. The collection thread is responsible for acquiring image data into memory and displaying it on the monitor, the analysis thread processes the images collected by the collection thread, and the storage thread saves the images collected into memory by the collection thread onto the hard disk. The three threads operate in three modes: in preview mode, only the collection thread works; in real-time analysis mode, the collection and analysis threads operate synchronously; in real-time storage mode, the collection and storage threads operate synchronously.

The camera's sampling frequency is set at 40 frames per second, enabling real-time sampling, analysis, and storage of the badminton match. To facilitate algorithm development and analysis using the collected and stored video images, an array of video images was collected and stored, with an average video storage frame rate of 38 frames per second. Using a computer with a higher CPU configuration and larger memory capacity will increase the storage speed.

(3) Court Calibration Before conducting image analysis processing, it is necessary to manually select several feature points on the badminton court in the image, as shown by the 8 arrow positions in Fig. 11.1, which are the four corners of the net and the four intersection points on the court below the net. These positional information parameters are not only crucial for determining the type of badminton but also for defining the processing range and excluding interference from off-court moving objects. Figure 11.2 shows the image after actual court calibration. The 8 numbers on Fig. 11.2 represent the 8 feature points of the badminton court obtained by clicking, and the white frame line indicates the range of the badminton court calculated based on the 8 feature points.

(4) Motion Target Extraction

To track and identify the trajectory of the badminton, it is first necessary to extract the badminton target from the complex background on-site and locate the badminton target, which is a crucial step in badminton trajectory tracking analysis. The method involves subtracting two consecutive images in the sequence to extract the moving target in the current image, followed by binarisation of the differential image by setting a threshold. Figure 11.3 shows the continuous frames 1 and 2 and their differential binarised image, with the binarisation threshold set at 5. The white pixels in the binarised image represent the detected moving parts of the badminton and players.

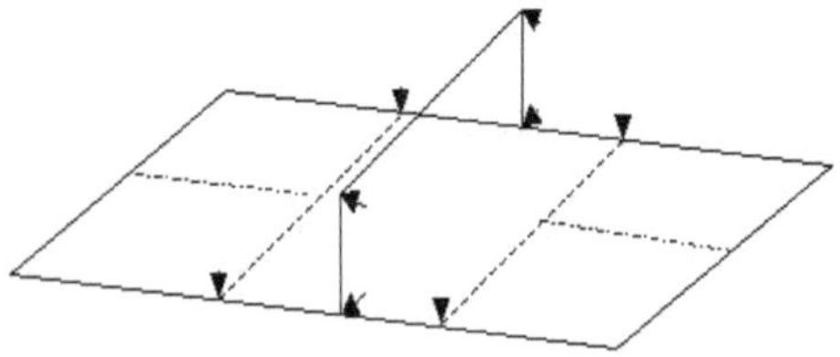

Fig. 11.1 Schematic diagram of characteristic points on badminton court

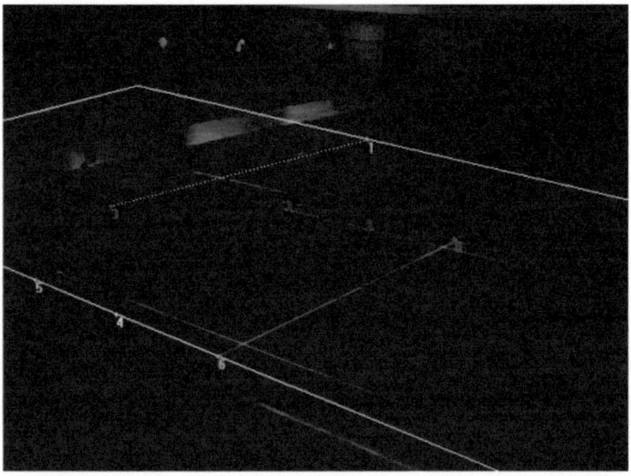

Fig. 11.2 Image of badminton court

(a) Frame 1 (b) Frame 2 (c) Difference binarization result

Fig. 11.3 Sequence frames and their difference binarization results

(5) Trajectory Classification and Connection

① The concept of direction number. To determine the flight direction and type of the badminton, the concept of direction number is introduced. When the coordinates of a point on a trajectory increase in the horizontal direction, the direction number of the trajectory is defined as "+", and for each additional trajectory point in that direction, the direction number increases by 1; when the coordinates of a point on a trajectory decrease in the horizontal direction, the direction number is defined as "−", and for each additional trajectory point in that direction, the direction number decreases by 1. The initial direction number of the trajectory is set to zero, and when a trajectory ends, its direction and approximate length can be determined based on the sign and magnitude of its direction number.

In the analysis and statistics of badminton types, the statistical data of each player is required. When a trajectory is determined to be a badminton trajectory, the direction of the badminton can be judged by the sign of the direction number at the endpoint of the trajectory. If the direction number is positive, it can be concluded that the badminton is moving from the left to the right of the image, and this data should be included in the statistics of the player on the left side of the image.

If the direction number is negative, it can be concluded that the badminton is moving from the right to the left of the image, and this data should be included in the statistics of the player on the right side of the image. The direction number can not only be used to determine the direction of the badminton but also, the magnitude of the direction number is one of the factors in determining the length of the badminton trajectory.

② Calculation of the target centroid. For the differential binarised image, the contour data of each region on the image is first detected, and then the average value of the contour pixel coordinates is stored as its centroid in a linked list for subsequent trajectory matching.

Figure 11.4 shows the calculation results of the centroids of the white regions on the binarised image. The larger "+" symbol on the net line of the court indicates the centre of the net line. The two larger "+" symbols above the players on the left and right sides respectively indicate the centroids of the players on both sides. The small "+" on each white region block indicates its respective centroid. The centroid of the player is calculated from the centroid of each white region block in the measurement area. To visually see the situation of the measurement area, the boundary line of the measurement area is marked with a white line on the binarised image.

(6) Motion Trajectory Extraction

① Recording the motion trajectory of the point target. Assume the motion trajectory of the point target is *Tra*, record the following information of *Tra* on each frame of the image:

- The position (*cx*, *cy*) of the point target *k* on the current frame of *Tra*; In the current frame, the angle *Ang* between the line connecting point target *k* and point target *k*-1 in the previous frame and the *x*-axis;

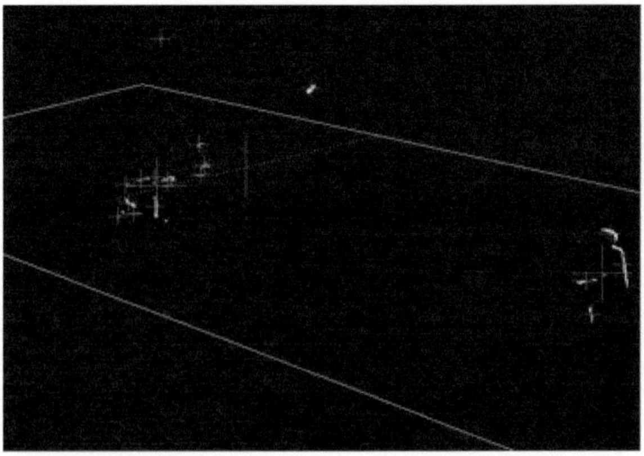

Fig. 11.4 Center of gravity for each area and pitch boundary

- The distance *Len* between point target k and k -1;
- The number of directions *Dir* of *Tra* in the current frame.

② Trajectory matching connection. Represent the centroid coordinates of the i-th white area in the m-th frame as $Ce\ n_m^i\ (x, y)$, and match it with the centroid coordinates of all white areas in the m-1 frame. Calculate the distance L and angle A between $Ce\ n_m^i\ (x, y)$ and the point $Ce\ n_{m-1}^{\max}\ (x, y)$ on the longest trajectory $Tra_{\max}$ in the m-1 frame. If the distance L is between the set minimum distance $Len_{\min}$ and maximum distance $Len_{\max}$, and the angle A is between the set minimum angle $A_{\min}$ and maximum angle $A_{\max}$, then the point $Ce\ n_m^i\ (x, y)$ is a point on the longest trajectory $Tra_{\max}$, and the information of this point is recorded in the longest trajectory, rewriting the corresponding direction number *Dir*. If the point $Ce\ n_m^i\ (x, y)$ cannot be matched with the longest trajectory, then match this point with other trajectory points in the m-1 frame, and record the information of this point in the linked list of the matched trajectory. If the point $Ce\ n_m^i\ (x, y)$ cannot be matched with any trajectory points in the m-1 frame, then treat this point as the starting point of a new trajectory and record the information of this point in a new trajectory linked list.

During the trajectory matching process, it is possible for multiple points in the same frame to meet the matching conditions with the same trajectory. In this case, choose the nearest point target for connection. If none of the centroid points of all regions in the m frame can be connected with the points in the m-1 frame, begin analysing the longest trajectory generated before the m frame to determine whether it is the trajectory of a shuttlecock.

Figure 11.5 shows the distribution of the centroid on a single frame image and the image after accumulating the centroids over multiple frames. It is difficult to determine the centroid of the shuttlecock on a single frame image, but by overlaying the centroid points on a sequence of continuous images, the trajectory of the shuttlecock's movement can be accurately distinguished. The movement trajectories of various moving regions in the image sequence are different, such as the trajectories of the player and the racket compared to the shuttlecock's trajectory, which differ significantly in terms of trajectory length, curvature, distance between points on the trajectory, number of directions, and other aspects. These trajectory features are important bases for extracting the shuttlecock's movement trajectory from the group of trajectories.

(7) Shuttlecock Trajectory Extraction

Let $Tra_{\max}$ represent the longest trajectory record in the group of trajectories, and let $\lambda_1, \lambda_2, \ldots, \lambda_m$ represent all the attributes for evaluating whether a trajectory is a shuttlecock movement trajectory. These attributes include the length of the trajectory, the curvature angle of the trajectory, the number of directions of the trajectory, the number of points on the trajectory, etc. The credibility $\Omega(Tra_{\max})$ of the maximum trajectory being a shuttlecock movement trajectory is defined by the following formula:

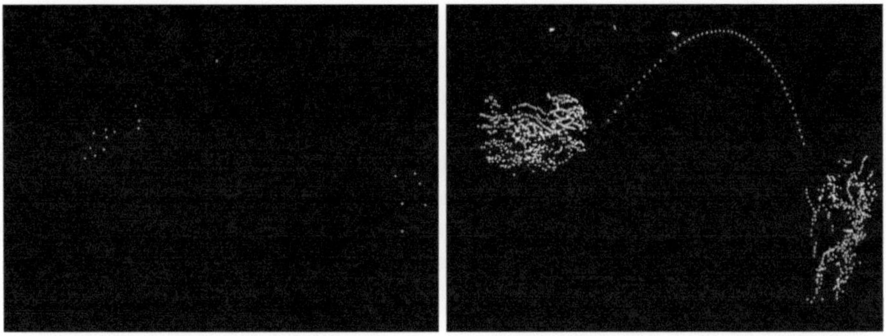

Fig. 11.5 Distribution of center points

$$\Omega(\mathrm{Tra}_{\max}) = \sum_{i}^{m} \Omega_i(\lambda_i) \tag{11.3}$$

If the credibility is greater than the set threshold, the trajectory is considered to be the shuttlecock's movement trajectory, and further analysis of the shuttlecock's type is conducted. After determining the shuttlecock's trajectory and type, all trajectory data is eliminated, and tracking measurements are restarted. If the trajectory $Tra_{\max}$ is not the shuttlecock's movement trajectory, all trajectory data is retained, and the connection judgment for the next frame continues. When no shuttlecock trajectory is detected within the set frame range, all trajectory data is cleared, and tracking measurements are restarted.

When capturing images of a badminton match, the following situation may occur: the shuttlecock is hit and flies out of the camera's field of view, and then falls back into the camera's field of view after a moment. At this time, it is necessary to determine whether the movement trajectory of the shuttlecock after flying out and falling back is the same trajectory. The distance is used to determine whether it is the same trajectory; when the distance (interval frame count) between the endpoint of the flying-out trajectory and the starting point of the falling-back trajectory is within the set range, it is considered the same trajectory, otherwise, it is considered a different trajectory.

Figure 11.6 shows the shuttlecock's movement trajectory extracted from the group of trajectories. Figure 11.6a shows the shuttlecock's movement trajectory entirely within the camera's field of view. Figure 11.6b shows the shuttlecock flying out of the camera's field of view and then falling back into the camera's field of view.

(8) Shuttlecock Type Determination

After extracting the shuttlecock's movement trajectory, according to the definition of shuttlecock types, it is determined which category of shuttlecock technique the obtained shuttlecock trajectory belongs to. According to the definition of shuttlecock types, the determined shuttlecock types include clears, smashes, drops, blocks,

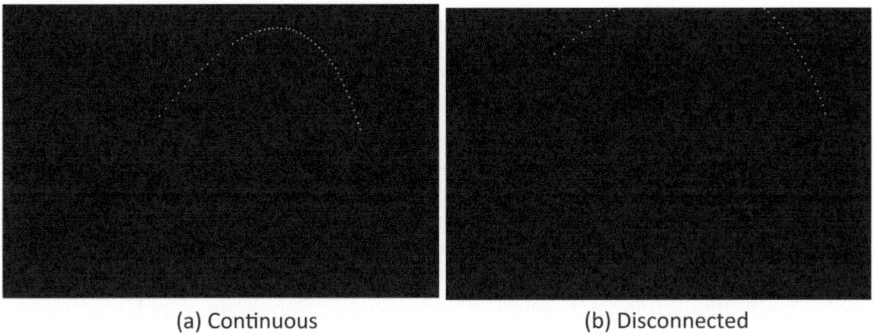

Fig. 11.6 Badminton track results

etc. The main basis for judgment is the technical parameters in the definition of shuttlecock types. These parameters include: the starting point and endpoint of the shuttlecock trajectory relative to the court, the flight angle of the shuttlecock, the flight speed of the shuttlecock, the position of the shuttlecock trajectory relative to the player, the distance between the shuttlecock trajectory and the net, etc.

Based on the characteristics of the shuttlecock trajectory in the definition of shuttlecock types and the relative position relationship between the shuttlecock trajectory and the court, the shuttlecock movement trajectory extracted using the above method is analysed to obtain statistical results of shuttlecock technique types. The analysis and processing of video sequence images with a total frame count close to 10,000 frames in four selected groups were conducted, and the obtained statistical results were compared with the results of manual judgment, as shown in Table 11.1. Group I had a total of 394 frames, Group II had a total of 1204 frames, Group III had a total of 3108 frames, and Group IV had a total of 6020 frames. The results show that the accuracy of shuttlecock type judgment is close to 100%. Only in Group 4 did a misjudgment occur once for a lift, and the reason for the misjudgment was that the shuttlecock's trajectory within the camera's field of view was too short, causing the shuttlecock's trajectory not to be the longest trajectory in the group and thus being ignored.

Table 11.1 Statistical results of shuttlecock types

Group	Lob		Toe		Smash		Block		Drop		Drive	
	Al	Ar	Al	Ar	Al	Ar	Al	Ar	Al	Ar	Al	Ar
I			2	2	2	2			1	1	1	1
II	2	2	9	9	3	3	1	1	6	6	1	1
III	9	9	9	9	2	2	3	3	8	8	20	20
IV	6	6	28	29	16	16	6	6	13	13	4	4

11.3.2 Analysis of Bee Dance Behaviour [2]

(1) Technical Objectives and Key Points

Studies have shown that the duration of a bee's waggle dance correlates with the distance to the nectar source, while the angle of the waggle dance indicates the direction of the nectar source. Currently, researchers generally obtain bee waggle dance data through manual marking methods, although there are cases of data acquisition through image tracking methods. Traditional image tracking methods typically involve applying luminescent materials to the observation target, supplemented by a light source to track and measure the luminescent points. This method is undoubtedly challenging when applied to small bees and can affect their behaviour.

This system aims to track and detect unmarked multiple-target bees through image tracking, performing statistical analysis on their movement trajectories to determine the interval of the bee waggle dance. This allows for the acquisition of information such as the waggle time and angle, providing raw data for analysing the information conveyed by the bee waggle dance. The technical points are as follows:

① Method for selecting tracking targets;
② Method for unmarked image tracking of targets;
③ Method for determining the bee waggle dance;
④ Method for calculating the duration of the bee waggle dance;
⑤ Method for calculating the direction of the bee waggle dance.

(2) Experimental Setup and Video Image Acquisition Fig. 11.7 is a schematic diagram of the experimental setup of this system. The experimental video is captured by a digital camera, with an image resolution of 640 × 480 pixels and a frame rate of 30 frames per second. The video is saved in AVI format. The bees crawl on a vertical plane, and the camera lens's optical axis is perpendicular to the vertical plane for shooting. The image processing is performed on a PC configured with a Pentium(R) Dual-Core processor, a main frequency of 2.6 GHz, and 2.00 GB of memory. Algorithm development was conducted using Microsoft Visual Studio 2010.

(3) Bee Movement Trajectory Tracking

① Selection of target bees. The top-left corner of the image is the origin, with the horizontal direction to the right as the positive direction of the x-coordinate, and the vertical direction downward as the positive direction of the y-coordinate. In the first frame of the video, the head point P_s and the tail point P_e of the target bee are manually clicked with the mouse. The length of the line $P_s P_e$ is recorded as d, and a square processing area with a side length of 1.5 times d is set for the bee. Each point on $P_s P_e$ is scanned to find the point closest to P_s with the largest $2R$-B value (where R and B are the red and blue components of the target pixel, respectively), defining this point as the bee target point P(position of "+" on the image).

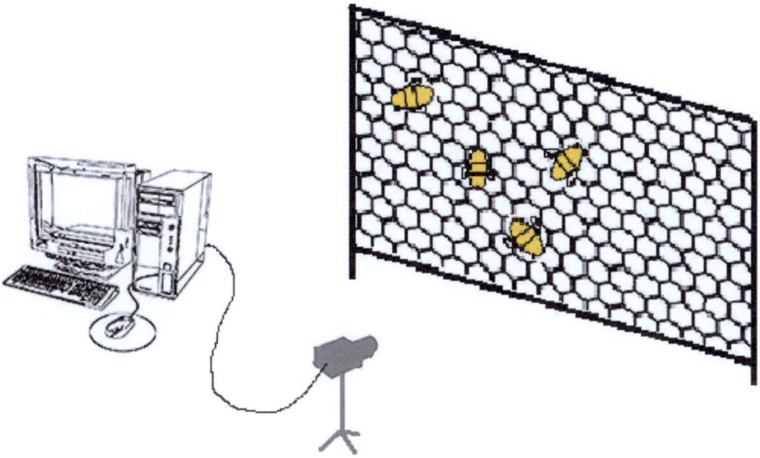

Fig. 11.7 Honeybee dance behavior analysis device

Figure 11.8 shows the initial frame image of the processed video. It can be seen from the figure that the background colour of the hive is very similar to the colour of the bees. Figure 11.9 is a $2R$-B grayscale image, where the bee target is enhanced.

Figure 11.10 is a line profile of the straight line on the target bee within the rectangular frame in Fig. 11.9. Although the brightness value of the target and the background fluctuates without a specific pattern, overall, the brightness value of the background is less than that of the target, and the point with the maximum brightness value A is certainly on the target bee. Therefore, taking the point closest to the bee's head with the maximum brightness value as the bee target point P is effective.

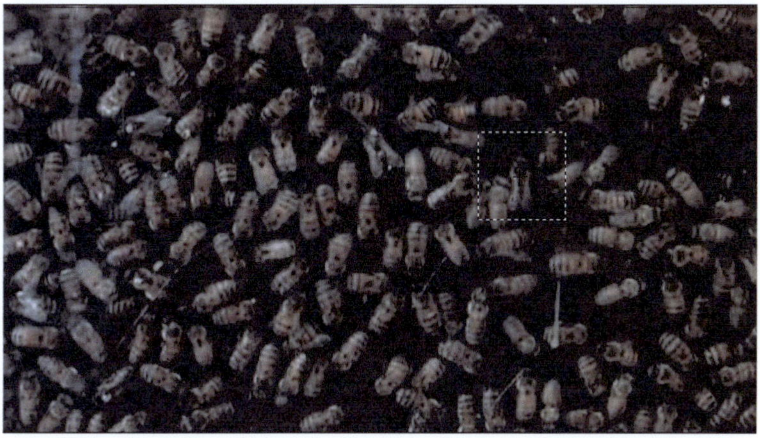

Fig. 11.8 Original image of honeycomb

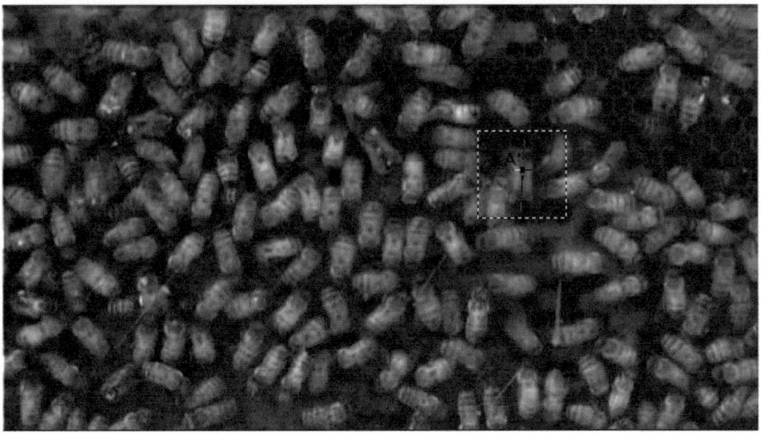

Fig. 11.9 2*R*-*B* gray scale image

Fig. 11.10 Line profile analysis diagram

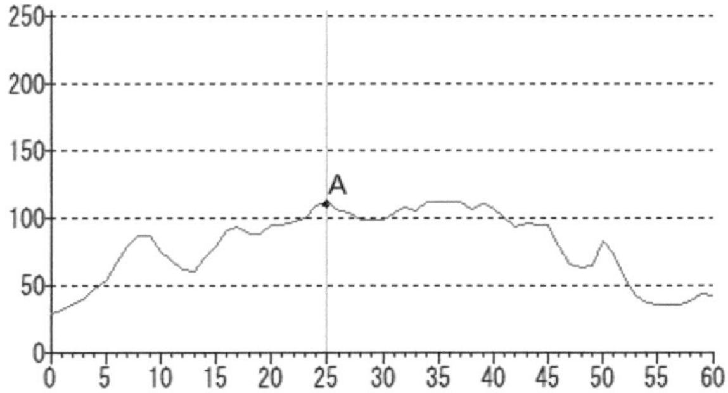

② Target point tracking. Starting from the second frame image, target point tracking detection is achieved through template matching with the previous frame image. A 9 × 9 pixel area template is established with the target point of the previous frame as the centre point. Template matching is performed on the current frame, and the matching area is referred to as sub-image $P(n)$ (where n is the sub-image number, $0 \leq n \leq 8$). The specific steps are as follows.

a. Establish the template. With the target point P of the previous frame as the centre, in a spiral manner as shown in Fig. 11.11, read its own and surrounding 80 pixels of R and B component values in a clockwise direction, and store them separately into arrays $R[k]$ and $B[k]$ ($0 \leq k \leq 80$). Sort the values in $R[]$ and $B[]$ as follows: find the maximum value of the R component in the outermost layer (i.e., $49 \leq k \leq 80$, a total of 32 points), and use this pixel as the starting point, with its preceding pixel as the endpoint, to rearrange the pixels in order. The newly arranged pixel R and B component values are then sequentially stored into arrays $SR[]$ and $SB[]$, respectively, to be used as a matching template.

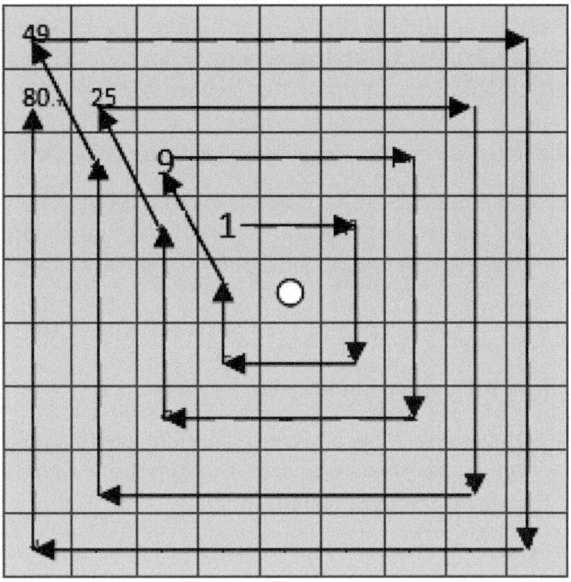

Fig. 11.11 Pixel reading sequence

b. Perform template matching on the current frame. As shown in Fig. 11.12, 0 represents the template target point P at the corresponding position on the current frame. On the current frame, place the template centre sequentially at positions 0 to 8 to obtain the corresponding sub-image $P\ (n)$. Use the method from step a to obtain the R and B component value arrays $R'[]$ and $B'[]$ for each pixel in sub-image $P\ (n)$, as well as the rearranged arrays $SR'[]$ and $SB'[]$. Calculate the matching degree DF for each sub-image using Eq. (11.4). The smaller this value, the higher the degree of matching.

$$DF = \sum_{k=0}^{80} \left| SR[k] - SR'[k] \right| + \sum_{k=0}^{80} \left| SB[k] - SB'[k] \right| \tag{11.4}$$

Fig. 11.12 Template movement sequence

8	1	2
7	0	3
6	5	4

Find the position N where the matching degree is highest, i.e., where DF is smallest (DF_{m}). If $N = 0$, stop searching; point 0 is the quasi-target point, and record the matching degree of this sub-image and the template as DF_{m}. If N is not equal to 0, move the template centre to point N, using this point as the new initial position 0 for the template centre, and continue searching for the quasi-target point.

c. Determine the target point. If Eq. (11.5) holds, the quasi-target point is considered the tracked target point; otherwise, the quasi-target point is not considered the tracked target point, and further target searching is required.

$$DF_{\min} < 5AR \tag{11.5}$$

where AR is the template area, i.e., $AR = 81$.

d. Target searching. Repeat steps b and c until a point satisfying Eq. (11.5) is found within the processing area, or the point with the smallest DF within the area is taken as the target point.

During the target point tracking process, the position of the target point in each frame is recorded, and the horizontal and vertical coordinates of the target point are sequentially stored into arrays X[] and Y[].

Figure 11.13 illustrates a set of examples of the template colour feature parameters R[], B[], SR[], and SB[]. Figure 11.13a and b show the R component array R[] and the B component array B[] of each pixel in the original template. In the figure, point b represents the point with the maximum R component in the outermost layer of the template (i.e., $49 \le k \le 80$, a total of 32 points), and points a and c are the starting and ending points of the original template, respectively. Figure 11.13c and d show the R component array SR[] of each pixel in the rearranged template. B component array SB[], as shown in the figure, rearranges the template in order starting from point b, with point a appended after point c.

This system reorders the template pixels starting from the point with the maximum value of the R component in the outermost layer of the template, then performs a matching operation once. This achieves the effect of multiple rotations and matching calculations of the template in traditional methods, reducing the computational load of template rotation and matching, significantly shortening processing time, and ensuring precise matching results.

Figure 11.14 shows the tracking result images of the target bee (within the dashed box) in the 2nd, 31st, 56th, and 125th frames from Fig. 11.8. The results indicate that the aforementioned method can accurately track the movement trajectory of the target bee.

(4) Bee Dance Determination

As shown in Fig. 11.15, the movement trajectory of the bee's waggle dance is in the shape of an 8. Point F is the starting point of the waggle, and point E is the end point. The direction FE is the straight-line direction of the bee's waggle dance, referred to as the crawling direction. The perpendicular direction to FE is the waggle

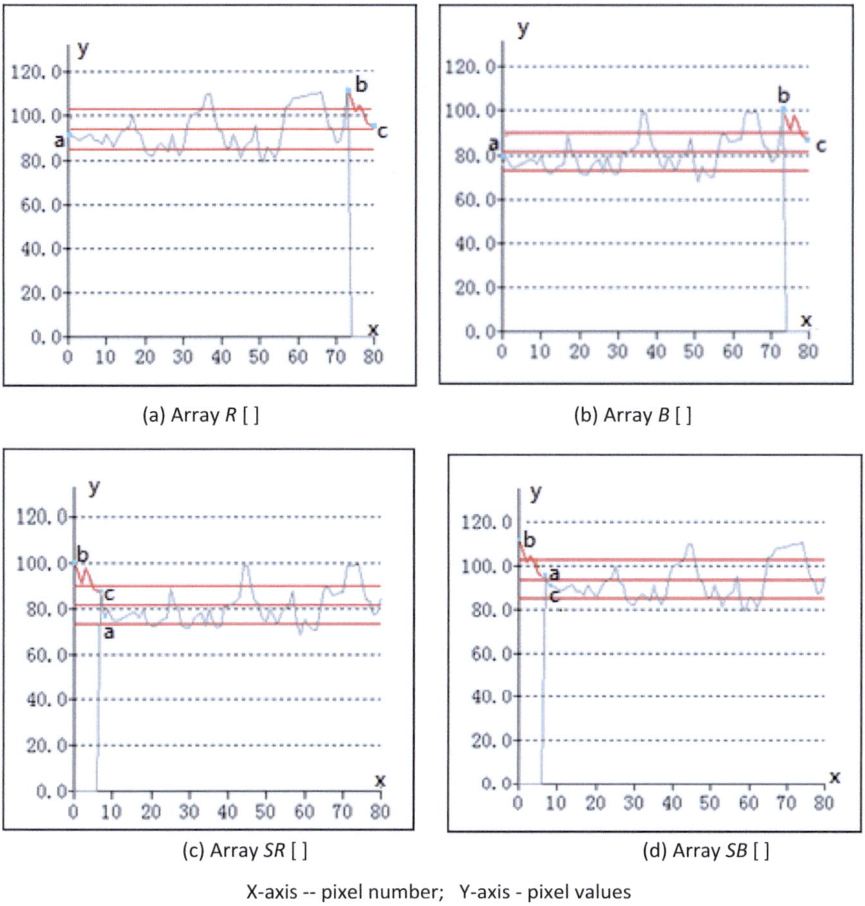

Fig. 11.13 Waveform diagram of template colour parameters

direction, and the inflection points of the waggle direction's coordinates are called waggle feature points. The bee repeatedly performs the same waggle dance several times near a location.

The absolute differences between the *x* and *y* coordinates of the aforementioned bee target point in each frame and its *x* and *y* coordinates in the first frame are sequentially stored in arrays D_x[] and D_y[] respectively. That is, arrays D_x[] and D_y[] represent the distances in the *x* and *y* directions from each point on the bee's movement trajectory to the starting point. Figure 11.16a and b are waveform diagrams of arrays D_x[] and D_y[] respectively, with the horizontal axis representing frame numbers and the vertical axis representing distance values (in pixels). By analysing the waveforms of arrays D_x[] and D_y[], the interval of the bee's waggle dance can be determined, thereby obtaining information such as the waggle time and waggle angle.

Fig. 11.14 Image of target bee tracking results

Fig. 11.15 How the bee waggle dance works

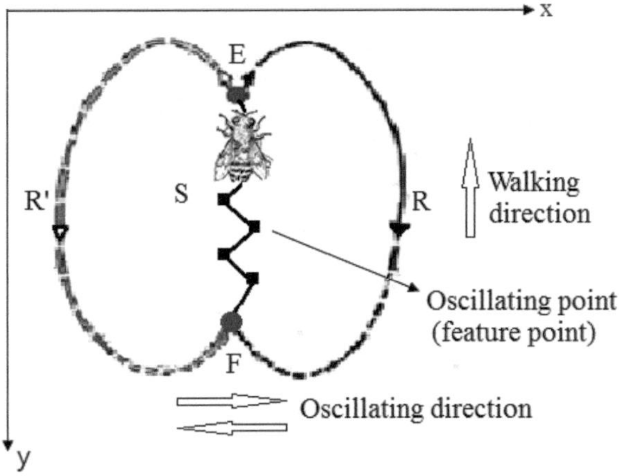

Taking $D_x[]$ as an example, the analysis process is as follows. Assume the size of array $D_x[]$ is M, and set the initial values of the peak position I_t, peak value t, trough position I_b, and trough value b all to 0. Scan the array $D_x[]$ from the start, comparing $D_x[k]$ $(1 \le k \le M)$ with $D_x[k\text{-}1]$ until the entire array is scanned.

① Find the peak. If the current point satisfies $D_x[k] > D_x[k\text{-}1]$, then compare $D_x[k]$ with… t in size, if $D_x[k] > t$, then record $I_t = k$, $t = D_x[k]$.

Otherwise, use Eq. (11.6) to judge. If the condition is met, then take the current I_t as the peak position, the current t as the peak value, and set $b = t$. Then proceed to step

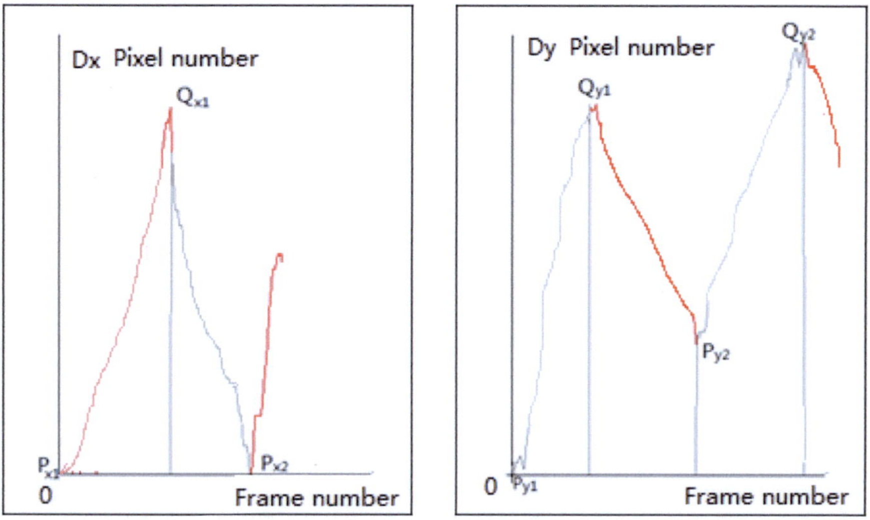

(a) Waveform diagram of array D_x [ ] (b) Waveform diagram of array D_y [ ]

Fig. 11.16 Waveform diagram of array D_x and D_y

② for trough search; otherwise, continue to repeat this step until a peak is found.

$$t - D_x[k] > D \text{ and } t - b > D \tag{11.6}$$

In the formula, $D = d/5.0$, where d is the length of the bee.

② Trough search. If the current point satisfies $D_x[k] < D_x[k\text{-}1]$, then compare $D_x[k]$ with b in size. If $D_x\ [k] < b$, then record $I_b = k$, $b = D_x[k]$.

Otherwise, use Eq. (11.7) to judge. If the condition is met, then take the current I_b as the trough position, the current b as the trough value, and set $t = b$. Then proceed to step ① for peak search; otherwise, continue to repeat this step until a trough is found.

$$D_x[k] - b > D \text{ and } t - b > D \tag{11.7}$$

After completing the scan of D_x, the positions of each peak and trough in the array are found. As shown in Fig. 11.16a, the troughs in D_x are determined as P_{x1}, P_{x2}, and the peak as Q_{x1}.

Then, determine the crawling direction of each interval (the area between adjacent peaks and troughs). Denote the starting point (left end) of the interval as I_s, and the endpoint (right end) as I_e. If $X[I_e] > X[I_s]$, then classify this interval direction as rightward; otherwise, classify it as leftward. Accordingly, the interval $P_{x1}\ Q_{x1}$ is classified as rightward, $Q_{x1}\ P_{x2}$ interval is classified as leftward.

For each point k within the interval, compare the y coordinate values with its adjacent points. If it satisfies Eq. (11.8) or Eq. (11.9), then k is considered a point with oscillation characteristics (inflection point of the coordinates), i.e., an oscillation characteristic point (see Fig. 11.15), and record its count N (defined as the number of oscillation characteristics). N The larger the value, the more pronounced the oscillation characteristics.

$$Y[k] < Y[k+1], Y[k] < Y[k-1] \tag{11.8}$$

$$Y[k] > Y[k+1], Y[k] > Y[k-1] \tag{11.9}$$

where, $I_s \leq k < I_e$.

Accordingly, obtain the characteristic parameter N_{R1} for the rightward interval P_{x1} Q_{x1} and the characteristic parameter N_{L1} for the leftward interval Q_{x1} P_{x2}.

Similarly, analyse the array $D_y[]$ as shown in Fig. 11.16b, to determine the troughs P_{y1}, P_{y2} and peaks Q_{y1}, Q_{y2} of the array $D_y[]$, obtaining the characteristic parameters N_{U1}, N_{U2} for the upward intervals P_{y1} Q_{y1}, P_{y2} Q_{y2} respectively, and the characteristic parameter N $_{D1}$ for the downward interval Q_{y1} P_{y2}.

Calculate the average values of the oscillation characteristic parameters for the upward, downward, leftward, and rightward intervals N_U, N_D, N_L, N_R, and identify the maximum value among them, which indicates the crawling direction of the bee waggle dance, with the waggle direction being perpendicular to the crawling direction.

If the oscillation characteristic parameter is greater than 20, the segment is considered an oscillation interval; otherwise, it is regarded as a non-oscillation interval. For the oscillation interval, set the starting frame and ending frame as DS_s and DS_e.

The swing time T can be obtained from Eq. (11.10).

$$T = (DS_e - DS_s)/R_f \tag{11.10}$$

In the formula, R_f represents the frame rate of the video file or real-time acquisition.

The swing angle is obtained by the following method. Calculate the average values of the horizontal and vertical coordinates of each point within the swing interval, X_a and Y_a, respectively. Using the point (X_a, Y_a) as a known point, perform a line fitting on the points within the swing interval using the Hough transform through the known point, and denote the resulting line as L. Define the crawling direction of the bee as the direction of L, and the angle between the line L and the vertically upward direction as the swing angle. It is stipulated that clockwise from vertically upward to vertically downward is 0 ~ 180°, and counterclockwise from vertically upward to vertically downward is 0 ~ −180°, as shown in Fig. 11.17, θ is the swing angle.

For bees that swing multiple times, calculate the swing time and angle for each instance, and take their average as the final parameter.

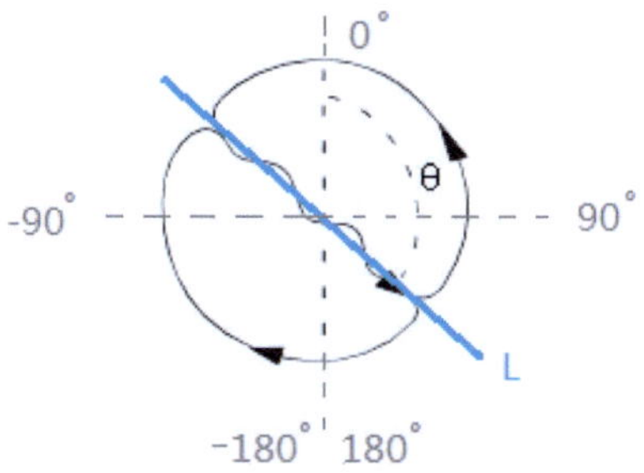

Fig. 11.17 Swing angle

Multiple targets can be selected simultaneously, and the above processing can be performed separately to achieve tracking and analysis of the motion trajectories of multiple target bees.

Figure 11.18 is a waveform diagram of the distance $D_y[]$ and $D_x[]$ from each frame to the initial frame of the target bee in Fig. 11.8. The horizontal axis x represents the frame number, and the vertical axis y represents the distance (in pixels) from the target point on each frame to the target point on the initial frame.

Figure 11.18a detected 6 upward intervals $P_{y1}\ Q_{y1}$, $P_{y2}\ Q_{y2}$, $P_{y3}\ Q_{y3}$, $P_{y4}\ Q_{y4}$, $P_{y5}\ Q_{y5}$, $P_{y6}\ Q_{y6}$ and 6 downward intervals $Q_{y1}\ P_{y2}$, $Q_{y2}\ P_{y3}$, $Q_{y3}\ P_{y4}$, $Q_{y4}\ P_{y5}$, $Q_{y5}\ P_{y6}$, $Q_{y6}\ P_{y7}$. Figure 11.18b detected $P_{x1}\ Q_{x1}$, $P_{x2}\ Q_{x2}$, $P_{x3}\ Q_{x3}$, $P_{x4}\ Q_{x4}$, $P_{x5}\ Q_{x5}$, $P_{x6}\ Q_{x6}$, and other 6 leftward intervals and $Q_{x1}\ P_{x2}$, $Q_{x2}\ P_{x3}$, $Q_{x3}\ P_{x4}$, $Q_{x4}\ P_{x5}$, $Q_{x5}\ P_{x6}$,

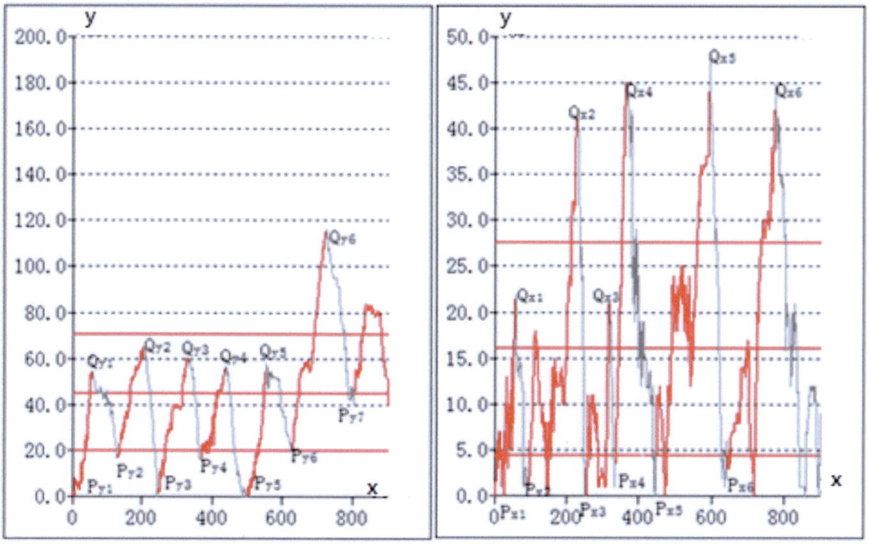

(a) Array Dy[] (b) array Dx[]

X-axis - frame number; Y-axis - distance (number of pixels)

Fig. 11.18 Distance change curve between each frame of the target bee and the initial frame

and other 5 rightward intervals. It can be seen that this algorithm effectively detects the peaks and troughs of the distance change curve.

From Fig. 11.18, the direction of the bee's movement can be determined. For example, $P_{y1}\ Q_{y1}$ indicates the bee moving away from the initial position in the y direction, and $Q_{y1}\ P_{y2}$ indicates returning to the initial position. Whether the bee is dancing needs to be determined by detecting the number of swing features in the direction perpendicular to the direction of movement, that is, to determine whether the bee is dancing in the y direction intervals $P_{y1}\ Q_{y1}$ and $Q_{y1}\ P_{y2}$, it is necessary to use the number of swing features in the perpendicular x direction. Figure 11.19 shows the change curve of the bee's x coordinate in the interval $P_{y1}\ Q_{y1}$, with the horizontal axis representing the frame number and the vertical axis representing the target bee's x coordinate on the image. It can be seen that the number of swing features (curve inflection points) is 32. Similarly, the number of swing features for other intervals moving away from the starting point (upward) $P_{y2}\ Q_{y2}$, $P_{y3}\ Q_{y3}$, $P_{y4}\ Q_{y4}$, $P_{y5}\ Q_{y5}$, $P_{y6}\ Q_{y6}$ are 28, 26, 30, 19, 12 respectively, with an average of 25 for the 6 feature numbers, i.e., the swing feature number in the upward direction N_U is 25.

Similarly, the downward interval $Q_{y1}\ P_{y2}$, Q_{y2}, P_{y3}, Q_{y3}, P_{y4}, Q_{y4}, P_{y5}, Q_{y5}, P_{y6}, Q_{y6}, P_{y7} have swing characteristic numbers of 10, 15, 2, 10, 8, 20 respectively, and the final swing characteristic number N_D is their average, 11. For Fig. 11.18b, the measured left and right swing characteristic numbers are $N_L = 15$ and $N_R = 6$. Among N_U, N_D, N_L, N_R, the largest is $N_U = 25$, indicating that the bee is performing a waggle dance in the y direction. In the y direction, there are four intervals, $P_{y1}\ Q_{y1}$, $P_{y2}\ Q_{y2}$, $P_{y3}\ Q_{y3}$, $P_{y4}\ Q_{y4}$, where the swing characteristic numbers exceed the threshold of 20, thus these four intervals are the bee's waggle dance intervals. The specific crawling angle is obtained through the Hough transform of the trajectory coordinates. Figure 11.20 shows the result image of tracking analysis conducted simultaneously on five target bees, processing a total of 1000 frames. The behaviours of the five target bees were correctly tracked and analysed, with only the target bee shown in Fig. 11.8

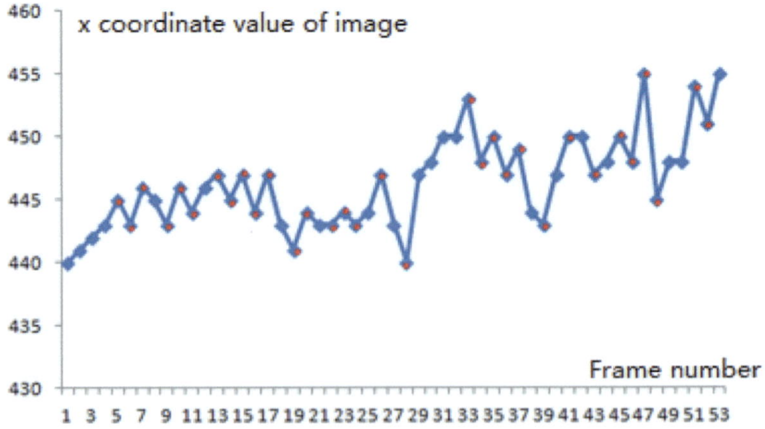

Fig. 11.19 Waveform of $P_{y1}\ Q_{y1}$ segment $X[]$

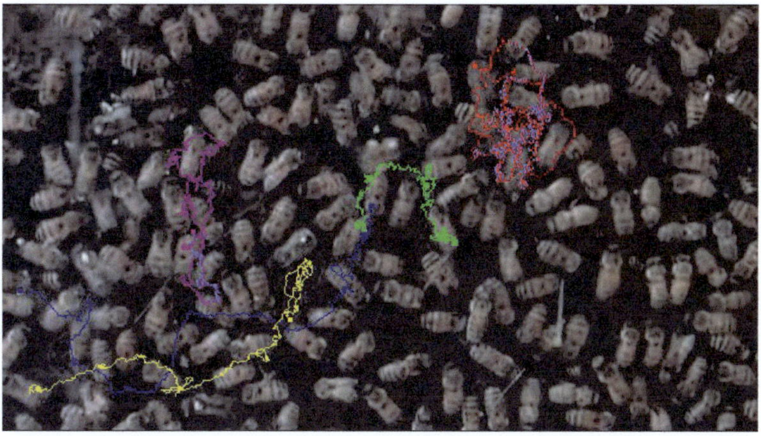

Fig. 11.20 Movement trajectory and waggle dance information of target bee

(the rightmost trajectory in Fig. 11.20) exhibiting waggle dance behaviour, and the colour of its waggle dance trajectory differs from that of the crawling trajectory.

References

1. Bingqi Chen, Zhiqiang Wang: A Statistical Method for Technical Data of a Badminton Match based on 2D Seriate Images [J]. Tsinghua Science and Technology. 2007, 12(5): 594–601.
2. Ming Xiaoqiang. Research on Unmarked Image Tracking and Analysis Method of Bee Waggle Dance [D]. Beijing: China Agricultural University, 2013.

Chapter 12
Fourier Transform

12.1 The World of Frequency [1]

The theme of this chapter is completely different from the image processing methods and perspectives introduced so far. Previously, many image processing methods were introduced, all of which are visually easy to understand. This is because they utilise the visual properties of images. However, frequency-based methods appear to process images using concepts that are unrelated to their visual properties.

When we think of frequency, we often associate it with the realm of sound. Therefore, let us use sound to analogously explain the frequency of images. As shown in Fig. 12.1, the low frequencies of an image correspond to the general parts, or the smooth areas of overall grayscale. The high frequencies of an image represent the subtle parts, namely the edges and noise. So, what is the goal of using frequency to process them? Let's use sound as a comparison to explain. Frequency processing of sound is something we commonly experience. For example, by adjusting the tone controller on a stereo sound system, reducing the TREBLE (high frequency) results in a muffled sound, whereas reducing the BASS (low frequency) results in a sharp sound. Images can also undergo frequency processing. Figure 12.2 shows an example of processing; if the high-frequency components are removed, the subtle parts disappear, making the image blurry. Conversely, if the low-frequency components are removed, the general parts disappear, leaving only the edges.

To process images using frequency, the image must first be transformed into the frequency world (frequency domain). This transformation is accomplished using the Fourier transform. Fourier transform is a specialised subject in mathematics, and frequency processing alone can be considered a research field.

The aim of this book is to explain in a simple and easy-to-understand manner, attempting to elucidate these complex contents in the simplest way possible. Firstly, we will discuss the transformation of one-dimensional signals into the frequency domain, and subsequently, the frequency transformation of two-dimensional signals, such as images.

© Chemical Industry Press 2026

B. Chen and S. Chen, *Machine Vision Technology*, Advanced and Intelligent Manufacturing in China, https://doi.org/10.1007/978-981-95-1184-6_12

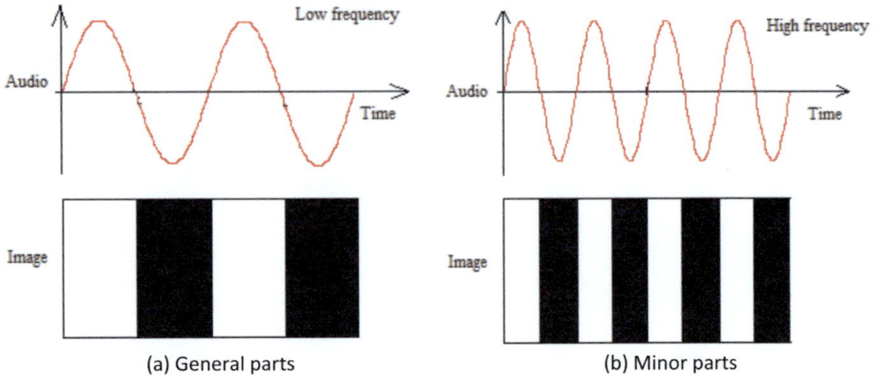

Fig. 12.1 Frequencies of sound and images

12.2 Frequency Transformation

The principle of frequency transformation is that any waveform can be decomposed into a sum of simple sine waves. For example, the waveform shown in Fig. 12.3a can be decomposed into four sine waves with different frequencies, as shown in Fig. 12.3b–e.

Taking the waveform shown in Fig. 12.3d as an example, see Fig. 12.4. If the basic sine wave passing through the origin with a magnitude of 1 is represented by a dashed line, the solid line wave can be determined by the magnitude (magnitude or amplitude) A and phase ϕ. Thus, the four waveforms in Fig. 12.3 b–e can be plotted as graphs with frequency f on the horizontal axis and magnitude A on the vertical axis, as well as frequency f on the horizontal axis and phase ϕ on the vertical axis, as shown in Fig. 12.5. This type of graph reflecting the relationship between frequency, magnitude, and phase is called a Fourier spectrum. In this way, the waveform of Fig. 12.3a is transformed into the frequency domain of Fig. 12.5.

It can be seen that no matter how complex the waveform is in the spatial domain, it can be transformed into the frequency domain. Generally, in the frequency domain, it is also in a continuous form, as shown in Fig. 12.6.

Expressed in formula form:

$$f(t) \underset{\text{Inverse Fourier transform}}{\overset{\text{Fourier transform}}{\longleftrightarrow}} A(f), \varphi(f) \tag{12.1}$$

This transformation is called the Fourier transform, which is a type of orthogonal transformation.

In general, in the Fourier transform, to simultaneously represent magnitude A and phase ϕ, a complex number form can be used. A complex number is a number represented by a combination of a real part a and an imaginary part b, expressed by the following formula:

(a) Original image (b) Remove high-frequency image

(b) Remove low-frequency image

Fig. 12.2 Example of frequency-based processing

$$a + jb \qquad \text{Here } (j = \sqrt{-1}) \tag{12.2}$$

Using this formula allows the concepts of magnitude and phase to be handled with a single complex number. Thus, the Fourier transform of Eq. (12.1) can be expressed using the complex function $F(f)$ or $F(\omega)$:

$$f(t) \underset{\text{Inverse Fourier transform}}{\overset{\text{Fourier transform}}{\longleftrightarrow}} F(f), F(\omega) \tag{12.3}$$

The process of deriving $F(f)$ or $F(\omega)$ from $f(t)$ is quite complex and will not be introduced here. The result is shown in Eq. (12.4):

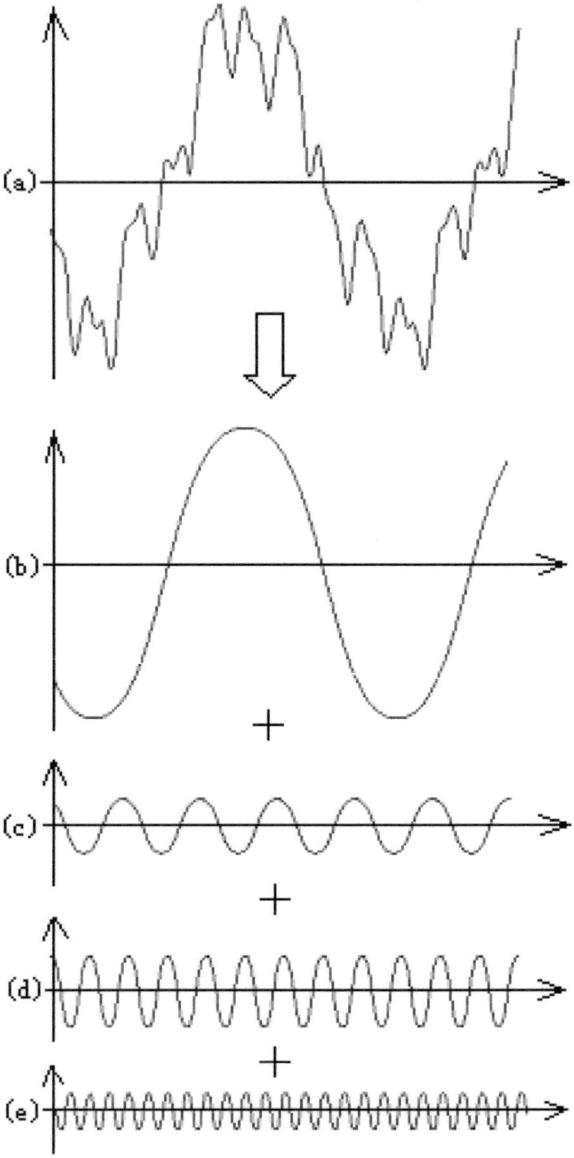

Fig. 12.3 Waveform decomposition into sine waves

$$F(f) = \int_{-\infty}^{\infty} f(t)\mathrm{e}^{-j2\pi ft}\mathrm{d}t \qquad \text{Fourier transform}$$

$$f(t) = \int_{-\infty}^{\infty} F(f)\mathrm{e}^{j2\pi fx}\mathrm{d}x \qquad \text{Inverse Fourier transform}$$

or

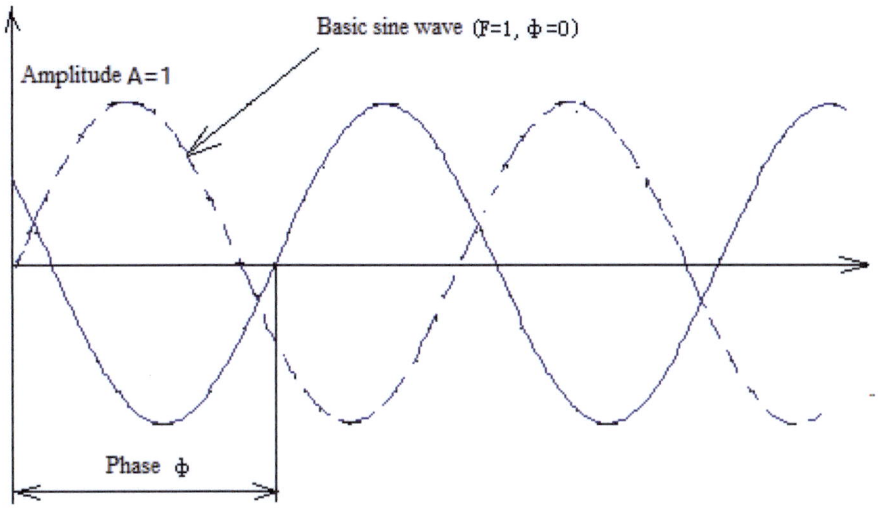

Fig. 12.4 Amplitude and phase of sine wave

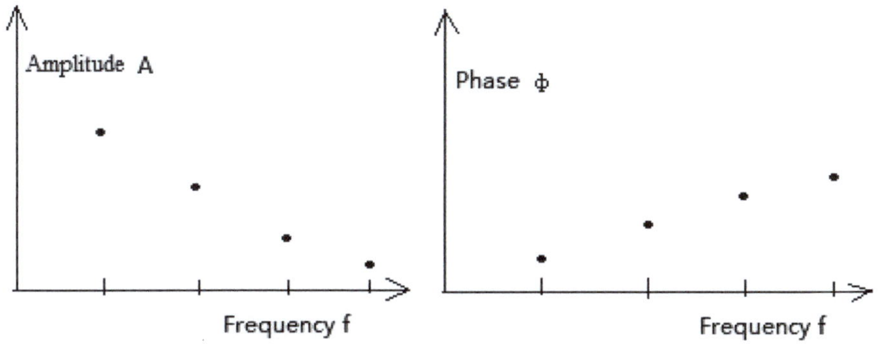

Fig. 12.5 Spectrum diagram of the waveform shown in Fig. 12.3a

$$F(\omega) = \int_{-\infty}^{\infty} f(t)\mathrm{e}^{-j\omega t}\mathrm{d}t \qquad \text{Fourier transform}$$

$$f(t) = \int_{-\infty}^{\infty} F(\omega)\mathrm{e}^{j\omega t}\mathrm{d}t \qquad \text{Inverse Fourier transform} \tag{12.4}$$

where the angular frequency $\omega = 2\pi f$. This is a very important fundamental formula used in all frequency processing. The purpose of this book is to process digital images using computers, without delving deeply into this seemingly complex formula. The difference between the field of computing and the field of mathematics lies in the following two points: one point is that, so far, the signals involved $f(t)$ are continuous signals (analogue signals) as shown in Fig. 12.7a, whereas the signals processed in the field of computing are sampled digital signals as shown in Fig. 12.7b. Another

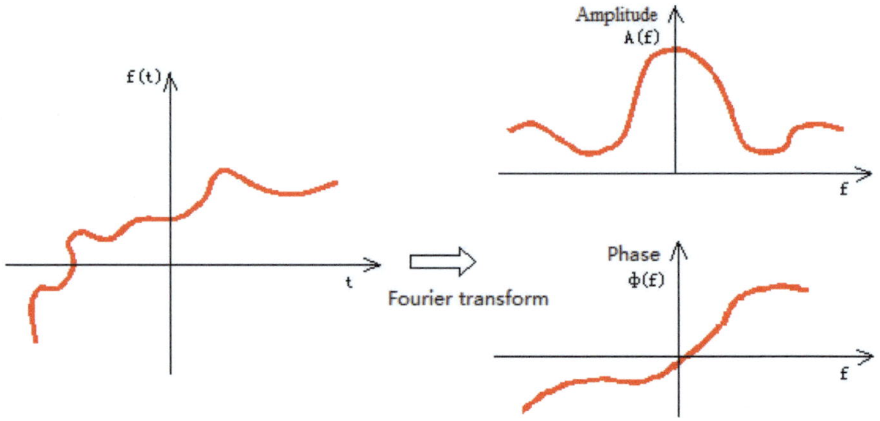

Fig. 12.6 Fourier transform

Fig. 12.7 Analog signal and digital signal

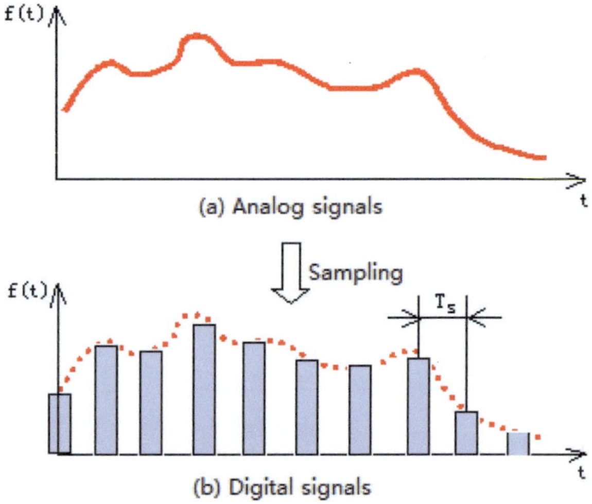

point is that considering infinity is common in mathematics, but computers must perform a finite number of operations. The Fourier transform, considering the above limitations, is called the discrete Fourier transform (DFT).

12.3 Discrete Fourier Transform

The discrete Fourier transform (DFT) can be derived by converting the Fourier transform in Eq. (12.4) into discrete values. Assume the input signal comprises $x(0)$, $x(1)$, $x(2)$, …, $x(\mathrm{N}-1)$, a total of N discrete points, then the result of the transformation

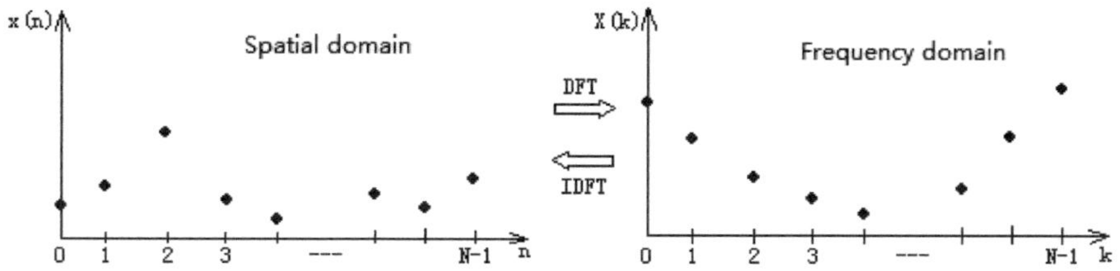

Fig. 12.8 N discrete signals become N frequency signals after DFT transformation

to the frequency domain (complex numbers) as shown in Fig. 12.8 is also N discrete values X (0), X (1), X (2), ..., X (N − 1).

The relational expression is shown as follows:

$$\begin{aligned} X(k) &= \frac{1}{\sqrt{N}} \sum_{n=0}^{N-1} x(n) W^{kn} \text{ DFT} \\ x(n) &= \frac{1}{\sqrt{N}} \sum_{k=0}^{N-1} X(k) W^{-kn} \text{ IDFT} \end{aligned} \tag{12.5}$$

where $k = 0, 1, 2, \ldots, N - 1$; $n = 0, 1, 2, \ldots, N - 1$; $W = \mathrm{e}^{-j\frac{2\pi}{N}}$; IDFT is the inverse discrete Fourier transform (Inverse Discrete Fourier Transform). This is the basic operational formula of the DFT. The integration operation is replaced by a summation operation, and W is called the twiddle factor.

In the field of complex numbers, there is Euler's formula, as shown in Fig. 12.9 and Eq. (12.6).

$$e^{jt} = \cos t + j \sin t \tag{12.6}$$

The twiddle factor can be substituted using Euler's formula as follows:

$$W^{kn} = \mathrm{e}^{-j\frac{2\pi}{N}kn} = \cos\left(\frac{2\pi}{N}kn\right) - j\sin\left(\frac{2\pi}{N}kn\right) \tag{12.7}$$

Fig. 12.9 Complex coordinates of magnitude 1 and declination angle t

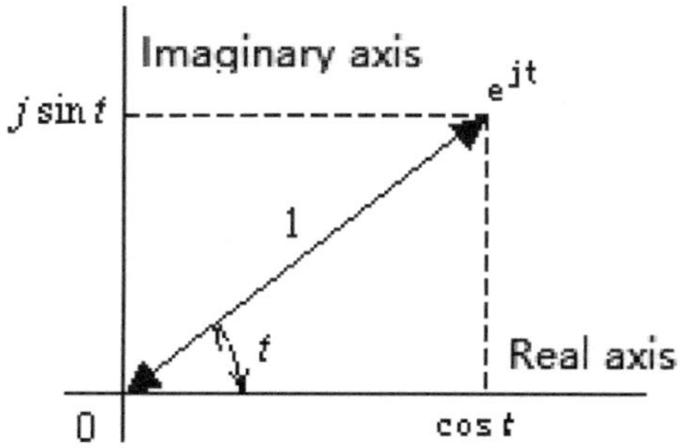

By substituting Eq. (12.7) into Eq. (12.5), only trigonometric functions and summation operations remain, allowing computation by computer, although the computational load is quite large. Therefore, the Fast Fourier Transform (FFT) algorithm was proposed, which can save a considerable amount of computation when the data is a power of 2.

When performing a Fast Fourier transform, the actual signal needs to be input as the real part, and the output is the real part of the complex number (denoted by *a*_ rl) and the imaginary part (denoted by *a*_im represents). If you want to understand the amplitude characteristic *A* (Amplitude Characteristic) and phase characteristic ϕ (Phase Characteristic), the following transformation can be performed:

$$\begin{aligned} A &= \sqrt{a\_\mathrm{rl}^2 + a\_\mathrm{im}^2} \\ \varphi &= \tan^{-1}\left(\frac{a\_\mathrm{im}}{a\_\mathrm{rl}}\right) \end{aligned} \tag{12.8}$$

What frequency components do the *N* sequences obtained in this way represent? Refer to Fig. 12.10, in fact, the far left is the DC component, and the far right is the sampling frequency component. Additionally, there is a notable feature that the amplitude characteristic is symmetrical about the point at half the sampling frequency, and the phase characteristic is point-symmetrical about the centre. What does this indicate?

First, let's understand the concepts of sampling frequency and sampling theorem. Refer to Fig. 12.7, after sampling an analogue image at a certain time interval *T* to obtain a digital image, 1/*T* (Hz) is called the sampling frequency. According to the sampling theorem, a digital signal can only represent an analogue signal with a frequency up to half the sampling frequency. For example, a CD uses a sampling frequency of 44.1 kHz, theoretically only representing sound signals from 0 to 22.05kHz. Therefore, when the sampling frequency is f_s, the meaning of replacing an analogue signal with a digital signal is essentially that it only has values between 0 and $f_s/2$.

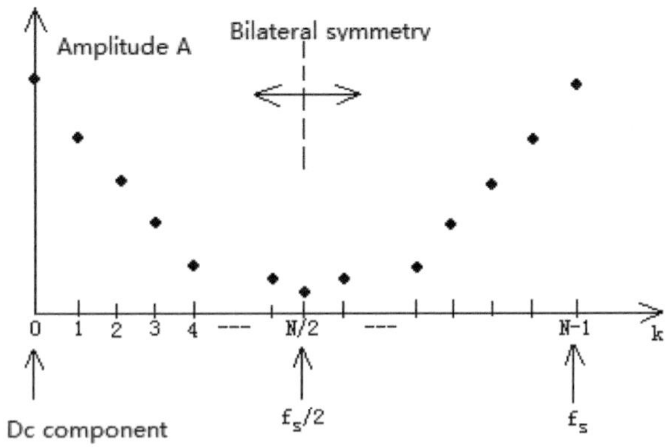

Fig. 12.10 Amplitude A and phase φ are obtained from DFT

12.4 Two-Dimensional Fourier Transform of Images

This section begins the main topic. All the signals introduced so far are one-dimensional signals, and since images are planar, they are two-dimensional signals with frequencies in both horizontal and vertical directions. Furthermore, in the frequency spectrum of an image, the center of the frequency plane is commonly regarded as the DC component.

Figure 12.11 shows the situation corresponding to an actual image when the horizontal frequency is u and the vertical frequency is v. Similarly, the amplitude characteristic of a two-dimensional spectrum is symmetrical about the amplitude A axis, and the phase characteristic is point-symmetrical about the origin.

So, how is the two-dimensional frequency calculated? A relatively simple method is to perform a one-dimensional FFT in the horizontal direction and a one-dimensional FFT in the vertical direction, as shown in the processing block diagram in Fig. 12.12.

By visualising the amplitude characteristic as a greyscale value, the result is shown in Fig. 12.13. Comparing Fig. 12.13a and b, it can be seen that images with fewer

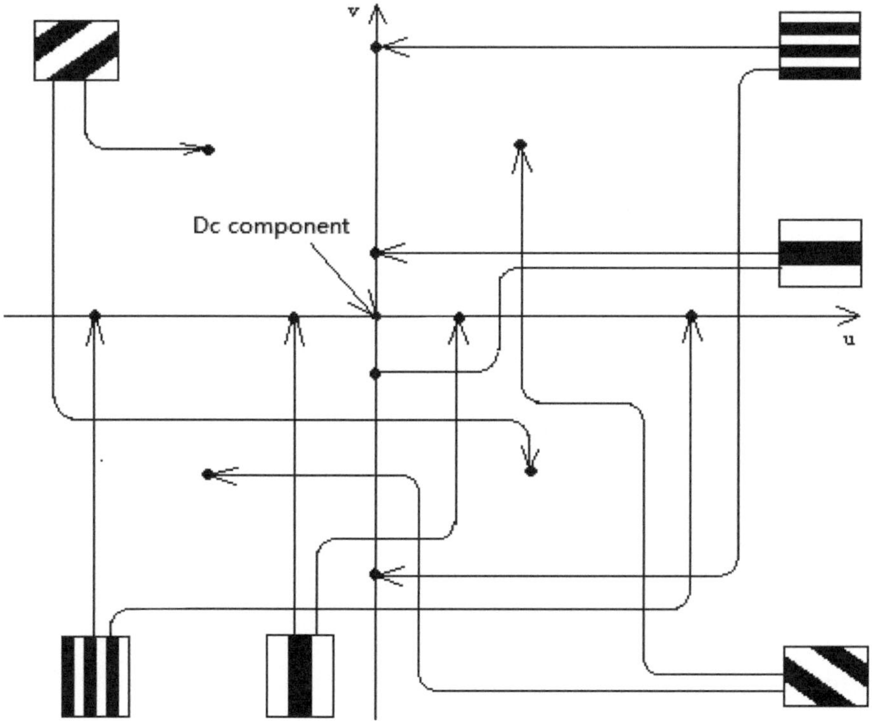

Fig. 12.11 Relation between two-dimensional frequency and image

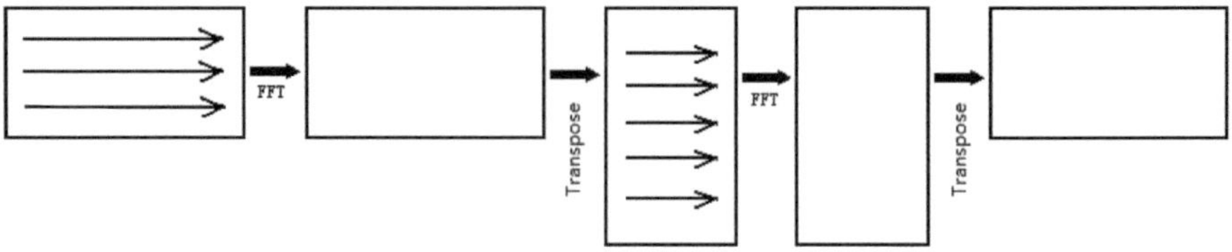

Fig. 12.12 Processing block diagram of 2D FFT

details have more low frequencies, while images with more details have more high frequencies.

12.5 Filtering Processing

The function of a filter is to allow certain things to pass through while blocking others. In the frequency domain, a filter allows certain frequencies to pass through while blocking others. As shown in Fig. 12.14, by setting the values of parameters a and b, frequencies above a and below b (frequencies indicated by the slanted lines) are allowed to pass, while other frequencies are blocked for filtering processing. Figure 12.15 shows the result of setting the high-frequency components of the frequency spectrum, obtained through DFT processing of an image, to zero, followed by IDFT processing to transform back to the image. It can be seen that the high-frequency components (detail parts) of the image have disappeared, making it blurred. Next, let's look at the result of setting the low-frequency components to zero, as shown in Fig. 12.16, where the edges have been extracted. This is because many high-frequency components are contained in the edges.

This filtering process can be considered as the multiplication of the frequency of the filter and the frequency of the image. In fact, by changing the frequency characteristics of this filter, various processing can be achieved. Assuming the input image is $f(i, j)$, the frequency of the image $F(u, v)$ becomes:

$$F(u, v) = D\left[f(i, j)\right] \quad \text{where } D[] \text{ represents DFT} \tag{12.9}$$

If the frequency characteristic of the filter is represented as $S(u, v)$, then the processed image $g(i, j)$ is expressed as:

$$g(i, j) = D^{-1}[F(u, v)S(u, v)] \quad \text{where } D^{-1}[] \text{ represents IDFT} \tag{12.10}$$

Here, assuming $S(u, v)$ is obtained by IDFT as $s(i, j)$, then Eq. (12.10) is transformed into:

$$g(i, j) = D^{-1}[F(u, v)S(u, v)]$$

(a) Images with little detail

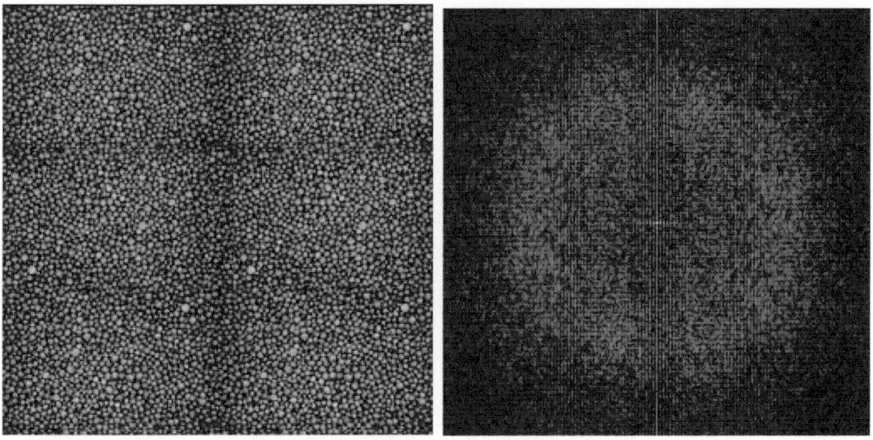

(b) More detailed images

Fig. 12.13 Example of image FFT transformation

$$
\begin{aligned}
&= D^{-1}[F(u,v)]D^{-1}[S(u,v)] \\
&= f(i,j)s(i,j)
\end{aligned}
\tag{12.11}
$$

The symbol $\otimes$ is called convolution operation, which has actually appeared many times in image processing so far. The differential operations using differential operators are convolution operations, for example, the Laplacian operator. From Eq. (12.11), we can derive the following very important property: convolution operation in the image (spatial domain) is exactly the same as the product operation in the frequency domain. From this result, it is evident that the Laplacian operator is essentially a filtering process that allows the high-frequency components of the image to pass through, thereby enhancing the high-frequency components. Similarly,

Fig. 12.14 Filter shape

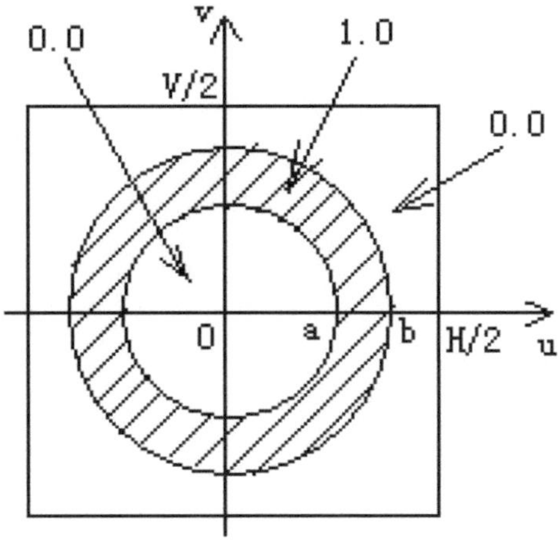

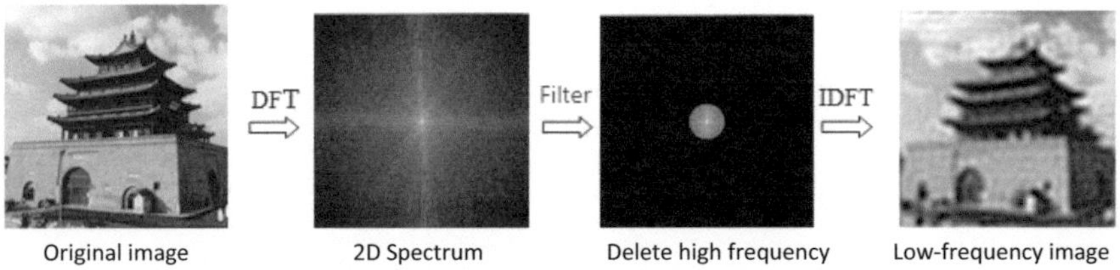

Fig. 12.15 Processing to remove image high frequency

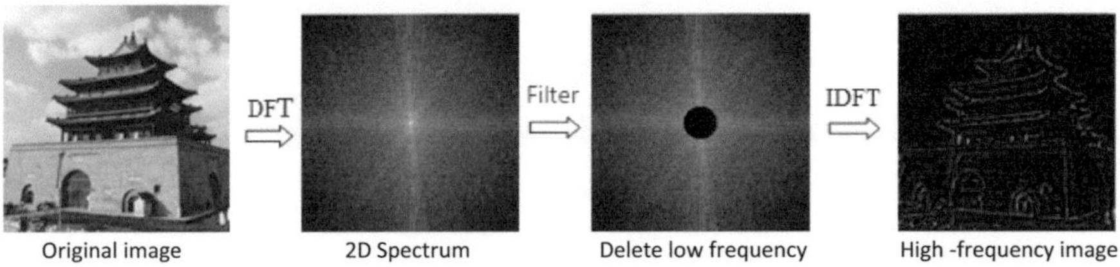

Fig. 12.16 Processing to remove image low frequency

smoothing (moving average method) is a filtering process that allows low-frequency components to pass through.

Reference

1. Chen Bingqi. Practical Digital Image Processing and Analysis [M]. 2nd Edition. Beijing: China Agricultural University Press, 2014.

Chapter 13
Wavelet Transform

13.1 Overview of Wavelet Transform [1]

Wavelet analysis is a new analytical method developed in the late 1980s, which serves as an outstanding example of the convergence of pure mathematics and applied mathematics, following the principles of Fourier analysis. The generation, development, and application of wavelet transform have benefitted from the collective efforts of experts, scholars, and engineers in numerous fields, including computer science, signal processing, image processing, applied mathematics, and earth sciences. Theoretically, the main contributors to the systematic framework of wavelet transform are mathematician Y. Meyer, geophysicist J. Morlet, and theoretical physicist A. Grossman. I. Daubechies and S. Mallat have played an extremely important role in applying this theory to the engineering field. Wavelet analysis has now become a hot topic with a very wide range of applications in scientific research and engineering technology. Different fields have different perspectives on wavelet analysis:

① Mathematicians say that wavelets are a beautiful representation of function spaces;
② Signal processing experts believe that wavelet analysis is a new theory for time–frequency analysis of non-stationary signals;
③ Image processing experts consider wavelet analysis an effective tool for space-scale analysis and multiresolution analysis in digital image processing;
④ Scholars in earth sciences and fault diagnosis regard wavelet analysis as a new technique for position-scale analysis in singularity detection;
⑤ Microlocal analysts view wavelet analysis as a new approach to time-scale analysis in fine-local analysis.

In summary, wavelet transform has a multiresolution characteristic, also known as a multiscale characteristic, allowing for the gradual observation of signals from coarse to fine. It can also be seen as filtering a signal with a set of band-pass filters. By appropriately selecting scale and translation factors, a scalable window can be obtained. By

© Chemical Industry Press 2026

B. Chen and S. Chen, *Machine Vision Technology*, Advanced and Intelligent Manufacturing in China, https://doi.org/10.1007/978-981-95-1184-6_13

suitably choosing the basic wavelet, wavelet transform can represent local features of signals in both time and frequency domains. The combination of multiresolution analysis and filter banks enriches the theoretical foundation of wavelet analysis and broadens its application scope. All this illustrates a simple fact: wavelet analysis has deeply rooted itself in many fields of scientific research and engineering technology that we are interested in. An era of researching and using wavelet transform theory and wavelet analysis has arrived.

13.2 Wavelets and Wavelet Transform

So far, Fourier transform, introduced in Chap. 12, is often used in general signal analysis and synthesis. However, because the Fourier basis uses infinite continuous trigonometric functions without local properties, time information is completely lost in the frequency domain after Fourier transform. In contrast, the wavelet transform introduced in this chapter can obtain local frequency information, making effective time–frequency analysis possible.

So, what are wavelets and wavelet transform? A musical score can be seen as a description of a two-dimensional time–frequency space, as shown in Fig. 13.1. Frequency (pitch) increases from the bottom upwards, while time (beat) progresses to the right; each note in a musical piece corresponds to a wavelet component (pitch burst) that will appear in the performance record of the piece; each wavelet's duration is encoded by the type of note (quarter note, half note, etc.) rather than by their horizontal extension. Suppose you want to analyse a record of a musical performance and write the corresponding score; this process can be called wavelet transform. Similarly, a musician's recording of a piece can be seen as an inverse wavelet transform because it reconstructs the signal using time–frequency representation.

Wavelet means "small wave" or "fine wave," which is a wave with an average value of 0 and an effectively finite duration. Specifically, wavelet is a function or signal $\psi(t)$ that in the space square integrable function $L^2(R)$ (where R represents real numbers) satisfies the following conditions.

$$\int_R |\psi(t)|^2 dt < \infty \tag{13.1}$$

$$\int_{R^*} \frac{|\Psi(\omega)|^2}{|\omega|} d\omega < \infty \tag{13.2}$$

ψ(t) is also called a basic wavelet or mother wavelet, and Eq. (13.2) is known as the admissibility condition.

$$\psi_{a,b}(t) = \frac{1}{\sqrt{a}} \psi\left(\frac{t-b}{a}\right) \tag{13.3}$$

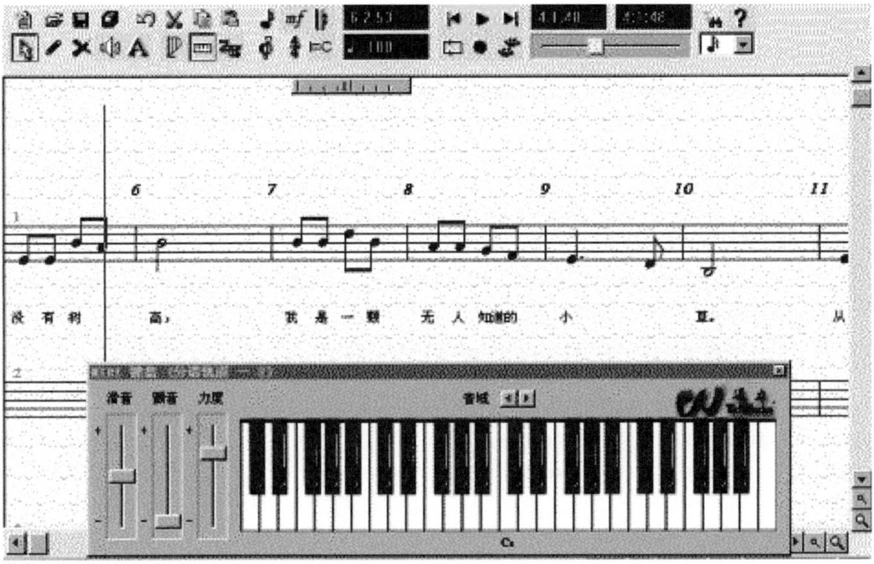

Fig. 13.1 Music score

The function defined in Eq. (13.3) represents a continuous wavelet function (continuous wavelet transform, CWT), which is generated by the mother wavelet and is dependent on the parameters a and b. It is commonly referred to as the wavelet function, as shown in Fig. 13.2. It is obtained by scaling the wavelet $\psi(t)$ horizontally by a factor of a and translating it by a distance b. The factor $1/\sqrt{a}$ is a normalization coefficient. Here, a is the scale parameter, and b is the shift parameter. Since a represents the time amplitude of the wavelet, 1/a is equivalent to frequency.

For any function or signal $f(t) \in L^2(R)$, its continuous wavelet transform is given by

$$W(a, b) = \frac{1}{\sqrt{a}} \int_R f(t) \psi^* \left(\frac{t - b}{a} \right) dt \tag{13.4}$$

Fig. 13.2 Wavelet and wavelet function

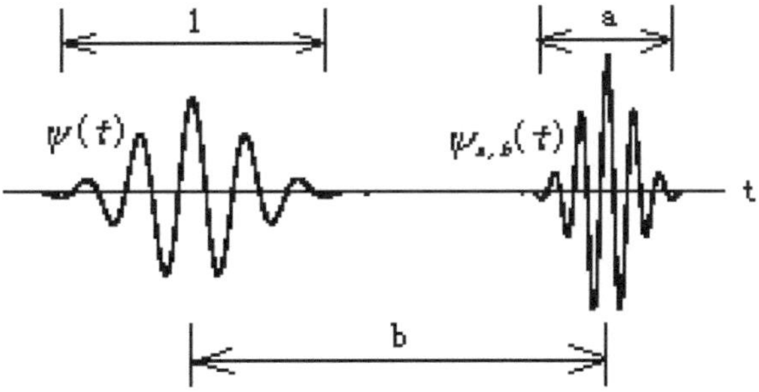

Wavelet functions are generally complex, and their inner product uses the complex conjugate. $W(a, b)$ is analogous to the Fourier coefficients in a Fourier transform, where $\psi^*(\cdot)$ is the complex conjugate of $\psi(\cdot)$. When $t = b$, it indicates how much of the component $\psi_{a,b}(t)$ is contained in the signal $f(t)$. Since wavelet bases differ from Fourier bases, wavelet transforms also differ from Fourier transforms, particularly because wavelet transforms have two parameters: the scale factor a and the shift factor b. As a increases, the time window stretches, the frequency window contracts, the bandwidth narrows, the central frequency decreases, and the frequency resolution increases. Conversely, as a decreases, the time window contracts, the frequency window stretches, the bandwidth widens, the central frequency increases, and the frequency resolution decreases. This precisely aligns with the natural rule in practical problems where high-frequency signals have short durations and low-frequency signals have long durations.

If the wavelet satisfies the condition shown in Eq. (13.5), then its inverse transform exists, and its expression is shown in Eq. (13.6).

$$C_\psi = \int_{-\infty}^{\infty} \frac{|\psi(\omega)|^2}{|\omega|} d\omega < \infty \tag{13.5}$$

$$f(t) = \frac{2}{C_\psi} \int_{0}^{\infty} \left[\int_{-\infty}^{\infty} W(a, b)\psi_{a,b}(t)db \right] \frac{da}{a^2} \tag{13.6}$$

It is evident that through the wavelet basis $\psi_{a,b}(t)$, the signal $f(t)$ can be represented. However, this representation requires an infinite integral based on a and b for signal reconstruction, which is impractical. When performing wavelet transforms and inverse transforms of signals based on numerical calculations, discrete wavelet transforms need to be used.

13.3 Discrete Wavelet Transform

According to the definition of continuous wavelet transform, it is known that under continuously varying scale a and translation b, wavelet bases have significant correlation, thus the information content of the continuous wavelet transform coefficients of a signal is redundant. It is necessary to restrict the values of wavelet bases $\psi_{a,b}(t)$ to some discrete points. Generally, a and b are taken as binary partition according to Eq. (13.7), thus discretising the continuous wavelet:

$$a = 2^j$$

$$b = k2^j \tag{13.7}$$

When $j = 0, \pm 1, \pm 2, \ldots$ is discretised, it is equivalent to halving the width of the wavelet function, then halving it again, or doubling it, then doubling it again, and so on, for scaling. Additionally, by $k = 0, \pm 1, \pm 2, \ldots$ it can cover all variable domains.

The wavelet function derived by substituting Eq. (13.7) into Eq. (13.3) is referred to as a dyadic wavelet, i.e.:

$$\psi_{j,k}(t) = \frac{1}{\sqrt{2^j}}\psi\left(\frac{t - k2^j}{2^j}\right) = 2^{-\frac{j}{2}}\psi\left(2^{-j}t - k\right) \tag{13.8}$$

The wavelet transform using this formula is called the discrete wavelet transform. This formula is the Daubechies representation, where the term before t 2^{-j} is equivalent to the angular frequency in Fourier transform, so when j is smaller, it represents high frequency. On the other hand, in the Meyer representation, there is no negative sign before j, so contrary to the Daubechies representation, the larger the j value, the higher the frequency. This j is called the level or resolution index.

By appropriately selecting ψ in Eq. (13.8), $\{\psi_{j,k}\}$ can become an orthogonal system. The orthogonal system includes translation orthogonality and scaling orthogonality.

13.4 Wavelet Families

The commonly used wavelet families are detailed below.

(1) Haar Wavelet

The Haar wavelet is the earliest and simplest wavelet, satisfying the scaling orthogonality condition. Any discussion of wavelets begins with the Haar wavelet. The Haar wavelet is represented by the formula shown in Eq. (13.9) and graphically in Fig. 13.3.

$$\psi(t) = \begin{cases} 1 & (0 \le t < 1/2) \\ -1 & (1/2 \le t < 1) \\ 0 & (other) \end{cases} \tag{13.9}$$

(2) Daubechies Wavelets

Ingrid Daubechies is one of the pioneers in wavelet research, having invented compactly supported orthogonal wavelets, thereby making discrete wavelet analysis practical. The Daubechies family of wavelets can be denoted as db N, where N is the order, and db is the wavelet name. The db1 wavelet is equivalent to the aforementioned Haar wavelet. Figure 13.4 shows the wavelet functions of the other nine members of the Daubechies family.

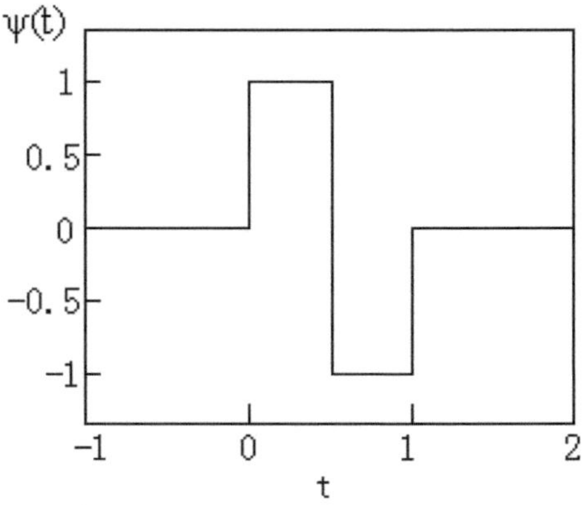

Fig. 13.3 Haar wavelet function

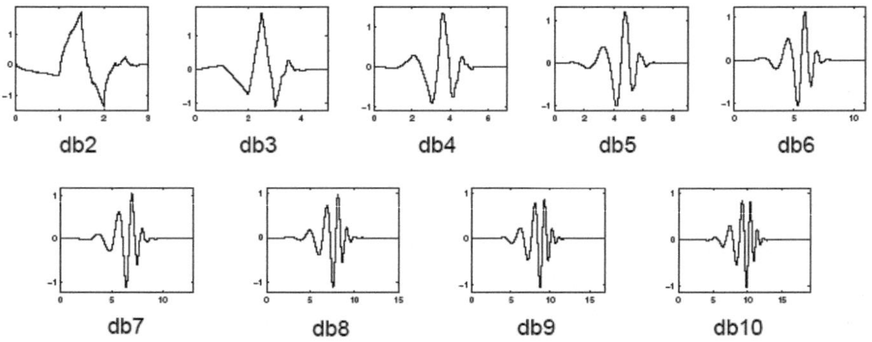

Fig. 13.4 Daubechies wavelets

Additionally, there are biorthogonal spline wavelets, Coiflets wavelets, Symlets wavelets, Morlet wavelets, Mexican Hat wavelets, Meyer wavelets, and others.

13.5 Decomposition and Reconstruction of Signals

The following uses wavelet coefficients to illustrate the methods of signal decomposition and reconstruction.

Firstly, the signal is approximately represented by a linear combination of functions called scaling functions. The linear combination of scaling functions is referred to as the approximated function. Furthermore, the accuracy of the approximation is called the level or resolution index, with level 0 being the most accurate approximation, and higher levels indicating coarser approximations. This section will demonstrate that the difference between the approximated function at any level j and the approximated function at the coarser level $j + 1$ is a linear combination of wavelets.

Ultimately, the signal can be represented by a linear combination of wavelets and scaling functions from level 1 to any level.

A rectangular pulse of width 1 is used as the scaling function $\varphi(t)$, and a linear combination of this function generates the approximated function $f_0(t)$ of any signal $f(t)$ as shown in Eq. (13.10):

$$f_0(t) = \sum_k s_k \phi(t - k) \tag{13.10}$$

where

$$\phi(t) = \begin{cases} 1 \ (0 \le t < 1) \\ 0 \quad (other) \end{cases} \tag{13.11}$$

The coefficient s_k is the average value of the signal $f(t)$ over the interval $[k, k + 1]$, given by Eq. (13.12):

$$s_k = \int_{-\infty}^{\infty} f(t)\phi^*(t - k)dt = \int_{k}^{k+1} f(t)dt \tag{13.12}$$

An example of the signal $f(t)$ and its approximated function f_0 is shown in Fig. 13.5. As shown in Fig. 13.6, the rectangular pulse of width 1 used to generate the approximated function $\varphi(t)$ is called the scaling function because its function is to scale the observed signal. Here, it is specifically referred to as the Haar scaling function.

Similar to wavelets, consider the integer translation and scaling of the scaling function, $\varphi_{j,k}$ as defined in Eq. (13.13):

$$\phi_{j,k}(t) = 2^{-\frac{j}{2}}\phi\left(2^{-j}t - k\right) \tag{13.13}$$

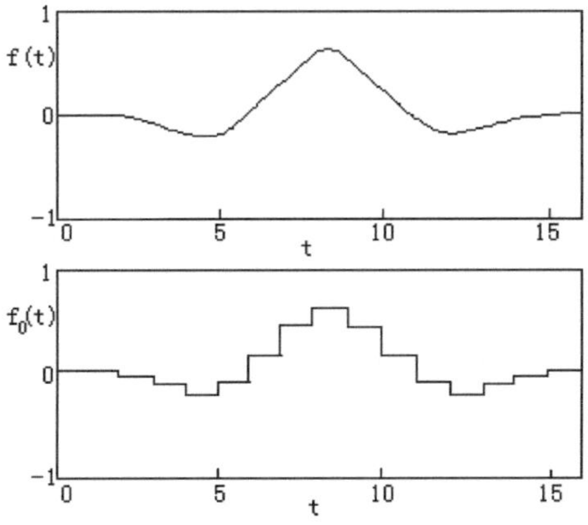

Fig. 13.5 Signal $f(t)$ and its approximate function $f_0(t)$

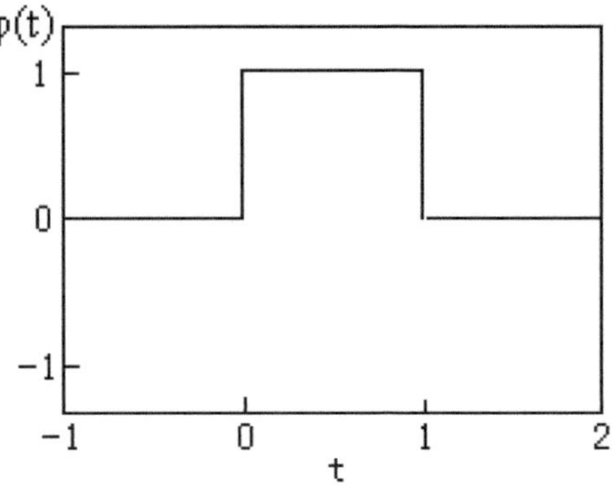

Fig. 13.6 Haar scaling function $\varphi(t)$

Next, using $\varphi_{j,k}$ Define the approximation function $f_j(t)$ of level j, as shown in Eq. (13.14):

$$f_j(t) = \sum_k s_k^{(j)} \phi_{j,k}(t) \tag{13.14}$$

where:

$$s_k^{(j)} = \int_{-\infty}^{\infty} f(t) \phi_{j,k}^*(t) dt \tag{13.15}$$

Additionally, since $\varphi_{j,k}(t)$ is orthonormal with respect to translation, therefore $s_k^{(j)}$ is obtained from the inner product of the approximation function of level jf_j and the scaling function $\varphi_{j,k}$, expressed by Eq. (13.16):

$$s_k^{(j)} = \int_{-\infty}^{\infty} f_j(t) \phi_{j,k}^*(t) dt \tag{13.16}$$

This $s_k^{(j)}$ is called the scaling coefficient. In Fig. 13.7, the signal $f(t)$ and its approximation functions $f_0(t)$ and $f_1(t)$ are shown.

Comparing $f_0(t)$ and $f_1(t)$, it is evident that the signal $f_1(t)$ is a coarser approximation. $f_1(t)$ in Eq. (13.13) is the case where $j = 1$, and the coefficient in front of t is 2^{-1}, which is half of the coefficient when $j = 0$. This coefficient is equivalent to the angular frequency in Fourier transform, so the width of the scaling function φ becomes twice that when $j = 0$. Therefore, the case of $f_1(t)$ is an attempt to approximate the signal $f(t)$ with a wider rectangular signal, which results in a decrease in signal resolution due to the inability to represent detailed information.

Since there is information loss when using f_1 to approximate (or approach) f_0, it is only by compensating $f_1(t)$ with the lost information $g_1(t)$ that $f_0(t)$ can be restored. That is:

$$f_0(t) = f_1(t) + g_1(t) \tag{13.17}$$

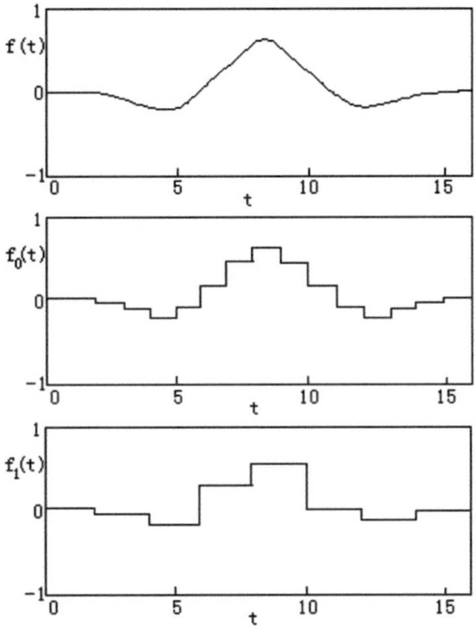

Fig. 13.7 Signal $f(t)$ and its approximate functions $f_0(t)$ and $f_1(t)$

$g_1(t)$ is the difference obtained by subtracting f_1 from f_0 in Fig. 13.7, as shown in Fig. 13.8.

The function $g_1(t)$ is referred to as the first-level wavelet component.

From Fig. 13.8, it is evident that the interval with a width of 1 is symmetric about the centre and oscillates vertically. Therefore, the components of $g_1(t)$ must be the function shown below:

$$\psi\left(\frac{t}{2}\right)=\begin{cases}1 & (0\leq t<1)\\ -1 & (1\leq t<2)\\ 0 & (other)\end{cases} \tag{13.18}$$

It is apparent that this $\psi(t)$ can only be the Haar wavelet shown in Eq. (13.9) from the previous section. This Haar wavelet can generate a family of functions $\psi_{\mathrm{j},k}$ through scaling and translation as described in the previous two sections using

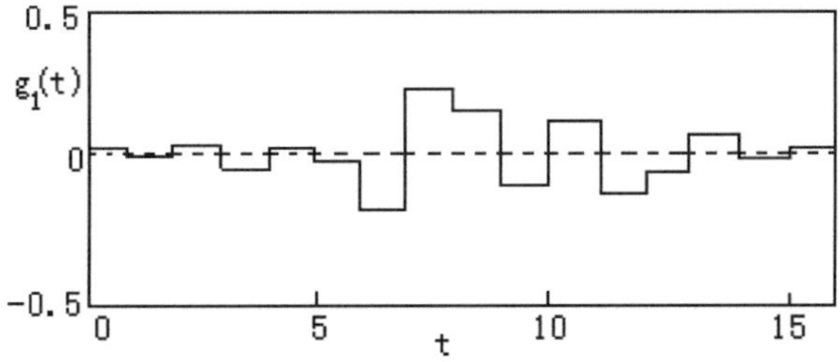

Fig. 13.8 Wavelet component $g_1(t)$

Eq. (13.8). At the first level ($j = 1$), $g_1(t)$ is expressed as a linear combination of $\psi_{j,k}$ as follows:

$$\psi\left(\frac{t}{2}\right) = \begin{cases} 1(0 \leq t < 1) \\ -1\,(1 \leq t < 2) \\ 0(other) \end{cases} \tag{13.19}$$

where $w_k^{(1)}$ is the first-level ($j = 1$) wavelet coefficient.

In summary, the approximation function at level 0 can be decomposed into a linear combination of the scaling function at level 1, $f_1(t)$, and a linear combination of the first-level wavelet, $g_1(t)$, as shown in Eq. (13.20):

$$f_0(t) = f_1(t) + g_1(t) = \sum_k s_k^{(1)} \phi_{1,k}(t) + \sum_k w_k^{(1)} \psi_{1,k}(t) \tag{13.20}$$

Extend this relationship to the general case at level j. That is, to generate a higher precision approximation function at level $j - 1$ from the approximation function at level j, f_{j-1}, it is only necessary to sum the approximation function at level j, $f_j(t)$, and the wavelet component $g_j(t)$:

$$f_{j-1}(t) = f_j(t) + g_j(t) \tag{13.21}$$

where

$$f_j(t) = \sum_k s_k^{(j)} \phi_{j,k}(t)$$

$$g_j(t) = \sum_k w_k^{(j)} \psi_{j,k}(t) \tag{13.22}$$

Next, consider representing the approximation function at level 0, $f_0(t)$, with decreasing precision down to the approximation function at level J. Substitute j(=1, 2, …, J) into Eq. (13.21) yields:

$$\begin{aligned} f_0(t) &= f_1(t) + g_1(t) \\ f_1(t) &= f_2(t) + g_2(t) \\ &\cdots \\ f_{J-1}(t) &= f_J(t) + g_J(t) \end{aligned} \tag{13.23}$$

In the above equations, by substituting the lowest $f_{J-1}(t)$ into the adjacent upper equation, and then substituting the result into its adjacent upper equation, and repeating this iterative process until $f_0(t)$, it can be seen that $f_0(t)$ can be expressed as the sum of the set of $f_J(t)$ and $g_j(t)$, as shown in Eq. (13.24):

$$f_0(t) = g_1(t) + g_2(t) + \cdots + g_J(t) + f_J(t) = \sum_{j=1}^{J} g_j(t) + f_J(t) \tag{13.24}$$

The meaning of this formula is: when the signal $f_0(t)$ is approximately represented by the Jth level approximation function $f_J(t)$, if the components lost in the rough approximation are sequentially added back, $f_0(t)$ can be restored. That is, the signal $f_0(t)$ can be expressed as the sum of the approximation function $f_J(t)$ at any rough level and the wavelet components from level 0 to level J. Therefore, it can be said that the signal $f_0(t)$ can be represented by J resolutions from level 1 to level J, i.e., multiresolution wavelets. This type of signal analysis is called multiresolution analysis.

Figure 13.9 illustrates an example of multiresolution analysis when $J = 2$. In this example, the relationship $f_0 = g_1 + g_2 + f_2$ holds. Thus, f_2 is rectangular in shape, but if J is increased, the width of the rectangle will be stretched beyond f_2. Therefore, when the signal contains a DC component (non-zero average value), it is necessary to represent this DC component with f_J, while the oscillating part coinciding with the DC is represented by wavelets. Since the linear combination of wavelets with an average value of zero still results in an average value of zero, it is impossible to represent the DC component with a finite number of wavelets.

Up to now, the example of the Haar wavelet has shown that once the scaling function is determined, the wavelet can be derived according to formula (13.17), i.e., $g_1(t) = f_0(t) - f_1(t)$. So, let us extend this relationship to wavelets beyond the Haar wavelet. The question is whether any function can obtain a scaling function and then derive a wavelet from this scaling function. According to the definition of multiresolution analysis, the answer is negative. A necessary condition for constituting multiresolution analysis is that the scaling function of level j, $\varphi_{j,k}$, can be expanded using the scaling function of level $j - 1$ with higher precision, $\varphi_{j-1,k}$. Mathematically, this is expressed as shown in Eq. (13.25):

$$\phi_{j,k}(t) = \sum_{n} p_n \phi_{j-1,2k+n}(t) = \sum_{n} p_{n-2k} \phi_{j-1,n}(t) \tag{13.25}$$

where the sequence p_n is the expansion coefficient. In the final equation above, n is replaced with $n - 2k$.

From the above equation, it is known that φ on both sides is a function of j, but p_n does not depend on j. In other words, the same sequence p_n is used in the expansion, independent of the series of j. It can be said that the sequence p_n is the inherent sequence connecting the scaling function of level j, $\varphi_{j,k}(t)$, with the scaling function of level $j - 1$ with higher precision, $\varphi_{j-1,k}(t)$.

However, according to the definition of multiresolution analysis, similar to the scaling function, the wavelet of level j, $\psi_{j,k}$, must also be expandable using the scaling function of level $j - 1$. Thus, similar to the case of the scaling function, the following mathematical expression holds:

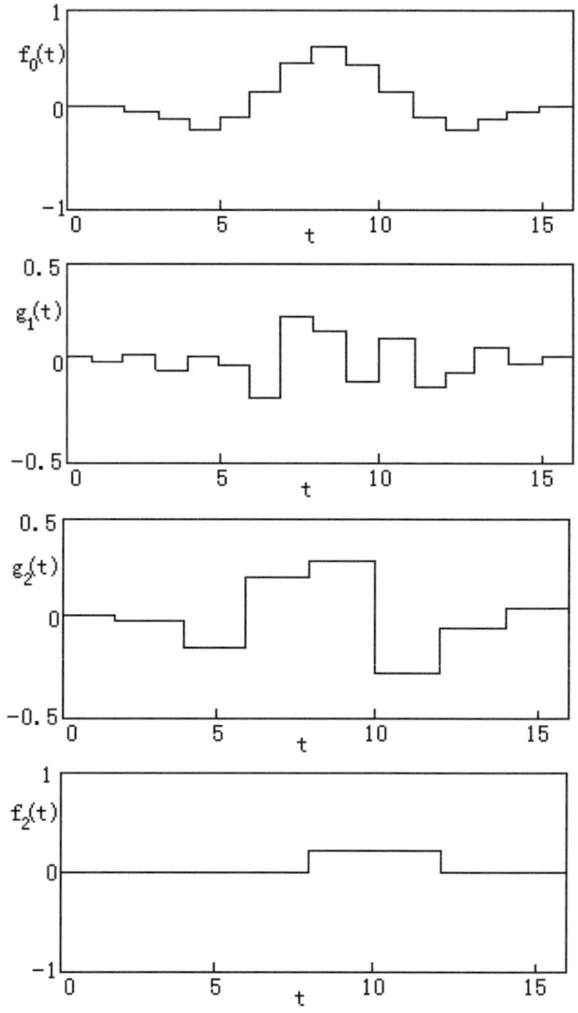

Fig. 13.9 Example of multi-resolution analysis $f_0 = g_1 + g_2 + f_2$

$$\psi_{j,k}(t) = \sum_n q_{n-2k}\phi_{j-1,n}(t) \tag{13.26}$$

where the sequence q_n is the expansion coefficient. This situation can also be described as the sequence q_n being the connection between the wavelet of level j, $\psi_{j,k}(t)$, and the wavelet of level $j-1$ with higher precision. The inherent sequence of the j − 1 level scaling function $\varphi_{j-1,k}(t)$. Since Eqs. (13.25) and (13.26) express the relationship between the scaling functions of levels j and $j-1$, as well as the relationship between the scaling function and the wavelet, they are referred to as the two-scale relation.

It is evident that the scaling function is essential for multi-resolution analysis. After satisfying the conditions of the two-scale relation in Eq. (13.25), the corresponding wavelet for this scaling function can be determined according to another two-scale relation in Eq. (13.26).

For orthogonal wavelets like Daubechies, due to the complexity of the shape of the function itself and its scaling function, it is difficult to represent them with known functions. Therefore, in 1989, Mallat proposed a method to represent orthogonal wavelets and their scaling functions using discrete sequences. In Daubechies wavelets, a natural number N is used to characterise the wavelet. The sequence representing the scaling function of Daubechies wavelets p_k is given in Table 13.1. The sequence representing the wavelet q_k is obtained by reversing p_k along the time axis and then reversing the sign of its coefficients, i.e.:

$$q_k = (-1)^k p_{-k} \tag{13.27}$$

Table 13.1 Daubechies sequence p_k

$N = 2$	$N = 3$	$N = 4$
0.48296291314453	0.33267055295008	0.23037781330889
0.83651630373780	0.80689150931109	0.71484657055291
0.22414386804201	0.45987750211849	0.63088076792986
− 0.12940952255126	− 0.13501102001025	− 0.02798376941686
	− 0.08544127388203	− 0.18703481171909
	0.03522629188571	0.03084138183556
		0.03288301166689
		− 0.01059740178507
$N = 6$	$N = 8$	$N = 10$
0.11154074335011	0.05441584224311	0.02667005790055
0.49462389039845	0.31287159091432	0.18817680007763
0.75113390802110	0.67563073629732	0.52720118893158
0.31525035170920	0.58535468365422	0.68845903945344
− 0.22626469396544	− 0.01582910525638	0.28117234366057
− 0.12976686756727	− 0.28401554296158	− 0.24984642432716
0.09750160558732	0.00047248457391	− 0.19594627437729
0.02752286553031	0.12874742662049	0.12736934033575
− 0.03158203931749	− 0.01736930100181	0.09305736460355
0.00055384220116	− 0.04408825393080	− 0.07139414716635
0.00477725751095	0.01398102791740	− 0.02945753682184
− 0.00107730108531	0.00874609404741	0.03321267405936
	− 0.00487035299345	0.00360655356699
	− 0.00039174037338	− 0.01073317548330
	0.00067544940645	0.00139535174707
	− 0.00011747678412	0.00199240529519
		− 0.00068585669496
		− 0.00011646685513
		0.00009358867032
		− 0.00001326420289

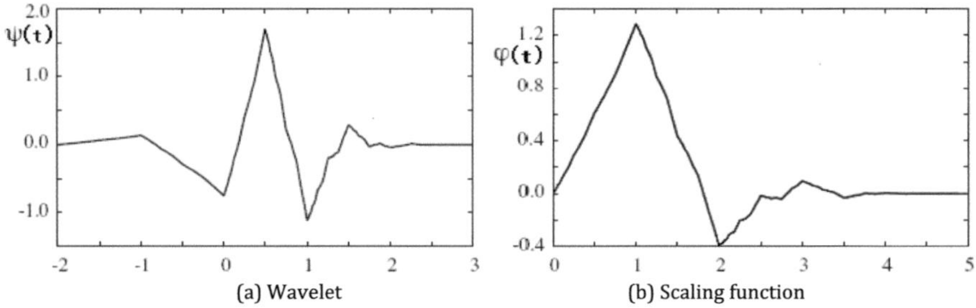

Fig. 13.10 Daubechies wavelet of N = 3 and its scaling function

Figure 13.10 illustrates the Daubechies wavelet and its scaling function for $N = 3$. Comparing Fig. 13.10 with the $N = 3$ entry in Table 13.1, it is found that p_k in the table only defines 6 values, while the figure represents a rather complex function shape. Although this book does not discuss why only 2 N values can represent such a complex function, it is possible to sequentially obtain high-precision functions through repeated iterative calculations starting from 2 N values. The next section will explain the method of directly obtaining expansion coefficients from 2 N discrete sequences.

13.6 Wavelet Transform in Image Processing

13.6.1 Two-Dimensional Discrete Wavelet Transform

From the above discussion, it is known that the orthogonal wavelet represented by Daubechies wavelets and its scaling function can be expressed using discrete sequences. In this section, the method discovered by Mallat for obtaining wavelet expansion coefficients using this discrete sequence is introduced.

As described in Sect. 13.3, the 0th level approximation function $f(t)$ of a continuous signal $f(t)$ is expanded by the 0th level scaling function according to Eq. (13.28):

$$f(t) \approx f_0(t) = \sum_k s_k^{(0)} \phi(t-k) \tag{13.28}$$

where

$$s_k^{(0)} = \int_{-\infty}^{\infty} f(t) \phi_{0,k}^*(t) dt \tag{13.29}$$

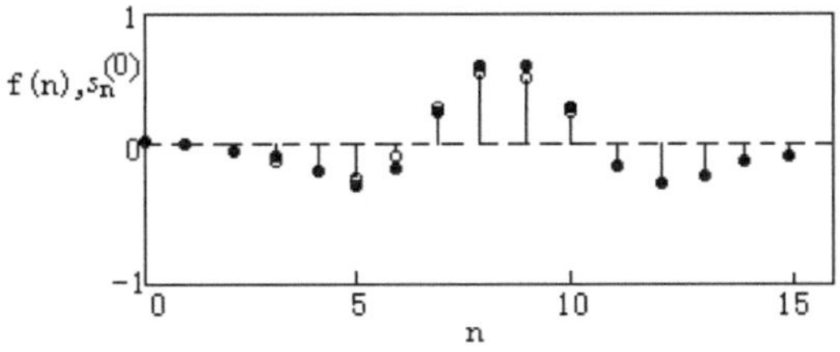

Where, the white dot represents the signal sampling value f(n), and the black dot represents the scale coefficient of the 0th level of Daubechies based on numerical calculation (N=2).

Fig. 13.11 Comparison of signal sampling value $f(n)$ and Daubechies scale coefficient $s_k^{(0)}$ $(N = 2)$

However, in Daubechies wavelets, although 2 N discrete sequences are given, since the scaling function $\varphi_{0,k}(t)$ is not provided, there is a problem of not being able to calculate $s_k{}^{(0)}$ using the above formula.

To overcome this problem, the method proposed by Mallat is to sample the signal to obtain the sequence $f(n)$ is regarded as $s_k^{(0)}$. Mallat discovered that because $\varphi_{0,k}$ changes k on a rectangular or triangular window and translates the time axis, for a certain k value, $s_k^{(0)}$ provides the median value of the signal visible from the window. This median value of the signal corresponds to $f(k)$. This means that $\varphi_{0,k}$ is like a delta function $\delta_k(t)$ (δ Function). Mallat believes that $\varphi_{0,\mathrm{k}}(t)$ has the same function of reconstruction based on the delta function $\delta_k(t)$. In terms of applications in image processing, $f(n)$ regarded as $s_k^{(0)}$ has been proven to be practically unproblematic.

As an example, the Daubechies scale coefficients obtained by numerical calculation $s_k^{(0)}$ $(N = 2)$ compared with $f(n)$ results, as shown in Fig. 13.11, show that $f(n)$ and $s_k^{(0)}$ are almost indistinguishable.

After obtaining $s_k^{(0)}$, it is possible to calculate the scale coefficients $s_k^{(j)}$ and wavelet coefficients $\omega_k^{(j)}$ for levels other than level 0 based on $s_k^{(0)}$. Using Eq. (13.30), it is possible to sequentially calculate the scale coefficients of higher orders (low resolution) from the level 0 scale coefficients $s_k^{(0)}$.

$$s_k^{(j)} = \sum_n p_{n-2k}^* s_n^{(j-1)} \tag{13.30}$$

By using Eq. (13.31), it is possible to sequentially calculate the wavelet coefficients of higher orders (low resolution) from the level 0 scale coefficients $s_k^{(0)}$.

$$w_k^{(j)} = \sum_n q_{n-2k}^* s_n^{(j-1)} \tag{13.31}$$

Below is an explanation of the wavelet transform for two-dimensional image data using discrete wavelets. The image data is given as two-dimensional discrete data, represented by $f(m, n)$. Similar to the case of two-dimensional discrete Fourier

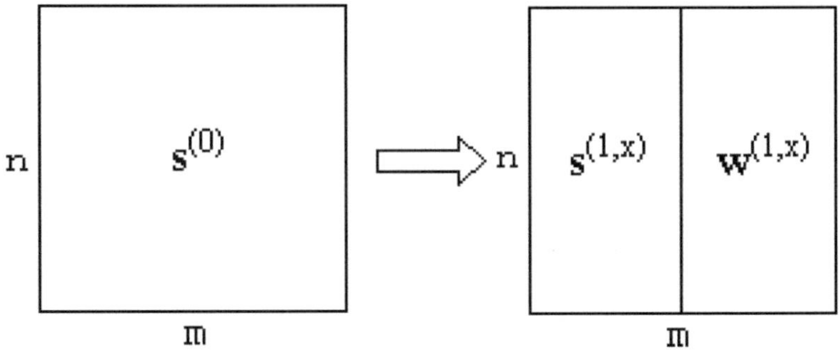

Fig. 13.12 Decomposition of $s_{m,n}^{(0)}$

transform, the discrete wavelet transform is first performed in the horizontal direction, and then its coefficients are transformed in the vertical direction. The image data $f(m, n)$ is regarded as the level 0 scale coefficients $s_{m,n}^{(0)}$.

First, perform the discrete wavelet transform in the horizontal direction.

$$s_{m,n}^{(j+1,x)} = \sum_k p_{k-2m}^* s_{k,n}^{(j)}$$

$$w_{m,n}^{(j+1,x)} = \sum_k q_{k-2m}^* s_{k,n}^{(j)} \tag{13.32}$$

where, $s_{m,n}^{(j+1,x)}$ and $w_{m,n}^{(j+1,x)}$ respectively represent the scale coefficients and wavelet coefficients in the horizontal direction. $j = 0$ is as shown in Fig. 13.12.

Next, perform the discrete wavelet transform in the vertical direction for the coefficients.

$$s_{m,n}^{(j+1)} = \sum_l p_{l-2n}^* s_{m,l}^{(j+1,x)}$$

$$w_{m,n}^{(j+1,h)} = \sum_l q_{l-2n}^* s_{m,l}^{(j+1,x)}$$

$$w_{m,n}^{(j+1,v)} = \sum_l p_{l-2n}^* w_{m,l}^{(j+1,x)}$$

$$w_{m,n}^{(j+1,d)} = \sum_l q_{l-2n}^* w_{m,l}^{(j+1,x)} \tag{13.33}$$

where, $w_{m,n}^{(j+1,h)}$ represents the coefficients where the scale function acts in the horizontal direction and the wavelet acts in the vertical direction, $w_{m,n}^{(j+1,v)}$ represents the coefficients where the wavelet acts in the horizontal direction and the scale function acts in the vertical direction, additionally, $w_{m,n}^{(j+1,d)}$ represents the coefficients

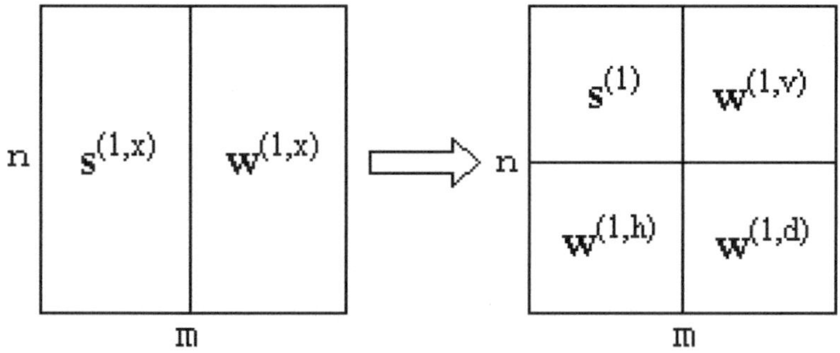

Fig. 13.13 Decomposition of $s_{m,n}^{(1,x)}$ and $w_{m,n}^{(1,x)}$

where the wavelet acts in both horizontal and vertical directions. $j = 0$ is as shown in Fig. 13.13.

Combining Eqs. (13.32) and (13.33) gives:

$$\begin{aligned}
s_{m,n}^{(j+1)} &= \sum_l \sum_k p_{k-2m}^* p_{l-2n}^* s_{k,l}^{(j)} \\
w_{m,n}^{(j+1,h)} &= \sum_l \sum_k p_{k-2m}^* q_{l-2n}^* s_{k,l}^{(j)} \\
w_{m,n}^{(j+1,v)} &= \sum_l \sum_k q_{k-2m}^* p_{l-2n}^* s_{k,l}^{(j)} \\
w_{m,n}^{(j+1,d)} &= \sum_l \sum_k q_{k-2m}^* q_{l-2n}^* s_{k,l}^{(j)}
\end{aligned} \tag{13.34}$$

In the above equation, only $s_{m,n}^{(j+1)}$ is further decomposed into 4 components, and by continuously iterating this process, multi-resolution decomposition is performed.

This reconstruction is the same as the one-dimensional case and is performed according to the following equation:

$$\begin{aligned}
s_{m,n}^{(j)} = \sum_k \sum_l [& p_{m-2k} p_{n-2l} s_{k,l}^{(j+1)} + p_{m-2k} q_{n-2l} w_{k,l}^{(j+1,h)} \\
&+ q_{m-2k} p_{n-2l} w_{k,l}^{(j+1,v)} + q_{m-2k} q_{n-2l} w_{k,l}^{(j+1,d)}]
\end{aligned} \tag{13.35}$$

13.6.2 Image Wavelet Transform Programming

Figure 13.14b shows the 4 components (or called 4 sub-images) after decomposing the original image signal (i.e., the first level wavelet decomposition). As can be seen

(a) Raw image (b) Level 1 wavelet decomposition (c) Level 2 wavelet decomposition

Fig. 13.14 Example of wavelet transform of image

from the figure, $w^{(1,v)}$ represents the high-frequency component in the vertical direction, $w^{(1,h)}$ represents the high-frequency component in the horizontal direction, $w^{(1,d)}$ represents the high-frequency component in the diagonal direction. Additionally, $s^{(1)}$ represents the low-frequency component averaged from $s^{(0)}$. The result of further decomposing $s^{(1)}$ into 4 components (i.e., the second level wavelet decomposition) is shown in Fig. 13.14c.

It can be seen that during the image decomposition process, the total data volume neither increases nor decreases. However, after a wavelet transform, an image results in a series of sub-images with different resolutions, which represent the low-frequency component sub-image and sub-images representing high-frequency components in different directions. Most values on the high-frequency component sub-images are close to 0, and this phenomenon becomes more apparent with higher frequencies. Therefore, for an image, the main information is contained in the low-frequency components, while the high-frequency components only contain detail information. Thus, the simplest image compression method is to retain the low-frequency components and discard the high-frequency components.

Figure 13.15 illustrates an example of wavelet decomposition and compression. Figure 13.15b shows the result of the first level wavelet decomposition of the original image shown in Fig. 13.15a, Fig. 13.15c shows the result of image recovery using only the low-frequency components (the sub-image in the top left corner) after the first level decomposition, and Fig. 13.15d shows the result of further decomposing the low-frequency component sub-image in Fig. 13.15b into the second level wavelet decomposition. Finally, the image recovered from the low-frequency components after the second level decomposition is shown in Fig. 13.15e.

It is evident that although the compression method that retains low-frequency components is simple, it lacks detailed information after image compression, affecting the image quality. This method can be used for transmitting images over the internet by first sending a low-resolution (high-level number) image, followed by images with progressively higher resolutions until the highest resolution image is transmitted. This approach can create a progressive effect, initially presenting the

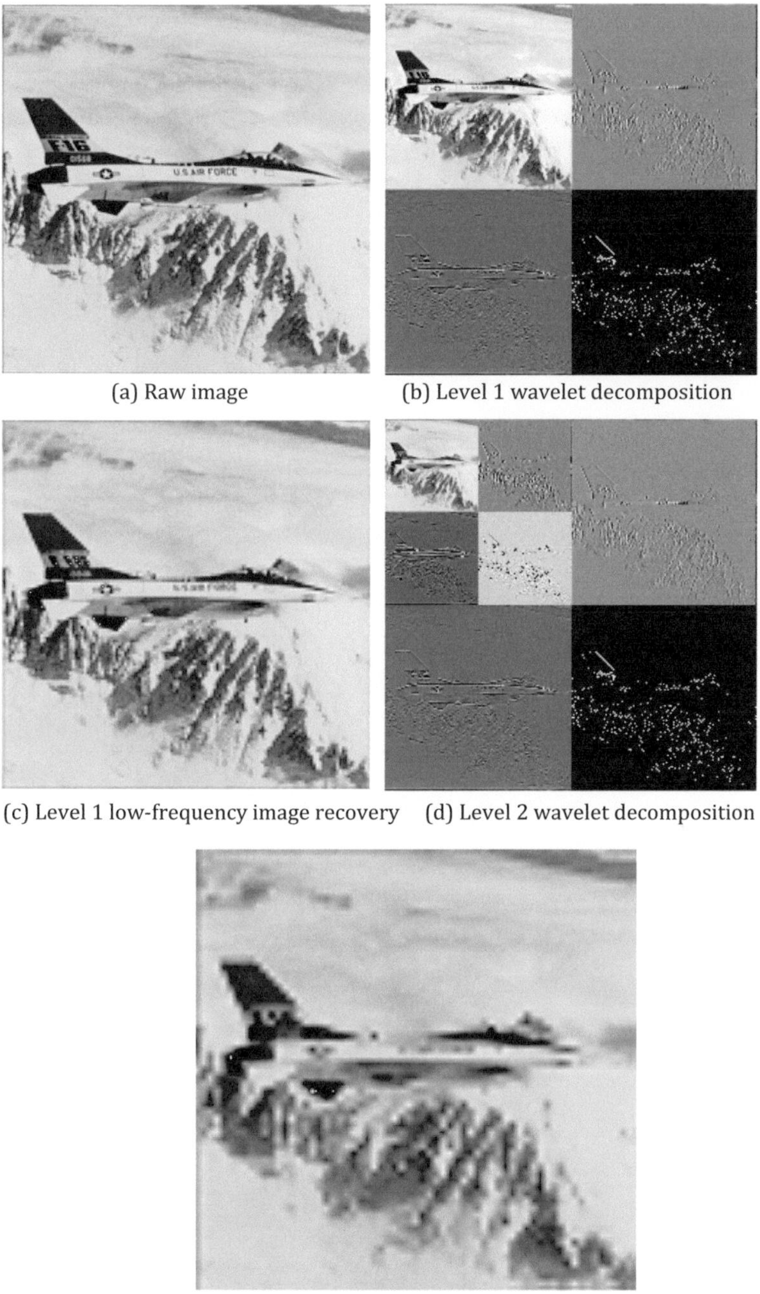

(a) Raw image
(b) Level 1 wavelet decomposition
(c) Level 1 low-frequency image recovery
(d) Level 2 wavelet decomposition
(e) Level 2 low-frequency image recovery

Fig. 13.15 Example of wavelet compression

general outline of the image and then gradually displaying more detailed aspects, as if the image is getting closer and clearer.

Another method of wavelet image compression involves using a wavelet tree, which will not be introduced here. Interested readers are encouraged to refer to other books.

Reference

1. Chen Bingqi. Practical Digital Image Processing and Analysis [M]. 2nd Edition. Beijing: China Agricultural University Press, 2014.

Chapter 14
Pattern Recognition

14.1 Concepts of Pattern Recognition and Image Recognition [1]

Pattern recognition is the process of classifying observed patterns into several categories when the objects of recognition can be categorized. Pattern classification can be considered as preprocessing or a part of pattern recognition. In our daily lives, we are constantly engaging in pattern recognition. Looking around, we can identify whether the surrounding objects are tables or chairs, and recognise whether the person opposite is Zhang San or Li Si; upon hearing sounds, we can distinguish whether it is a car passing by or glass breaking, whether it is a cat meowing or human speech, who is speaking, and what is being said; upon smelling odours, we can tell whether it is fried fish or stinky tofu. These pattern recognition abilities we possess seem very ordinary, and no one would be surprised by them. However, after the advent of computers, when people attempted to use computers to achieve the pattern recognition capabilities that humans possess, the difficulty of this task gradually became apparent.

What is a pattern? Broadly speaking, observable entities existing in time and space, if we can distinguish whether they are the same or similar, can be referred to as patterns.

The understanding of patterns should take note of the following points:

- A pattern does not refer to the entity itself, but to the information we obtain from the entity. Patterns often manifest as information with temporal or spatial distribution.
- When using computers for pattern recognition, the information with temporal and spatial distribution in the computer is represented as arrays.
- The indices of elements in an array can correspond to time and space, or to other identifiers. For example, in the process of pattern recognition where a doctor determines the type of disease based on various test indicators, the various test

© Chemical Industry Press 2026

B. Chen and S. Chen, *Machine Vision Technology*, Advanced and Intelligent Manufacturing in China, https://doi.org/10.1007/978-981-95-1184-6_14

items do not correspond to actual time and space. Thus, the aforementioned time and space should be understood in a broader and more abstract sense.

To grasp objective entities, people categorise them based on their degree of similarity. The function and purpose of pattern recognition lie in correctly categorising a specific entity into a particular category. For instance, when viewing a face from different angles, the images on the retina differ, but we can recognise who the person is and categorise all images from different angles into the category of that person. If each category is named and represented by a specific symbol, then pattern recognition can be seen as mapping information with temporal or spatial distribution to that symbol.

Typically, the information obtained from observing specific individual entities, which has temporal or spatial distribution, is referred to as a sample, while the category to which the sample belongs or the entirety of samples within the same category is referred to as a class.

Image recognition is a branch of pattern recognition, specifically referring to when the object of pattern recognition is an image. Specifically, it can be a photograph, image, handwritten character, remote sensing image, ultrasound signal, CT image, MRI image, radio photograph, etc.

The field of image recognition is very broad, including the recognition and classification of mechanical parts; identifying forests, lakes, cities, and military facilities from remote sensing images; determining and forecasting weather based on meteorological satellite observation data; examining human health conditions using ultrasound images, CT images, or MRI images; automatically sorting products in factories; and conducting security checks at airports and other locations based on facial photographs. Although these research topics in image recognition are diverse, their key issue is primarily classification.

14.2 Composition of Image Recognition Systems

An image recognition system is mainly composed of four parts: image data acquisition, preprocessing, feature extraction and selection, and classification decision, as shown in Fig. 14.1.

Below is a brief explanation of these parts.

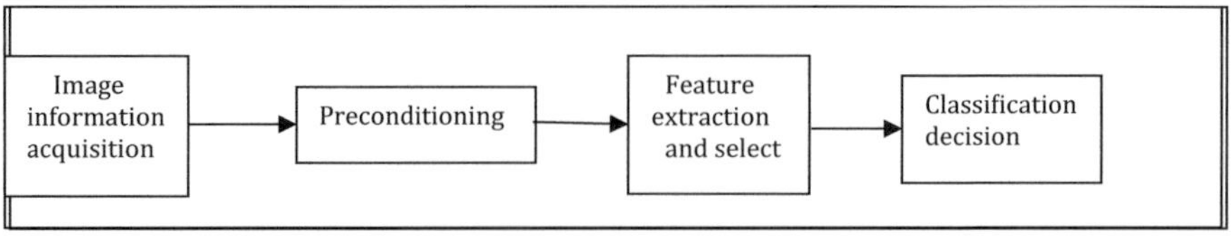

Fig. 14.1 Basic composition of image recognition system

- Image Information Acquisition: Through measurement, sampling, and quantification, a two-dimensional image can be represented by a matrix.
- Preprocessing: The purpose of preprocessing is to remove noise, enhance useful information, and restore degradation caused by measuring instruments or other factors.
- Feature Extraction and Selection: The amount of data obtained from an image is quite large. For example, a text image can have thousands of data points, and the data volume of satellite remote sensing data is even larger. To effectively achieve classification recognition, the original data must be transformed to obtain features that best reflect the essence of classification, which is the process of feature extraction and selection. Generally, the space composed of original data is called measurement space, and the space on which classification recognition relies is called feature space. Through transformation, samples represented in the higher-dimensional measurement space can be converted into samples represented in the lower-dimensional feature space. Samples in the feature space can often be represented as a vector, i.e., a point in the feature space.
- Classification Decision: Classification decision involves using statistical methods in the feature space to classify the identified objects into specific categories. There are mainly two methods: one is supervised classification, where the characteristics of the input object and its category are explained, allowing the machine to learn, and then for a new input, its characteristics are analysed to determine which category it belongs to. The other is unsupervised classification, also known as clustering, where only the characteristics of the input object are known, but not its category, and the computer automatically groups similar characteristics into one category based on certain criteria.
- There is no precise demarcation between classification decision and feature extraction and selection. An ideal feature extractor can simplify the work of the classifier, while an all-purpose classifier would not require a feature extractor. Generally speaking, feature extraction is more dependent on the object being identified than classification.

14.3 The Relationship Between Image Recognition, Image Processing, and Image Understanding

As can be seen from Fig. 14.2, the primary task of image recognition is to acquire images, but regardless of the acquisition method used, various interferences will be introduced during the acquisition process. Therefore, to improve the effectiveness of image recognition, the acquired images must be preprocessed before feature extraction, using methods from Chap. 3 for colour correction, methods from Chap. 5 to filter out interference and noise, enhance the image, and methods from Chap. 6 for geometric correction, etc. Sometimes, images also need to be transformed (such as frequency transformation in Chap. 12, wavelet transformation in Chap. 13, etc.) to facilitate computer analysis. Of course, to find the target we want to analyse in the

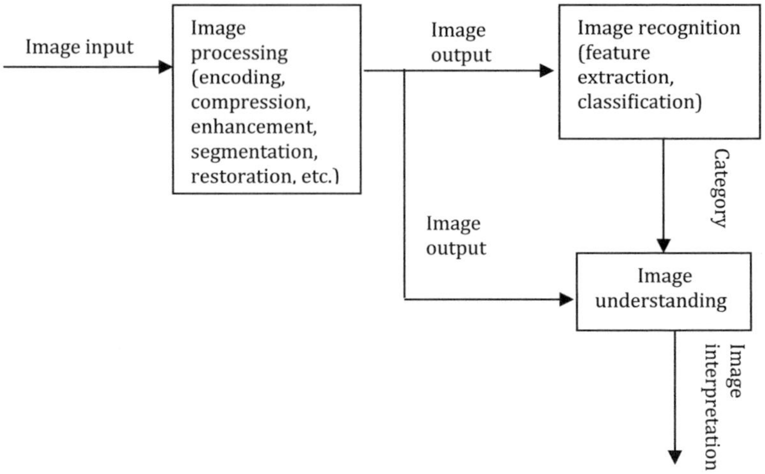

Fig. 14.2 Schematic diagram of image processing, image recognition and image understanding

image, methods from Chap. 3 are also needed to segment the image, i.e., target location and separation. If the acquired image is degraded, restoration processing is also required to improve the fidelity of the image. In practical applications, due to the large amount of information in images, compression is also necessary during transmission and storage. Therefore, the image processing part includes image encoding, image enhancement, image compression, image restoration, image segmentation, etc. The purposes of image processing are twofold: one is to determine whether there is needed information in the image, and the other is to segment out the needed information.

Image recognition involves classifying the processed images to determine the category name. It includes two processes: feature extraction and classification. For the content of feature extraction, please refer to Chap. 3 of this book. It should be noted here that image segmentation does not necessarily occur entirely during image processing; sometimes segmentation and recognition occur simultaneously. Therefore, image processing and image recognition can be carried out in an interleaved manner.

The ultimate goal of image processing and image recognition is to describe and interpret the image, so as to ultimately understand what the image is, i.e., image understanding. Therefore, image understanding is based on image processing and image recognition, performing structural syntactic analysis according to classification results, and describing and interpreting the image. Thus, it is a general term for image processing, image recognition, and structural analysis.

14.4 Image Recognition Methods

There are many image recognition methods, mainly including the following four types: template matching, statistical classification, syntactic or structural classification, and neural network methods. A brief description of these four methods can be found in Table 14.1.

14.4.1 Template Matching Method

Template matching is the earliest and relatively simple image recognition method, which is essentially a statistical recognition method. Matching is a general operation used to define the degree of similarity between a template and an input sample, often represented by a correlation coefficient. When using the template matching method, templates are first established through a training sample set, and then the sample to be recognized is matched against each template to obtain results. Of course, when defining templates and similarity functions, the posture and scale of the entity must be considered. This method works well in many situations, but its main drawback is that changes in perspective may lead to matching errors.

14.4.2 Statistical Pattern Recognition

If an image undergoes feature extraction and obtains an m-dimensional feature vector, then this sample can be regarded as a point in an m-dimensional feature space. The goal of pattern recognition is to select appropriate features so that samples of different classes occupy different regions in the m-dimensional feature space, and samples of the same class are as compact as possible in the m-dimensional feature space. After a training set is given, the decision boundary is determined in the feature space

Table 14.1 Common methods of image recognition

Method	Representation	Identification mode	Canonical criterion
Template matching	Sample, pixel, curve	Correlation coefficient, distance measurement	Classification error
Statistical identification	Trait	Classifier	Classification error
Syntactic or structural recognition	Construction language	Rule, grammar	Acceptable error
Neural network	Sample, pixel, feature	Network function	Minimum root mean square error

through training, dividing samples of different classes into different categories. In statistical decision theory, the decision boundary is determined by the probability density distribution function of each class, which must be known in advance or obtained through learning. Learning is divided into parametric and non-parametric; the former knows the form of the probability density distribution function and needs to estimate its parameters, while the latter does not know the form and requires direct inference of the probability density distribution function.

Statistical recognition methods are divided into geometric classification and probabilistic statistical classification.

(1) Geometric Classification

In statistical classification, samples are regarded as points in the feature space. To determine which category an input sample belongs to, it can be judged by which region of the feature space the sample point falls into. It can be divided into distance method, linearly separable, and non-linearly separable.

a. Distance Method

This is the simplest and most intuitive geometric classification method. Below, the nearest neighbor method is used as an example to introduce this type of method. Suppose there are c categories $\omega_1, \omega_2, \ldots, \omega_c$ in a pattern recognition problem, with N_i samples in each category, $i = 1, 2, \ldots, c$. We can define the discriminant function for class ω_i as

$$g_i(x) = \min_k x - x_i^k k = 1, 2, N_i \tag{14.1}$$

where the superscript i of x_i^k indicates class ω_i, and k indicates the k -th sample in class ω_i with N_i samples. According to the above formula, the decision rule can be written as:

If $g_j(x) = \min_i g_i(x), i = 1, 2, \ldots, c$, then the decision is $x \in \omega_j$.

The intuitive explanation is: for an unknown sample x, we only need to compare the Euclidean distance between x and the known samples, represented by $N = \sum_{i=1}^{c} N_i$, to decide that x is of the same class as the sample closest to it.

The K-nearest neighbor method is an extension of the nearest neighbor method. The K-nearest neighbor method involves taking the k nearest neighbors of an unknown sample x, and classifying x into the class to which the majority of these k neighbors belong. Specifically, it involves finding the k samples closest to x among N known samples. If $k_1, k_2, \ldots, k_c$ are the numbers of these k neighbors belonging to classes $\omega_1, \omega_2, \ldots, \omega_c$, then we can define the discriminant function as

$$g_i(x) = k_i, i = 1, 2, \ldots, c \tag{14.2}$$

The decision rule is:

If $g_j(x) = \max_i k_i$, then the decision is $x \in \omega_j$.

Fig. 14.3 Binary image of Fig. 6.1

The following example illustrates the processing procedure and results of the K-nearest neighbor method. Figure 14.3 is a binary image obtained by extracting brightness values above 30 and applying a 3-time median filter to Fig. 6.1 from Chap. 6.

For the binary image in Fig. 14.3, feature measurement is performed using the method from Chap. 6. The measured feature data include roundness, area, perimeter, and centroid coordinates. For example, the feature parameter of roundness is shown in Fig. 14.4a. These feature data are classified using the K-nearest neighbor method, and the classification results are shown in Fig. 14.4b. Based on the classification results, different classes of images are represented with different grey values, as shown in Fig. 14.5, where class 0 with larger roundness, such as oranges and pears, is represented with brighter grey values, and class 1 with smaller roundness, such as two bananas, is represented with darker grey values.

Classification can also be performed using the measured perimeter, area, and centroid coordinates. Different parameters may yield different classification results, and for different images, some parameters may not achieve good classification results. Figure 14.6 shows the Visual C++ interface for pattern recognition, which is integrated with the parameter measurement and display functions from Chap. 6 for ease of use. The "Display Parameters" and "Pattern Recognition" buttons can only be used after executing "Parameter Measurement".

b. Linearly Separable

Linearly separable essentially involves finding a linear discriminant function. The following example illustrates this with a two-class problem. Assume the discriminant function $g(x)$ is a linear function of x

$$g(x) = w^T x + \omega_0 \tag{14.3}$$

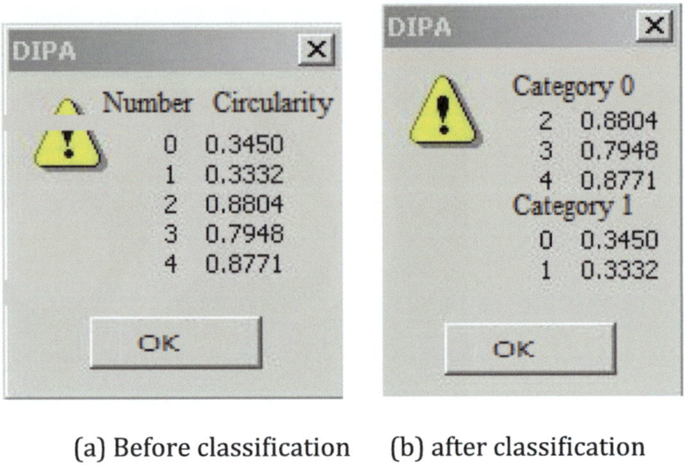

Fig. 14.4 Feature parameters of circularity in Fig. 14.3

Fig. 14.5 Image after circularity classification by k-nearest neighbor method

In this equation, x is d is the feature vector, and w is called the weight vector, represented as follows:

$$x = \begin{bmatrix} x_1 \\ x_2 \\ \vdots \\ x_d \end{bmatrix} w = \begin{bmatrix} w_1 \\ w_2 \\ \vdots \\ w_d \end{bmatrix} \tag{14.4}$$

ω_0 is a constant, referred to as the threshold.

The decision rule is:

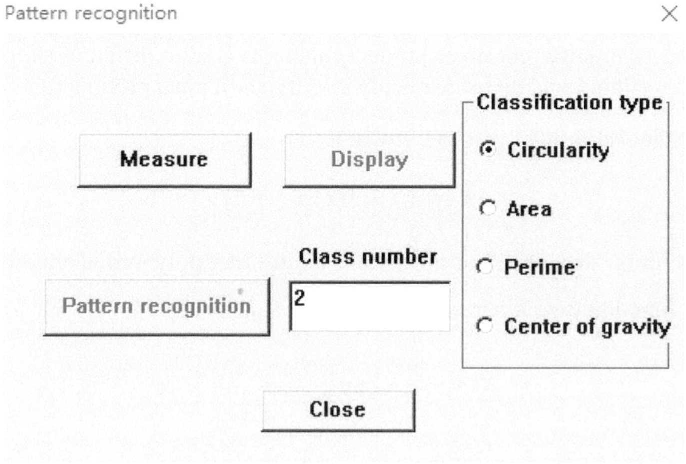

Fig. 14.6 Window interface of parameter measurement and K-nearest neighbor classification

$$g(x) = g_1(x) - g_2(x) \tag{14.5}$$

If

$$\begin{cases} g(x) > 0, & \textit{Make a decision } x \in \omega_1 \\ g(x) < 0, & \textit{Make a decision } x \in \omega_2 \\ g(x) = 0, & \textit{Then we can classify } x \textit{ any way we want} \end{cases}$$

The equation $g(x)$ defines a decision boundary, which separates points classified as belonging to class ω_1 from those classified as belonging to class ω_2. When $g(x)$ is a linear function, this decision boundary is a hyperplane.

Designing a linear classifier involves using a training sample set to establish a linear discriminant function, where the only unknowns are the weight vector w and the threshold ω_0. Thus, the problem of designing a linear classifier is transformed into finding the extremum points of the criterion function w^* and ω_0^* using the training sample set. This belongs to optimisation techniques, which will not be elaborated here.

c. Non-linearly separable

In practice, many pattern recognition problems are not linearly separable. For such problems, the most common method is to transform the non-linearly separable feature space into a linearly separable feature space through some mapping, and then use a linear classifier for classification. The support vector machine is used as an example below to illustrate this.

The basic idea of a support vector machine can be summarized as follows: first, transform the feature space into a higher-dimensional space through a non-linear transformation, and then find the optimal linear classification surface in this new space. This non-linear transformation is achieved by defining an appropriate inner product function. Different inner product functions lead to different support vector machine algorithms, and there are mainly three types of inner product function forms:

- Polynomial form inner product function

$$K(x, x_i) = [(x \cdot x_i) + 1]^q \tag{14.6}$$

The resulting support vector machine is a qth-order polynomial classifier.

- Kernel function type inner product

$$K(x, x_i) = \exp\left\{-\frac{|x - x_i|^2}{\sigma^2}\right\} \tag{14.7}$$

The resulting support vector machine is a radial basis function classifier.

- S-type function as inner product

$$K(x, x_i) = \tanh(v(x \cdot x_i) + c) \tag{14.8}$$

The obtained support vector machine is a two-layer perceptron neural network.

(2) Probabilistic Statistical Classification Method

The geometric classification method mentioned earlier is based on the premise that the pattern is geometrically separable, but such conditions are not often met. Pattern distribution is often not geometrically separable, meaning different patterns may appear in the same region. In this case, classification needs to use probabilistic statistical classification methods. The probabilistic statistical classification method mainly discusses three aspects: striving for optimal statistical decision-making, parameter estimation when the form of density distribution is known, and parameter estimation when the form of density distribution is unknown (or too complex). We will not elaborate on these here.

14.4.3 New Pattern Recognition Methods

The development of pattern recognition has a history of several decades, and many theories have been proposed. These theories and methods are based on statistical theory to find decision rules that can separate two types of samples. In these theories, pattern recognition is essentially pattern classification.

We know that humans focus on "recognition" when understanding things and only pay attention to "distinction" in finer details. For example, when humans recognize animals such as cows, sheep, horses, and dogs, they are actually recognizing the common features of all individuals of each animal, rather than seeking the differences between different species. Therefore, we can see that the focus of pattern recognition should not only be on "distinction" but also on "recognition." Traditional pattern recognition only pays attention to "distinction" and does not emphasize the concept of "recognition." Unlike traditional pattern recognition, Academician Wang Shoujue proposed the biomimetic pattern recognition (BPR) theory in 2002. It approaches pattern recognition from the perspective of "recognizing" patterns, rather than from the perspective of "dividing" patterns as traditional pattern recognition does. Because this method is closer to human cognition, this new pattern recognition method is called "biomimetic pattern recognition."

Biomimetic pattern recognition differs from traditional pattern recognition in the sense that it starts from the recognition of a class of samples to find similarities among similar samples. Biomimetic pattern recognition introduces certain universally existing rules among similar samples and seeks the optimal coverage of the distribution area of similar samples in the feature space from the perspective of understanding the distribution of similar samples in the feature space. This makes biomimetic pattern recognition completely different from traditional pattern recognition. Table 14.2 lists some main differences between biomimetic pattern recognition and traditional pattern recognition.

In the real world, if two similar samples are not completely identical, this difference must be a gradual process. That is, we can certainly find a gradual sequence that transforms from one of these homologous samples to the other, and all samples in this sequence belong to the same category. This rule of continuity among homologous samples is called the principle of homology-continuity (PHC). The mathematical description is as follows: In the feature space R^N, let the set formed by all sample points of class A be A. For any two samples $\vec{x}, \vec{y} \in A$ and $\vec{x} \neq \vec{y}$, if given $\varepsilon > 0$, there must exist a set B satisfying

$$B = \{\vec{x}_1 = \vec{x}, \vec{x}_2, \ldots, \vec{x}_{n-1}, \vec{x}_n = \vec{y} |$$

$$d(\vec{x}_m, \vec{x}_{m+1}) < \varepsilon, \forall m \in [1, n-1], m \in N\} \subset A \tag{14.9}$$

where $d(\vec{x}_m, \vec{x}_{m+1})$ is the distance between sample $\vec{x}_m$ and $\overrightarrow{x}_{m+1}$.

Table 14.2 Differences between biomimetic pattern recognition and traditional pattern recognition

Traditional pattern recognition	Bionic pattern recognition
The optimal partition process between multiple classes of samples	The cognitive process of a class of samples
The distinction between a class of samples and a finite class of known samples	The distinction between one class of samples and an infinite number of unknown samples
Otherness based on different types of samples	Similarity based on similar samples
Find the optimal interface between different classes	Find optimal coverage of the same type of sample

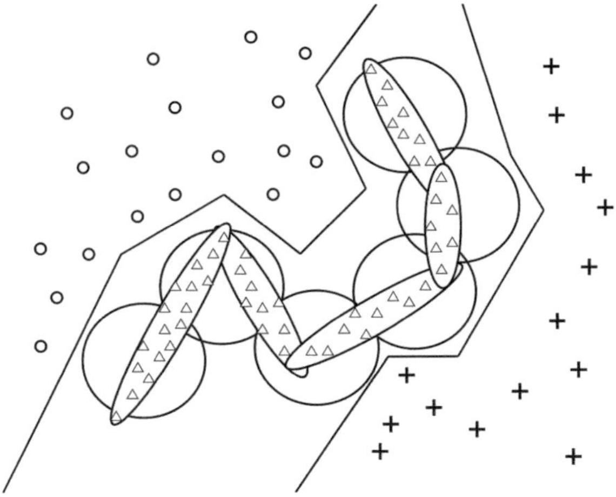

Fig. 14.7 Schematic diagram of bionic pattern recognition, traditional BP network and traditional radial basis function (RBF) network pattern recognition. *Note* Triangles represent samples to be identified. Circles and crosses are two different types of samples from triangles. The broken line is the division method of traditional BP network pattern recognition. The great circle is the division of RBF network

The principle of homogenous continuity is used as "prior knowledge" for the distribution of sample points in bionic pattern recognition. Therefore, bionic pattern recognition takes the analysis of the relationships between training sample points in the feature space as its foundation, and the principle of homogenous continuity provides the possibility for this. Traditional pattern recognition assumes that "all usable information is contained in the training set," yet it precisely overlooks the important rule of continuity existing between homogenous samples. Traditional pattern recognition aims for the optimal division of different class samples in the feature space, whereas bionic pattern recognition aims for the optimal coverage of a single class sample distribution in the feature space. Figure 14.7 is a schematic diagram of bionic pattern recognition, traditional BP network, and traditional radial basis function (RBF) network pattern recognition in two-dimensional space.

According to the principle of homogenous continuity, the mapping (which must be a continuous mapping) of any class of objects (such as class A) in the feature space R^N is certainly a continuous region, denoted as P. Considering the impact of random interference, all samples located near the set P should also belong to class A. We denote the distance between sample $\vec{x}$ and set P as:

$$d(\vec{x}, P) = \min_{\vec{y} \in P} d(\vec{x}, \vec{y}) \tag{14.10}$$

Thus, the optimal coverage P_{A} of class A samples in the feature space is:

$$P_A = \{\vec{x} | d(\vec{x}, P) \leq k\} \tag{14.11}$$

where k is the selected distance constant. In the R^N space, this optimal coverage is an N-dimensional complex shape, dividing the entire space into two parts, one belonging to class A and the other not. However, in practice, it is impossible to collect all samples of class A, so this optimal coverage P_{A} cannot actually be constructed. We can approximate this optimal coverage P_A by using a combination of many simpler covering units. In this case, using bionic pattern recognition to determine whether a sample belongs to this class is essentially determining whether the sample belongs to at least one of these simpler covering units.

14.5 Face Image Recognition System

The following example explains how to perform pattern recognition using face images. This example uses the ORL face database from the University of Cambridge, UK (http://www.cam-orl.co.uk/facedatabase.html). The database contains 40 individuals, each with 10 images. All photos are frontal portraits with a monochrome background. Each photo is a greyscale image of 92 × 112 pixels. Figure 14.8 shows some images from the database.

(1) Preprocessing

1. ①Determine the location of the face.
2. ②Correct the tilt of the face.
3. ③Determine the precise position of the eyes, using the left eye as $A'(x'_a, y'_a)$ point, with the right eye as $B'(x'_b, y'_b)$ point.

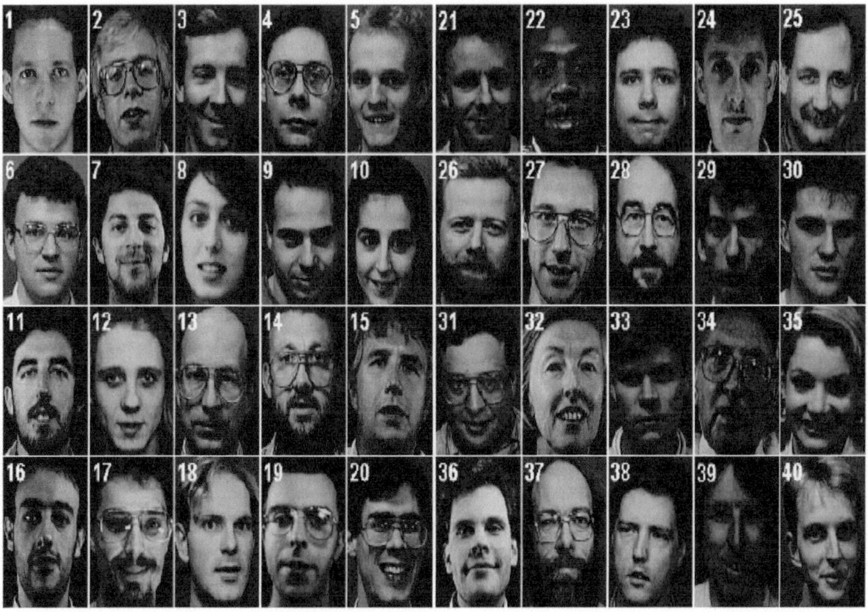

Fig. 14.8 Part of the images in ORL face library

4. ④Using P point as the centre, scale the image by a ratio of $|x'_a - y'_a|$: 30 to become a 255 × 255 image, and then apply a 3 × 3 mosaic to the image to obtain an 85 × 85 face image, where the coordinates of P point are determined by the following formula:

$$x_p = \frac{3}{2}(x'_a - x'_b) - 1$$

$$y_p = 3y'_a + 2|x'_a - x'_b - 1| \tag{14.12}$$

5. ⑤Using the points where the two eyes are located in the 85 × 85 face image obtained after scaling, $A(x_a, y_a)$, $B(x_b, y_b)$ as base points to determine points C, D, and E, with the formulas as follows:

$$x_c = \frac{1}{2}(x_a + x_b); \quad y_c = y_a + 25$$

$$x_d = x_c; \quad y_d = y_a + 40$$

$$x_e = x_c; \quad y_e = y_a \tag{14.13}$$

Thus, the five points A, B, C, D, and E as shown in Fig. 14.9 are obtained.

6. ⑥Using the five points A, B, C, D, and E shown in Fig. 14.9 as base points, perform feature extraction to obtain a 512-dimensional feature vector representing the face.
7. ⑦Use differential processing to reduce the impact of ambient light.

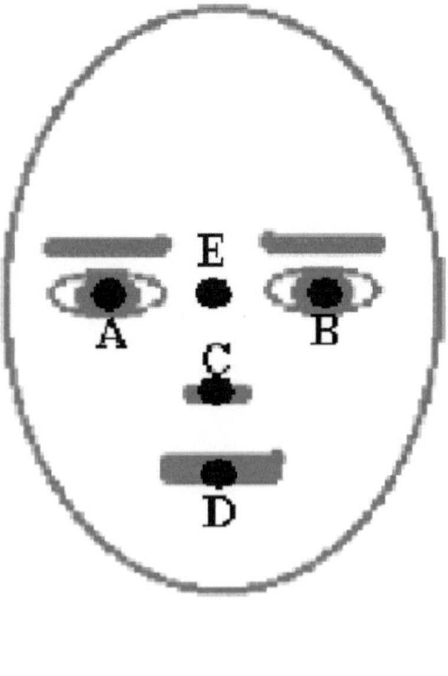

Fig. 14.9 Schematic diagram of feature extraction method

(2) Face Recognition Model

Since the process of a face turning from left to right (or from right to left) is a continuous change process, the change of its feature points mapped into the feature space must also be continuous. We assume that the face only changes in the left–right direction, so there is only one free variable, and the set of feature points should form a one-dimensional manifold distribution. The coverage shape of a certain type of face in the feature space should be a topological product of a one-dimensional manifold homotopic to a curve segment and a 512-dimensional hypersphere, thus forming a closed subspace of that type of sample. Assuming the curve segment is A, and the hypersphere radius is R, then the subspace of that type of sample P_{a} is

$$P_{\mathrm{a}} = \left\{x|\min\left[\rho(x, y)\right] < R, y \in A, x \in R^{512}\right\} \quad (14.14)$$

Assuming the number of training samples collected for each face is K, the training sample set S is represented as follows:

$$S = \{x|x = s_1, s_2, s_3, \ldots, s_K\} \quad (14.15)$$

The samples s_1, s_2, s_3, …, s_K are collected by sequentially rotating at different angles.

In order to achieve coverage of the subspace P_{a} using a limited number of neurons in the neural network, we can approximate the curve segment A with several line segments to form a polyline segment B. Then, by using a 512-dimensional hypersphere with radius R and the topological product of B, we can approximately cover P_{a}, resulting in P_{b}, which is the subspace of the actual obtained sample of this type. Since there are K training samples in total, $K - 1$ line segments can be used to approximate A, with each line segment represented by B_i ($i = 1, 2, \ldots, K - 1$), thus:

$$B_i = \{x|x = as_i + (1 - a)s_{i+1}, a \in [0, 1], s_i \in S, x \in R^{512}\}$$

$$B = \bigcup_{i=1}^{k-1} B_i \quad (14.16)$$

The range covered by each neuron is:

$$P_i = \{x|\min\left[\rho(x, y)\right] \le R, y \in B_i, x \in R^{512}\} \quad (14.17)$$

To achieve coverage of P_i, the neuron structure shown below is adopted:

$$y_i = f\left[\Phi(s_i, s_{i+1}, x)\right] \quad (14.18)$$

In the formula, s_i, s_{i+1} are the feature vectors of the i and $i + 1$ training samples; x is the input vector, i.e., the feature vector of the sample to be recognised; y_i is the i.

The output of a neuron. Φ is a computational function determined by a multi-weight vector neuron (multiple vector inputs, a scalar output), expressed as:

$$\Phi(s_i, s_{i+1}, x) = \min\left[\rho(x, y)\right], y \in \{z|z = \alpha s_i + (1 - \alpha)s_{i+1}, \alpha \in [0, 1]\} \quad (14.19)$$

f is a nonlinear transfer function, using the following step function:

$$f(x) = \begin{cases} 1, & x \leq R \\ 0, & x > R \end{cases} \quad (14.20)$$

The entire $K - 1$ neuron-covered sample subspace is:

$$P_b = \bigcup_{i=1}^{k-1} P_i \quad (14.21)$$

(3) Sample Training

Since the characteristic of bionic pattern recognition is to determine its own sample subspace based on the relationship of its own type of samples, the training process only requires samples of this type. When adding new sample types, there is no need to retrain the already trained samples of each type. For a certain type of sample (a specific person's face), the training process is as follows:

- Extract features from each face to obtain K feature vectors.
- Starting from the curve segment composed of the first feature vector and the second feature vector, train neurons covering the range of this segment in a 512-dimensional space until the training of all $K - 1$ line segments corresponding to $K - 1$ neurons is completed.
- Store the parameters of $K - 1$ neurons to complete the training of this type of sample.

(4) Sample Recognition

Each type of face feature subspace is composed of $K - 1$ neurons, with the neuron structure as follows [the meaning of each parameter is the same as described in Eq. (14.18)]:

$$y_i = f\left[\Phi(s_i, s_{i+1}, x)\right] \quad (14.22)$$

Then the discriminant function for this type is:

$$F_m(x) = F\left(\sum_{i=1}^{k-1} y_i\right) \quad (14.23)$$

where m is the identifier number for this type, F is a step function, as follows:

$$F(x) = \begin{cases} 1, \; x > 0 \\ 0, \; x \leq 0 \end{cases} \tag{14.24}$$

Therefore, when $F_{\mathrm{m}}(x)$ outputs 1, the sample x belongs to type m; otherwise, it does not belong to type m.

Reference

1. Chen Bingqi. Practical Digital Image Processing and Analysis [M]. 2nd Edition. Beijing: China Agricultural University Press, 2014.

Chapter 15
Neural Networks

15.1 Artificial Neural Networks [1]

Since ancient times, the mystery of the origin of human intelligence has consistently attracted the research enthusiasm of countless philosophers and natural scientists. Through long-term and relentless efforts, biologists and neurologists have observed and understood the human brain, concluding that the intelligent activities of the human brain are inseparable from its material foundation, including its physical structure and the various biological, chemical, and electrical activities occurring within it. This led to the establishment of neural network theory and the theory of the nervous system structure, with neural network theory forming the foundation for subsequent theories of neural conduction and brain function. Based on these theoretical foundations, scientists believe that it is possible to study human intelligent activities and cognitive phenomena by imitating the structure and function of the human brain's nervous system. On the other hand, before the nineteenth century, whether it was classical mathematics represented by Euclidean geometry and calculus or classical physics represented by Newtonian mechanics, these classical sciences were, in general, linear. However, the objective world is so complex and varied, with nonlinear situations everywhere, and the human brain's nervous system is no exception. Complexity and nonlinearity are interconnected, and thus, the study of nonlinear science is also key to understanding complex systems. To better understand the objective world, we must study nonlinear science. Artificial neural networks, as a nonlinear network model similar to brain intelligence, thus emerged. Therefore, the creation of artificial neural networks was not accidental but a product of the full development of science and technology in the early twentieth century.

An artificial neural network is a dynamic model of a distributed parallel information processing algorithm structure that mimics the behavioural characteristics of human neural networks. It uses components that receive multiple input stimuli and produce "excited" outputs when the weighted sum exceeds a certain threshold to mimic the working method of human neurons. Through the structure of these neuron

© Chemical Industry Press 2026

B. Chen and S. Chen, *Machine Vision Technology*, Advanced and Intelligent Manufacturing in China, https://doi.org/10.1007/978-981-95-1184-6_15

components and the weight coefficients reflecting the strength of associations, its "collective behaviour" possesses various complex information processing functions. Particularly, the formation of robust, fault-tolerant, anti-interference, adaptive, self-learning, and other flexible and powerful functions at the macro level is not due to the continuous improvement of component performance but is achieved through complex interconnections. Therefore, an artificial neural network is a connection mechanism model with many important characteristics of complex systems.

The essence of an artificial neural network is a mathematical expression of the transformation from input to output, and this mathematical relationship is determined by the network's structure, which must be designed and trained according to specific problems. It is precisely because of these characteristics of neural networks that they have been widely applied in pattern recognition technology. In a broad sense, a pattern is a certain characteristic category of an object, such as information like images, text, sonar signals, and the morphological forms of plant and animal species. Pattern recognition is the mapping of the characteristic category of the studied object into a "category number" to achieve recognition of the specific category of the object. Artificial neural networks are particularly suitable for solving such problems, forming a new pattern information processing technology. The main applications in this area include the recognition of graphic symbols, symbols, handwriting, and speech, the recognition of targets such as radar and sonar, robot vision and hearing, and various nearest neighbour pattern clustering and classification recognition.

15.1.1 The Biological Basis of Artificial Neural Networks

Artificial neural networks (ANN) are designed based on the research results of biological neural networks. They consist of a series of neurons and their corresponding connections, with a good mathematical description, which can be implemented not only with appropriate electronic circuits but also conveniently simulated with computer programs.

The human brain contains 10^{11} biological neurons, which are connected into a system through 10^{15} connections. Each neuron has the independent ability to receive, process, and transmit electrochemical signals. This transmission is completed via the neural pathways that constitute the brain's communication system, as shown in Fig. 15.1.

In this system, each neuron is connected to many other neurons in the system through synapses. Research suggests that the signals emitted by the same neuron through its extending branches are identical; however, their effects on different receiving neurons vary, primarily determined by the corresponding synapse. The greater the "connection strength" of the synapse, the stronger the received signal; conversely, the smaller the "connection strength" of the synapse, the weaker the received signal. The "connection strength" of the synapse can change with the training received by the system.

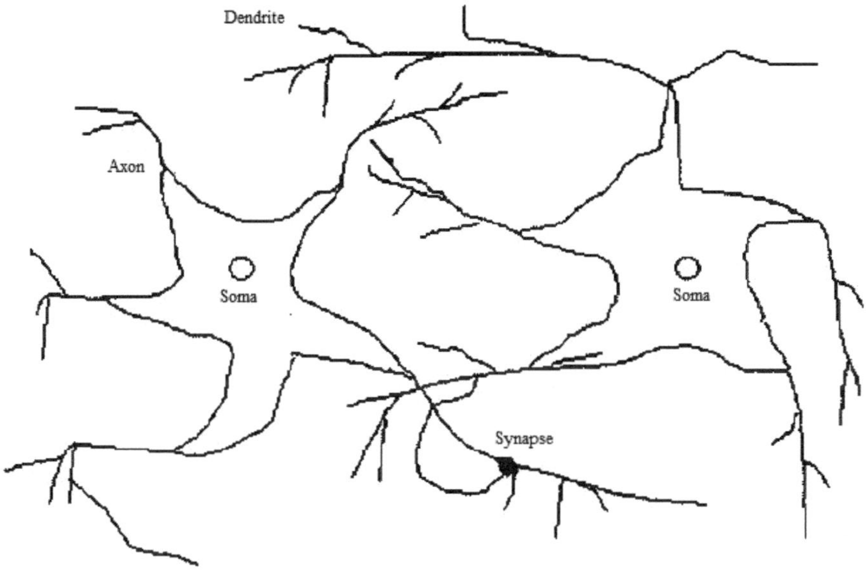

Fig. 15.1 Typical neuron

In summary, the biological nervous system has the following characteristics:

① Neurons and their connections;
② The connection strength between neurons can change with training;
③ Signals can either be excitatory or inhibitory;
④ The cumulative effect of signals received by a neuron determines the state of that neuron;
⑤ The connection strength between neurons determines the strength of signal transmission;
⑥ Each neuron can have a "threshold."

15.1.2 Artificial Neurons

From the above, it is known that neurons are the most basic units that constitute a neural network. Therefore, to construct an artificial neural network model, the primary task is to construct an artificial neuron model (as shown in Fig. 15.2).

Moreover, we hope that this model is not only a simple and easy-to-implement mathematical model but also possesses the six characteristics of biological neurons introduced in the previous section.

Each neuron consists of a cell body, an axon connecting to other neurons, and some other shorter branches extending outward—dendrites. The function of the axon is to transmit the output signal (excitation) of this neuron to other neurons. The many nerve endings at its terminal allow the excitation to be simultaneously transmitted

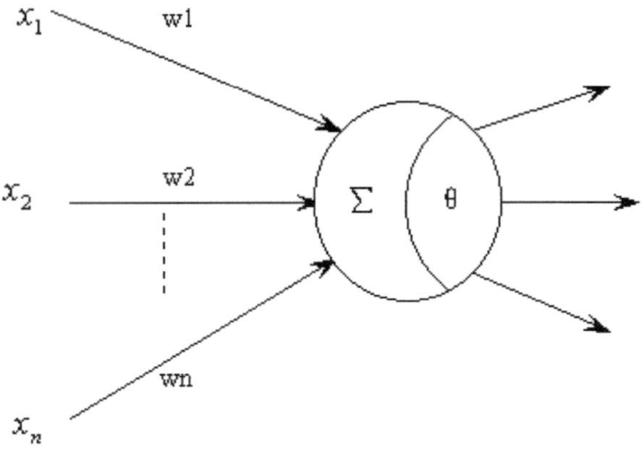

Fig. 15.2 Neuron without activation function. x_1, x_2, …, x_n are information from other artificial neurons as input to that artificial neuron. w_1, w_2, …, w_n in order of their corresponding join weights

to multiple neurons. The function of dendrites is to receive excitation from other neurons. The neuron cell body processes all received signals simply (e.g., weighted summation, considering all input signals with different emphasis reflected in the weights) and then outputs through the axon. The part where the dendrites of a neuron connect with the nerve endings of another neuron is called a synapse.

15.1.3 Learning of Artificial Neurons

Acquiring knowledge and improving performance through learning from the environment is an important feature of artificial neurons. According to the amount of information provided by the environment, the learning methods of the network can be divided into the following three types.

① Supervised Learning: This learning method requires the presence of an external "teacher" who can provide the expected output results (correct answers) for a given set of inputs. The learning system can adjust system parameters based on the difference (error signal) between the known output and the actual output.
② Unsupervised Learning: There is no external "teacher," and the learning system adjusts its parameters or structure entirely according to certain statistical rules provided by the environment (this is a self-organising process).
③ Reinforcement Learning: This type of learning is between the above two situations, where the external environment only provides an evaluation (reward or punishment) of the system's output results, rather than giving the correct answer. The learning system improves its performance by reinforcing those actions that are rewarded.

Learning algorithms can also be divided into three types:

① Error Correction Learning: Its ultimate goal is to minimise a target function based on the error signal, ensuring that the actual output of each output unit in the network closely approximates the expected output in some statistical sense. Once the form of the target function is selected, error correction learning becomes a typical optimisation problem. The most commonly used target function is the mean square error criterion.

② Hebbian Learning: In 1949, Canadian psychologist Hebb proposed the Hebbian learning rule, suggesting that changes occur in the synapses during the learning process, leading to enhanced synaptic connections and improved transmission efficiency. The Hebbian learning rule forms the basis of connection learning. The rule he proposed can be summarised as "when the activation of neurons at both ends of a synapse is synchronised, the strength of that connection should be increased, otherwise, it should be decreased."

③ Competitive Learning: In competitive learning, multiple output units in the network compete with each other, ultimately resulting in only one strongest activator.

15.1.4 Activation Function of Artificial Neurons

The artificial neuron model was proposed by psychologist McCulloch and mathematical logician Pitts as the M-P model (as shown in Fig. 15.3). They combined the basic model of artificial neurons and the activation function to form an artificial neuron, which can also be called a processing element (PE).

Activation Function

$$y = f\left(\sum_{\iota=0}^{n-1} \omega_\iota \chi_\iota - \theta\right) \tag{15.1}$$

f is called the activation function or action function, and The output is 1 if the sum of its inputs is greater than the internal threshold θ; otherwise, it is 0. Let

$$\sigma = \sum_{\iota=0}^{n-1} \omega_\iota \chi_\iota - \theta \tag{15.2}$$

The definition of the function f is as follows:

$$y = f(\sigma) = \begin{cases} 1, & \sigma > 0 \\ 0, & \sigma < 0 \end{cases} \tag{15.3}$$

When $\sigma > 0$, the neuron is activated and enters an excited state, $f(\sigma) = 1$; when $\sigma < 0$, the neuron is inhibited, $f(\sigma) = 0$. The activation function has nonlinear characteristics. Common nonlinear activation functions include step type, piecewise linear type, Sigmoid type (S-type), and hyperbolic tangent type, as shown in Fig. 15.4.

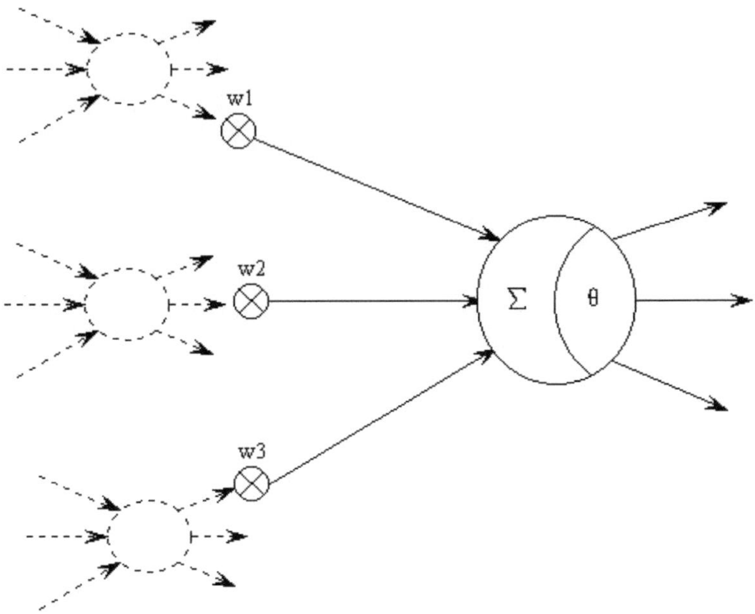

Fig. 15.3 M-P model

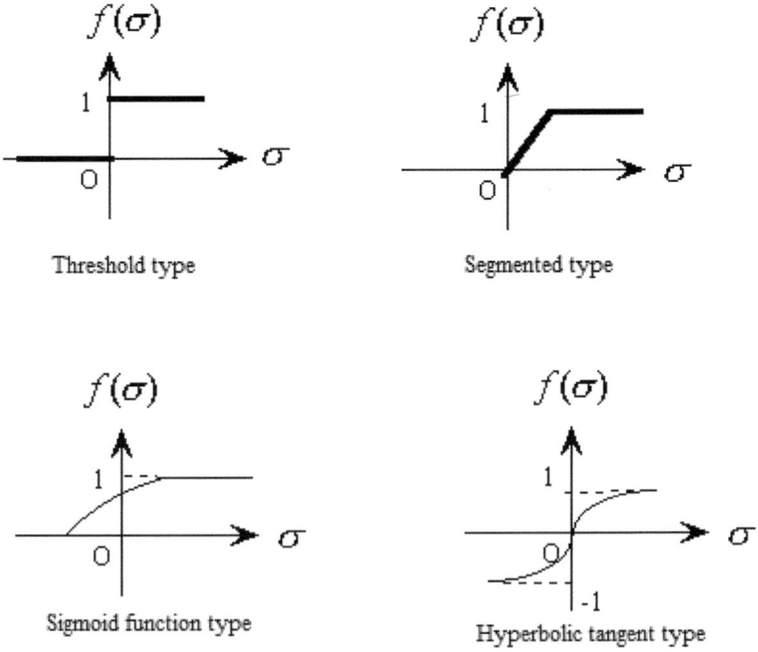

Fig. 15.4 Common nonlinear excitation function curve

① Step function:

$$f(\chi) = \begin{cases} 1, & \chi \geq 0 \\ 0, & \chi \leq 0 \end{cases} \quad \text{or} \quad f(\chi) = \begin{cases} 1, & \chi \geq 0 \\ -1, & \chi \leq 0 \end{cases} \tag{15.4}$$

② Sigmoid (S-type) function:

$$f(\chi) = 1/(1 + e^{-\chi}) \tag{15.5}$$

③ Hyperbolic tangent function:

$$f(\chi) = \tanh(\chi) = {(e^{\chi} - e^{-\chi})}/{(e^{\chi} + e^{-\chi})} \tag{15.6}$$

④ Gaussian function:

$$f(\mathbf{x}) = \exp\left(-\frac{1}{2\sigma_i^2}\sum_j (x_j - w_{ji})^2\right) \tag{15.7}$$

Among these, the step function is mostly used in discrete neural networks, the S-type function is often used in continuous neural networks, and the Gaussian function is used in radial basis function neural networks (radial basis function NN).

15.1.5 Characteristics of Artificial Neural Networks

Artificial neural networks are systems formed by a large number of neurons extensively interconnected, and this structural characteristic determines that artificial neural networks have the ability for high-speed information processing. Although the computational function of each neuron is very simple, and the signal transmission rate is relatively low (approximately 100 times/s), due to the extremely parallel interconnection function between neurons, it ultimately allows an ordinary human brain to complete tasks in about 1 s that would require at least billions of processing steps for a current computer to accomplish.

The knowledge storage capacity of artificial neural networks is very large. In neural networks, the storage of knowledge and information is manifested as the distributed physical connections between neurons. It is represented and stored in a distributed manner across the entire network within each neuron and its connections. Each neuron and its connections only represent a part of the information, rather than a complete specific concept. Only through the distributed comprehensive effect of each neuron can specific concepts and knowledge be expressed.

Artificial neural networks, with their large number of neurons and vast information storage capacity, possess a strong ability to process uncertain information. Even if the input information is incomplete, inaccurate, or ambiguous, neural networks can still associate and think of a complete image of things stored in memory. As long as the input pattern is close to the training samples, the system can provide correct reasoning conclusions. It is precisely because of the structural characteristics of artificial neural networks and their distributed information storage that they have another significant advantage over other judgment recognition systems, such as expert systems: robustness. Biological neural networks do not lose memory of original patterns due to the loss of individual neurons. The most compelling evidence is that when a person's brain suffers minor damage due to an accident, they do not lose all memory of original things. Artificial neural networks exhibit a similar situation. For various reasons, whether due to hardware or software implementation, if one or more neurons fail, the entire network can still continue to function.

Artificial neural networks, unlike current computers, are a type of nonlinear processing unit. Only when the comprehensive processing result of all input signals by the neurons exceeds a certain threshold does it output a signal. Therefore, neural networks are highly nonlinear, ultra-large-scale continuous-time dynamic systems. They break through the limitations of traditional digital electronic computers based on linear processing, marking a significant leap in human intelligent information processing capabilities and the ability to simulate human brain intelligent behaviour. The aforementioned functions and characteristics of neural networks make their application prospects very promising.

15.2 BP Neural Network

15.2.1 Introduction to BP Neural Network

The BP neural network (back-propagation neural network), also known as the error back-propagation neural network or multilayer feedforward neural network, is a unidirectional propagation multilayer feedforward neural network. The first layer is the input nodes, the last layer is the output nodes, and there is one or more hidden layers of nodes in between. The neurons in the hidden layer use a Sigmoid-type transformation function, while the neurons in the output layer use a pure linear transformation function. Figure 15.5 shows the topology of a three-layer feedforward neural network. The characteristics of this neural network model are that neurons in each layer are only connected to neurons in adjacent layers, there are no connections between neurons within each layer, and there are no feedback connections between neurons in each layer.

The input–output relationship of a BP neural network is a highly nonlinear mapping relationship. If the number of input nodes is n and the number of output nodes is m, then the network is a mapping from an n-dimensional Euclidean space to an m-dimensional Euclidean space (in 1989, Robert Hecht-Nielsen proved that any continuous function within a closed interval can be approximated by a BP

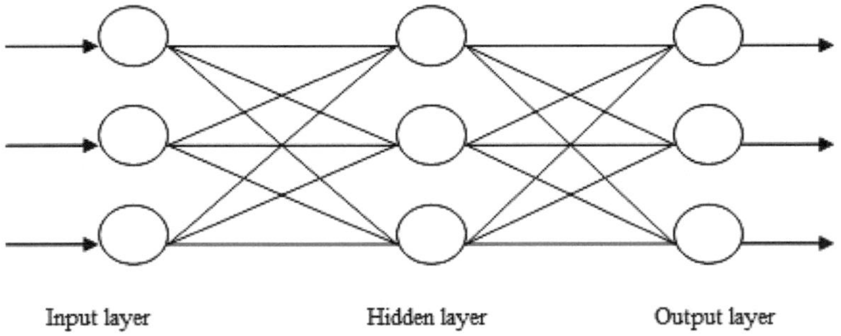

Fig. 15.5 Topology of BP neural network

network with a hidden layer, thus a three-layer BP network can perform arbitrary n-dimensional to m-dimensional mapping).

Two basic theorems have been proven regarding BP networks:

Theorem 1 (Kolmogorov Theorem) *Given any continuous function* $f{:}[0, 1]^n \to R^m$, *f can be implemented by a three-layer feedforward neural network, where the first layer, i.e., the input layer, has n neurons; the middle layer has* $2n + 1$ *neurons; the third layer, i.e., the output layer, has m neurons.*

Theorem 2 *Given any* $\varepsilon > 0$, *for any L2-type continuous function* $f{:}[0, 1]^n \to R^m$, *there exists a three-layer BP network that can approximate f within any ε squared error precision.*

Theorems 1 and 2 demonstrate that BP neural networks can approximate any nonlinear continuous function with arbitrary precision. Once the structure of the BP network is determined, it is trained using an input–output sample set, which involves learning and adjusting the network's weights and thresholds to achieve the given input–output mapping relationship.

Increasing the number of layers in a network can reduce errors and improve accuracy. However, as the number of layers increases, the network structure becomes more complex, and the number of network weights increases sharply, thereby increasing the training time of the network. Accuracy can also be improved by adjusting the number of nodes in the hidden layer, making the training results easier to observe and adjust. Therefore, a network structure with fewer hidden layers is usually preferred.

BP networks often adopt a structure with one hidden layer. Whether network training can converge and improve accuracy can be achieved by adjusting the number of neurons in the hidden layer. Compared to networks with multiple hidden layers, this method significantly reduces learning time and computational load. However, in specific problems, there is no clear theoretical guideline or method for determining the number of hidden layers and hidden layer nodes. In recent years, there has been much research on the optimization of BP neural network structures, which is a very important issue in the design of network topology. If there are too few hidden layer nodes, the learning process may not converge; however, if there are too many hidden layer nodes, it may result in long convergence times and overfitting, leading to a decrease in the network's fault tolerance and generalization ability. Each application problem requires a network structure suitable for itself, and the optimization of the neural network structure under a set of given performance criteria is a very complex issue.

The final performance of a BP neural network is not only determined by the network structure but also related to the initial point, the learning order of the training data, etc. Therefore, whether the chosen network topology has the best network performance is somewhat random. The choice of the number of hidden layer units has always been a complex issue in the application of neural networks. In fact, the application of ANN often translates into determining the network's structural parameters and obtaining the connection weights. Too few hidden layer units may result in a network that cannot be trained or is not "robust" enough to recognize previously unseen samples, with poor fault tolerance; but too many hidden layer units can lead to excessively long learning times, and the error may not be optimal. Therefore, there is a problem of determining the appropriate number of hidden layer units. In practical design, a more realistic approach is to train and compare different numbers of neurons and then appropriately add a little margin. A trained BP network can provide suitable outputs for inputs not in the sample set, a property known as generalization. From the perspective of function fitting, this indicates that the BP network has interpolation capabilities.

15.2.2 Training and Learning of BP Neural Networks

Assuming each layer of a BP network has N processing units, the input received by the ith neuron in the hidden layer is:

$$net_i = x_1 w_{1i} + x_2 w_{2i} + \cdots + x_n w_{ni} \tag{15.8}$$

In this equation, x_i represents the input sample received by the ith neuron in the input layer, and w_{ni} represents the connection weight between the ith neuron in the input layer and the nth neuron in the hidden layer.

The activation function usually selected is the Sigmoid function (Fig. 15.6) or the hyperbolic tangent function, which can reflect the nonlinear characteristics of biological neurons and satisfy the differentiable condition required by the BP algorithm. The output is:

$$o = f(net) = \frac{1}{1 + e^{-net}} \tag{15.9}$$

In this equation, net represents the input to the neuron, and o represents the output of the neuron.

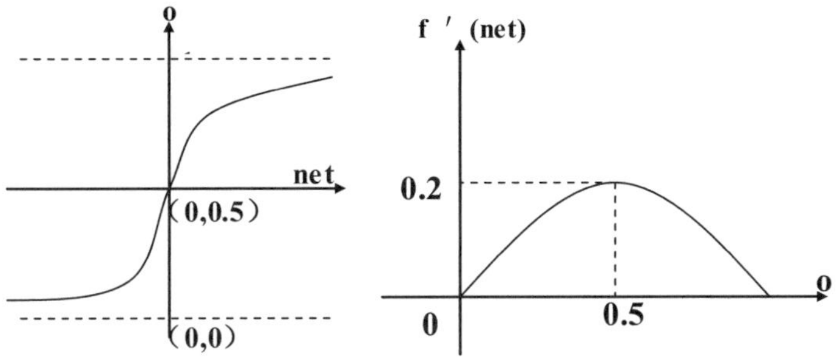

Fig. 15.6 Sigmoid function

$$f'(net) = -\frac{1}{(1+e^{-net})^2}(-e^{-net}) = o - o^2 = o(1-o) \tag{15.10}$$

(1) BP Neural Network Learning Process

① Forward Propagation Phase.

a. Take a sample from the sample set (x_p, y_p), and input x_p into the network;
b. Calculate the corresponding actual output o_p:

$$o_p = f_l(\ldots(f_2(f_1(x_p w_1)w_2)\ldots)w_l) \tag{15.11}$$

② Backward Propagation Phase—Error Propagation Phase.

a. Calculate the difference between the actual output o_p and the corresponding ideal output y_p;
b. Adjust the weight matrix to minimise the error;
c. Error measure of the network for the pth sample:

$$Ep = \frac{1}{2}\sum_{j=1}^{m}(y_{pj} - o_{pj})^2 \tag{15.12}$$

where, y_{pj} represents the expected output of the jth dimension for the pth sample. o_{pj} represents the actual output of the jth dimension for the pth sample.

d. Error measure of the network for the entire sample set:

$$E = \sum_p Ep \tag{15.13}$$

The following introduces the δ learning rule, the essence of which is to use the steepest descent method to change the weights in the direction of the negative gradient of the error function. If the change in weight w_{ji} is denoted as Δw_{ji}, then:

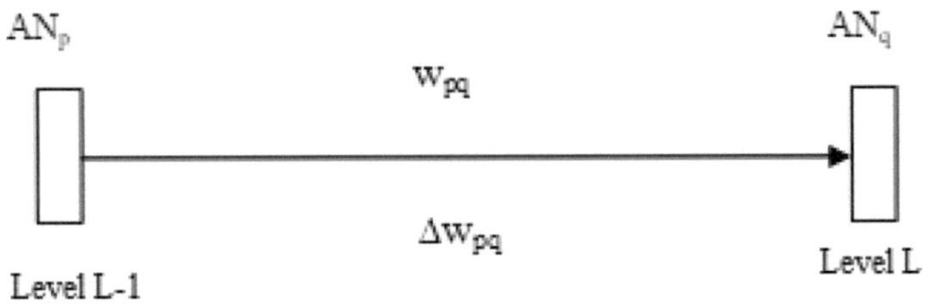

Fig. 15.7 Weight adjustment of output layer

$$\Delta w_{ji} \propto -\frac{\partial E_p}{\partial w_{ji}} \tag{15.14}$$

Because:

$$\frac{\partial E_p}{\partial w_{ji}} = \frac{\partial E_p}{\partial \alpha_{pji}} \frac{\partial \alpha_{pji}}{\partial w_{ji}} = \frac{\partial E_p}{\partial \alpha_{pji}} o_{pj} = -\delta_{pj} o_{pj} \tag{15.15}$$

Let:

$$\delta_{pj} = \frac{\partial E_p}{\partial \alpha_{pji}} \tag{15.16}$$

Thus:

$$w_{ji} = \eta \delta_{pj} o_{pj}, \eta > 0 \tag{15.17}$$

This is commonly referred to as the δ learning rule.

(2) Error Propagation Analysis

① Adjustment of Output Layer Weights (Fig. 15.7).

$$w_{pq} = w_{pq} + \Delta w_{pq} \tag{15.18}$$

In this formula, w_{pq} represents the connection weight between the qth neuron in the output layer and the pth neuron in the hidden layer.

$$\begin{aligned} \Delta w_{pq} &= \alpha \delta_q o_p \\ &= \alpha f_n{'}(net_q)(y_q - o_q) o_p \\ &= \alpha o_q (1 - o_q)(y_q - o_q) o_p \end{aligned} \tag{15.19}$$

② Adjustment of Hidden Layer Weights (Fig. 15.8).

$$\mathrm{v_{hp}} = \mathrm{v_{hp}} + \Delta \mathrm{v_{hp}} \tag{15.20}$$

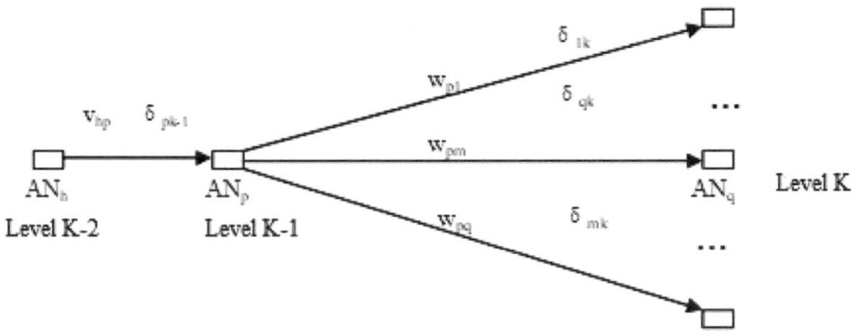

Fig. 15.8 Hidden layer weight adjustment

In this formula, v_{hp} represents the connection weight between the hth neuron in the $k-2$ layer and the pth neuron in the $k-1$ layer.

$$\begin{aligned}\Delta \mathrm{v_{hp}} &= \alpha * \delta * \mathrm{p_{k-1}o_{hk-2}} \\ &= \alpha \mathrm{f}'_{\mathrm{k-1}}\,(\mathrm{net_p})(\mathrm{w_{p1}}\delta_{1k} + \mathrm{w_{p2}}\delta_{2k} + \cdots + \mathrm{w_{pm}}\delta_{mk})\mathrm{oh_{k-2}} \\ &= \alpha o_{pk-1}\left(1 - o_{pk-1}\right)(\mathrm{w_{p1}}\delta_{1k} + \mathrm{w_{p2}}\delta_{2k} + \cdots + \mathrm{w_{pm}}\delta_{mk})\mathrm{oh_{k-2}}\end{aligned} \tag{15.21}$$

15.2.3 Improved BP Neural Network

The adjustment of the connection weights between the output layer and the hidden layer depends on three factors: α, d_l^k and y_j^k, and the weight adjustment between the hidden layer and the input layer also depends on three factors: β, e_j^k and x_i^k. It is evident that the adjustment is proportional to the correction error and also proportional to the output value of the hidden layer or the input signal value. That is, the higher the activation value of the neuron, the more active it is during this learning process, and the greater the adjustment of the weights connected to it. However, the adjustment of the threshold is only proportional to the correction error in form. The larger the learning coefficients α and β, the faster the learning speed, but it may cause oscillations in the learning process.

When using the BP network for target value prediction, the so-called "overfitting" phenomenon is often observed, meaning that the trained BP network fits the learning samples well, but there is a significant deviation in the prediction values for samples not involved in learning. When the size of the learning sample set is not large enough compared to the complexity of the network, "overfitting" is often more severe. This should be noted in the application of neural network models.

Disadvantages of the BP algorithm:

① Slow convergence speed, requiring thousands of iterations, and as the dimension of training examples increases, network performance deteriorates;

② There is no theoretical guidance for selecting the number of hidden nodes in the network;

③ From a mathematical perspective, the BP algorithm is a gradient steepest descent method, which may lead to local minima problems.

When local minima occur, the error appears to meet the requirements on the surface, but the solution obtained may not be the true solution to the problem, so the BP algorithm is incomplete. Therefore, many improved algorithms have emerged, such as using adaptive variable step size to accelerate the BP algorithm to improve convergence speed, and using simulated annealing (SA) to improve the BP algorithm to avoid local minima problems.

In the training of neural networks using the BP neural network model, the additional momentum method is adopted to prevent the network from falling into local minima. The additional momentum method allows the network to consider not only the effect of the error on the gradient when adjusting its weights but also the influence of the trend on the error surface. Its function is like a low-pass filter, allowing the network to ignore minor variations on the network. Without additional momentum, the network may fall into shallow local minima, but with the effect of additional momentum, it may slide over these minima. The additional momentum method reduces the network's sensitivity to local details of the error surface and can effectively prevent the network from falling into local minima. This method adds a value proportional to the previous weight change to each weight change based on the backpropagation method, and generates new weight changes according to the backpropagation method. The weight adjustment formula with an additional momentum factor is:

$$\Delta w_{ij}(k+1) = (1 - mc) * \eta * \delta_i * x_j + mc * \Delta w_{ij}(k) \tag{15.22}$$

where, k is the number of training iterations; Δw is the increment of the weight; η is the learning rate; δ is the error; x is the network input; mc. For the momentum factor, it is generally around 0.9.

The information of a BP network is dispersedly stored on the connection weights of neurons in adjacent layers. The evolutionary training process of the network is the process of continuously adjusting the connection weights during the network's learning of sample data. In the BP algorithm, we attempt to adjust the weights of a neural network so that the actual output of the training set can be as close as possible to the target output. When there is a nonlinear relationship between input and output, and the training samples are sufficient, this algorithm is very effective. However, in practice, the gradient-based BP algorithm has exposed its weaknesses, such as slow convergence speed and poor global search capability. Therefore, the adjustment and improvement of learning rules and learning rates are also important aspects of BP network design.

For a specific problem, choosing an appropriate learning rate is quite difficult. This is because a small learning rate can lead to a longer training time, while a large learning rate may cause system instability. Moreover, a learning rate that is effective at the beginning of training may not be suitable for later training. To solve this problem, an automatic adjustment learning rate method, namely the adaptive learning rate method, is used in network training. The criterion of the adaptive learning rate method is to check whether the correction value of the weight truly reduces the error function. If it does, it indicates that the selected learning rate is too small, and an increment can be added to it; if not, and over-adjustment occurs, then the learning rate should be reduced. Below is a formula for adjusting the adaptive learning rate.

$$\eta(k+1) = \begin{cases} 1.05 * \eta(k), SSE(k+1) < SSE(k) \\ 0.7 * \eta(k), SSE(k+1) > 1.04 * SSE(k) \end{cases} \tag{15.23}$$

$$SSE = \sum_{i=1}^{n} (y_i - y_i\prime)^2 \tag{15.24}$$

where η is the learning rate; k is the number of training iterations; SSE is the error function; y_i is the output value of the learning sample; y_i' is the actual output value of y_i after network training; n is the number of learning samples.

15.3 Application of BP Neural Network in Digital Character Recognition

Digital character recognition is increasingly widely used in modern daily life, such as automatic vehicle license plate recognition systems, online handwriting recognition systems, office automation, etc. With the development of office automation, printed digital character recognition technology has received increasing attention. Printed digital character recognition has a wide range of applications in different fields. If machines are used to recognize bank signatures, they can perform more work in the same amount of time, saving time and resources, improving work efficiency, and effectively reducing costs. With the rapid development of China's social economy and highway transportation, and the sharp increase in car ownership, the adoption of advanced, efficient, and accurate intelligent traffic management systems is urgent. The automation and intelligence of vehicle monitoring and management are of great significance in the traffic system. Vehicle automatic recognition systems can be widely used in highway and bridge toll stations, urban traffic monitoring systems, ports, airports, and parking lots for license plate authentication in actual traffic systems, to improve the automation level of vehicle monitoring and management in the traffic system. Automatic vehicle license plate recognition is one of the key technologies in intelligent traffic management systems, and the recognition of vehicle license plates is mainly the recognition of digital characters. The use of machine vision technology for license plate recognition has become widespread. Using BP neural networks to achieve digital character recognition is one of the common methods. In the practical application of neural networks, 80% to 90% of artificial neural network models use BP networks or their variations.

15.3.1 Principle of BP Neural Network Digital Character Recognition System

A typical neural network digital character recognition system consists of preprocessing, feature extraction, and a neural network classifier. Preprocessing involves removing useless information from the raw data, eliminating noise and other interference factors. Common methods used for preprocessing raw data images include gradient sharpening, smoothing, binarization, character segmentation, and amplitude normalization to extract useful information. The feature extraction part in a neural network digital character recognition system may not always be present, thus dividing the systems into two main categories: ① With feature extraction: This type of system is essentially a combination of traditional methods and neural network techniques. This approach can fully utilize human experience to obtain pattern features and the classification ability of neural networks to recognize characters. Feature extraction must reflect the characteristics of the entire character, but its anti-interference ability is not as strong as that of category ②. ② Without feature extraction: This omits feature extraction, with the entire character directly serving as the input to the neural network (some refer to this method as using character grid features). In this approach, the complexity of the neural network structure is greatly increased, and the increase in input pattern dimensions leads to a large network scale. Additionally, the neural network structure needs to completely eliminate the impact of pattern deformation by itself, but the network has good anti-interference performance and a high recognition rate.

In the BP neural network model, the input is the feature vector of the digital character, and the output nodes of the neural network model represent the number of characters. For 10 digital characters, the output layer has 10 neurons, each representing a digit. The number of hidden layers must be chosen carefully, and the number of neurons in each layer must be appropriate. Then, an appropriate learning algorithm must be selected to achieve good recognition results.

During the learning phase, a large number of samples should be used for training and learning. Through extensive learning of samples, the connection weights of each layer of the neural network are adjusted to ensure correct recognition results for the samples. This is akin to how humans memorize numbers, with the neurons in the network resembling human brain cells, and changes in connection weights akin to changes in the interactions between brain cells. In sample learning, the neural network adjusts the network weights, similar to how humans remember the image of each number. The network weights are the content remembered by the network, and the learning phase of the network is like the process of humans repeatedly learning to recognize numbers from not knowing them. The neural network memorizes numbers based on the overall feature vector, and as long as most features match the learned samples, it can be recognized as the same character. Therefore, even when samples have significant noise, the neural network model can still correctly recognize them. In the digital character recognition phase, as long as the preprocessed feature vector is used as the input to the neural network model, the output of the model after network computation is the recognition result. This principle can be used to establish a network model for digital character recognition.

15.3.2 Establishment of the Network Model

Firstly, designing and training a neural network to recognize 10 digits means that whenever an input representing a certain digit is given to the trained network, the network can correctly identify the digit at the output end. It is evident that the network has memorized all 10 digits. The training of the neural network should be supervised to train 10 groups of arrays at the input end, each representing digits 0 to 9, which can correspond to specific positions from 1 to 10 at the output end. Therefore, each digit's bitmap must be digitized to construct input samples. After processing such as grayscale image binarization, gradient sharpening, tilt adjustment, noise filtering, image segmentation, and size standard normalization, each training sample digital character is converted into an 8×16 matrix of Boolean values. For example, the digit 0 can be represented by a 0, 1 matrix as follows:

$$\text{Letter0} = \begin{bmatrix} 0&0&1&1&1&1&1&0 \\ 0&0&1&0&0&0&1&0 \\ 0&0&1&0&0&0&1&0 \\ 0&0&1&0&0&0&1&0 \\ 0&0&1&0&0&0&1&0 \\ 0&0&1&0&0&0&1&0 \\ 0&0&1&0&0&0&1&0 \\ 0&0&1&0&0&0&1&0 \\ 0&0&1&0&0&0&1&0 \\ 0&0&1&0&0&0&1&0 \\ 0&0&1&0&0&0&1&0 \\ 0&0&1&0&0&0&1&0 \\ 0&0&1&0&0&0&1&0 \\ 0&0&1&0&0&0&1&0 \\ 0&0&1&0&0&0&1&0 \\ 0&0&1&1&1&1&1&0 \end{bmatrix}$$

Additionally, the network must have fault tolerance. In practical situations, the network cannot receive an ideal Boolean vector as input. After digitizing the noise, when the noise mean is 0 and the standard deviation is less than or equal to 0.2, the system can correctly recognize the input vector, which is the fault tolerance capability of the network.

For the requirement of recognising numbers, the neural network is designed as a two-layer BP network with $8 \times 16 = 128$ input terminals, and the output layer has

10 neurons. The purpose of training the network is to ensure that its output vector correctly represents the number vector. However, due to the presence of noise signals, the network may produce imprecise outputs, but after training with a competitive transfer function, it can correctly identify number vectors with noise. The steps to establish the network model are as follows.

Initialisation: First, generate input sample data and output vectors, then establish a two-layer neural network. Network Training: To ensure that the generated network has a certain fault tolerance for input vectors, the best approach is to train the network using both ideal signals and signals with noise. The aim of training is to obtain good weight data so that it can recognise a sufficient number of samples that have never been learned before. This involves inputting a portion of the samples into the BP network and forming a weight file through multiple adjustments.

The basic training process is described as follows.

① Assign random values between [0, 1] to the connection weights between the input layer and the hidden layer w_{ij}, the connection weights between the hidden layer and the output layer v_{il}, and the threshold θ_j; and specify the learning coefficients α, β and the activation function of the neurons.

② Provide the image containing $n \times n$ pixel data as the input pattern $X_k = (x_{1k}, x_{2k}, \ldots, x_{nn})$ to the network, and randomly generate the output pattern pair $Z_k = (z_{1k}, z_{2k}, \ldots, x_{\mathrm{nn}})$.

③ Calculate the output y_j of each neuron in the hidden layer using the network settings:

$$\begin{cases} s_j = \sum\limits_{i=1}^{n} w_{ij} x_i - \theta_j \\ y_j = f(s_j) \end{cases} \tag{15.25}$$

④ Calculate the response C_l of the neurons in the output layer using the network settings:

$$\begin{cases} u_l = \sum\limits_{j=1}^{p} v_{jl} y_j - \gamma_l \\ C_l = f(u_l) \end{cases} \tag{15.26}$$

⑤ Use the given output data to calculate the generalised error $d_l{}^{\mathrm{k}}$ of the neurons in the output layer:

$$d_t^k = (z_l^k - C_l^k) f'(u_l) \tag{15.27}$$

⑥ Calculate the generalised error $e_j{}^{\mathrm{k}}$ of each neuron in the hidden layer:

$$e_j^k = \left[\sum_{l=1}^{q} v_{jl} d_l^k \right] f'(s_j^k) \tag{15.28}$$

⑦ Use the generalised error d_l^k of the neurons in the output layer and the output y_j of each neuron in the hidden layer, adjust the connection weights v_{jl} and the neuron threshold γ_l, between the hidden layer and the output layer:

$$\begin{cases} \Delta v_{jl} = \alpha d_l^k y_j^k, & (0 < \alpha < 1) \\ \Delta \gamma_l = -\alpha d_l^k \end{cases} \tag{15.29}$$

⑧ Using the generalised error of the hidden layer neurons e_j^k, and the input of each neuron in the input layer X^k, adjust the connection weights between the input layer and the hidden layer w_{ij} and the neuron threshold θ_j:

$$\begin{cases} \Delta w_{ij} = \beta e_j^k x_i^k, & (0 < \beta < 1) \\ \Delta \theta_j = -\beta e_j^k \end{cases} \tag{15.30}$$

⑨ Randomly select another input–output data set, return to step ③ for learning; repeat the learning process using all data sets. At this point, the network completes a learning process using the sample set.
⑩ Repeat the next learning process until the network's global error is less than the set value, or the number of learning iterations reaches the set limit.
⑪ Test the performance of the trained network to check if it meets the requirements.

Steps ③ and ④ in the above are the forward propagation process, while steps ⑤ to ⑧ are the error backpropagation process. Through repeated training and adjustments, the neural network eventually converges to weight factor values that correctly reflect the objective process. Train the network using ideal input signals until its mean square error reaches the desired accuracy.

15.3.3 Demonstration of Digital Character Recognition

For an image to be recognised, processes such as colour to greyscale conversion, binary segmentation, noise reduction, tilt adjustment, text segmentation, text width adjustment, text alignment, feature vector extraction, BP network training, reading the number of nodes in each layer, and text recognition are required. In practical applications, train on a digital image of the specification shown in Fig. 15.9 to verify the application of the BP neural network in image text recognition.

Fig. 15.9 Training sample

(1) Network Training

The following are the specific operational steps for network training.

Step 1: First, load Fig. 15.9 into the system.

Step 2: Open the binarisation window, set the low threshold to 200, select "Above", and perform binarisation as shown in Fig. 15.10.

Step 3: Close the binarisation processing window, apply median filtering three times to denoise the image.

Step 4: Open the "Text Recognition Based on BP Neural Network" window, as shown in Fig. 15.11.

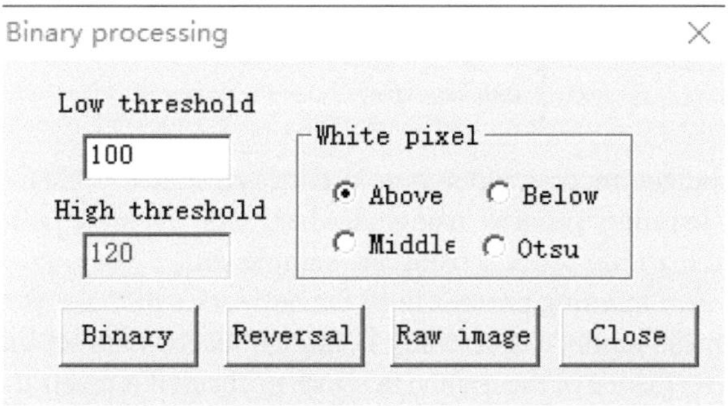

Fig. 15.10 Binarization processing window

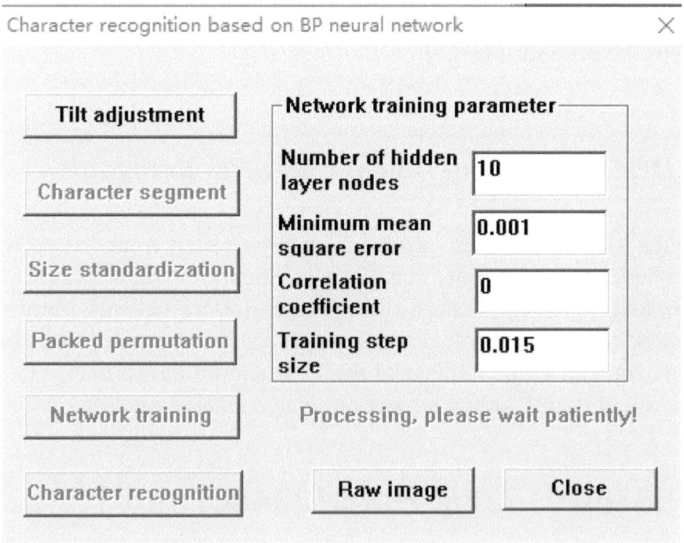

Fig. 15.11 Text recognition window

Fig. 15.12 Training samples after pretreatment

Step 5: Execute "Tilt Adjustment" to adjust the tilt of the text. After execution, the "Character Segmentation" button becomes active.

Step 6: Execute "Character Segmentation" to segment each character. After execution, the "Size Normalisation" button becomes active.

Step 7: Execute "Size Normalisation" to generate characters of standard size. After execution, the "Compact Arrangement" button becomes active. The system fixes the standardised size to a height of 8 pixels and a width of 16 pixels.

Step 8: Execute "Compact Arrangement" to arrange the standardised characters in order. After execution, the "Network Training" and "Text Recognition" buttons become active. Figure 15.12 shows the text sample image obtained after the above steps.

Step 9: Execute "Network Training" to internally generate a file storing the weights for future text recognition. "Network Training" requires some time, so patience is needed during execution.

After the above steps, a file storing the weights will be obtained, and there will be no need for further training in future text recognition. If the training results are unsatisfactory, retraining can be done by changing the network training parameters or the training image.

Since the pre-processed objects are 8×16 pixel characters, 128 neurons are used at the input end, with each input neuron representing a pixel value of the processed image. The output layer uses 10 neurons, corresponding to 16 digits. The transfer functions of the neurons in the hidden and output layers both use Sigmoid, as this function's output range of [0, 1] meets the requirement for Boolean output. The neural network parameters are set as shown in Fig. 15.11, with 10 hidden layer nodes, a minimum mean square error of 0.001, and a training step size of 0.015. The process of neural network training is to ensure that its output vector correctly represents the digital vector.

(2) Text Recognition

Figure 15.13 is the text image to be recognised, and this image is a colour image. The following introduces the specific steps for text recognition of this image.

Step 1: Read the image and convert the colour image into a grayscale image.

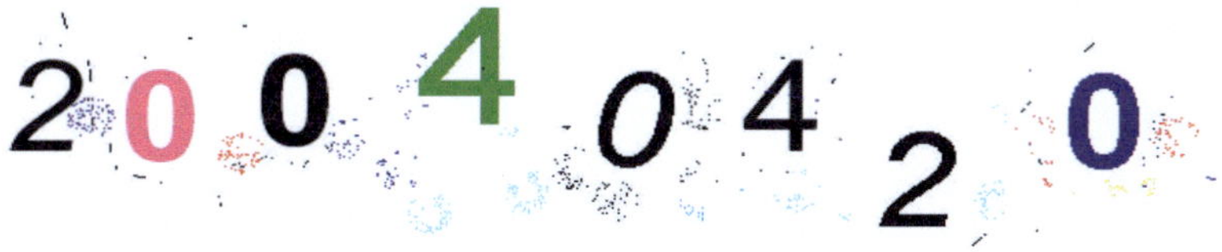

Fig. 15.13 Test sample

Fig. 15.14 Test sample after pretreatment

Fig. 15.15 Identification results

Steps 2 to 8: Exactly the same as steps 2 to 8 of "Network Training". Figure 15.14 is the test sample image after processing through the above steps.

Step 9: Execute the "Text Recognition" command to output the recognition result of Fig. 15.15. Note that this output result is a relatively ideal recognition result. Generally, the recognition accuracy rate is above 80%.

Reference

1. Chen Bingqi. Practical Digital Image Processing and Analysis [M]. 2nd Edition. Beijing: China Agricultural University Press, 2014.

Chapter 16
Deep Learning

16.1 The Development History of Deep Learning

In June 2012, The New York Times disclosed the Google Brain project, which attracted widespread public attention. Led jointly by the renowned machine learning professor Andrew Ng from Stanford University and the world-leading expert in large-scale computer systems, Jeff Dean, using a parallel computing platform with 16,000 CPU cores, they trained a machine learning model called "deep neural networks" (DNN), achieving great success in fields such as speech recognition and image recognition. This machine learning model contains a total of 1 billion nodes internally. However, this network cannot be compared to the human brain, which contains over 86 billion neurons.

In November 2012, Microsoft publicly demonstrated a fully automated simultaneous interpretation system at an event in Tianjin, China. The speaker delivered a speech in English, and the computer in the background automatically completed speech recognition, English-to-Chinese machine translation, and Chinese speech synthesis in one go, with very smooth results. It is reported that the key technology supporting this is also DNN, or deep learning (DL).

At Baidu's annual meeting in January 2013, founder and CEO Robin Li announced the establishment of Baidu Research Institute, with its first initiative being the "Institute of Deep Learning" (IDL).

Why do internet companies with big data compete to invest a large amount of resources in developing deep learning technology? What is deep learning? Why does deep learning exist? How did it come about? What can it do? What difficulties currently exist? This section will explain the above questions.

Machine learning is a discipline focused on studying how computers can simulate or replicate human learning behavior to acquire new knowledge or skills and reorganise existing knowledge structures to continuously improve their performance [1]. Can machines have learning abilities like humans? In 1959, Samuel in the United States designed a chess program with learning ability, which could improve its chess

© Chemical Industry Press 2026

B. Chen and S. Chen, *Machine Vision Technology*, Advanced and Intelligent Manufacturing in China, https://doi.org/10.1007/978-981-95-1184-6_16

skills through continuous games. Four years later, this program defeated its designer. Three years later, it defeated a champion in the United States who had held the title for eight years. This program demonstrated the capabilities of machine learning and raised many thought-provoking social and philosophical questions.

For example, image recognition, speech recognition, natural language understanding, weather prediction, gene expression, content recommendation, etc. Currently, the approach we use to solve these problems through machine learning is as follows (using visual perception as an example): starting with data acquisition through sensors (cameras); then pre-processing, feature extraction, feature selection, followed by inference, prediction, or recognition; the last part, which is the machine learning part, is where most of the work is done, and there is a lot of research.

The middle three parts, summarised as feature representation, play a crucial role in the accuracy of the final algorithm, and the main computation and testing work of the system are spent on this large part. Manually selecting features is very laborious, and whether it can be done well largely depends on experience and luck. Deep learning achieves the goal of automatically selecting features by learning some features. Its alias is unsupervised feature learning, meaning that humans do not participate in the feature selection process.

Machine learning is a discipline focused on studying how computers can simulate or replicate human learning behavior. The human visual mechanism is as follows: starting from the intake of raw signals, followed by preliminary processing (certain cells in the cerebral cortex detect edges and directions), then abstraction (the brain determines the shape of the object in front, for example, it is round), and then further abstraction (the brain further determines what specific object it is, for example, it is a balloon).

In general, the information processing of the human visual system is hierarchical. High-level features are combinations of low-level features, and as you move from low-level to high-level, the feature representation becomes increasingly abstract, better able to express semantics or intentions. The higher the level of abstraction, the fewer possible guesses exist, which is more conducive to classification. For example, the correspondence between a set of words and a sentence is many-to-one, the correspondence between a sentence and semantics is many-to-one, and the correspondence between semantics and intention is also many-to-one; this is a hierarchical system. The "deep" in deep learning represents this layered system. So how does deep learning draw on this process? After all, since it is handled by computers, one issue faced is how to model this process.

Features are the raw materials of machine learning systems, and their impact on the final model is undeniable. When data is well represented by features, a linear model can typically achieve satisfactory accuracy. So, what do we need to consider regarding features?

At what granularity of feature representation can a learning algorithm be effective? For an image, pixel-level features are of no value. For example, in a photo of a car, nothing can be gleaned from the pixel level, and it cannot distinguish between a car and a non-car. However, if the features are structured (or meaningful), such as whether it has headlights or tyres, it becomes easy to distinguish between a car and a non-car, allowing the learning algorithm to be effective. Complex graphics are often composed of some basic structures. This rule applies not only to images but also to sounds.

Small blocks of graphics can be composed of basic edges, while more structured, complex, and conceptual graphics require higher-level feature representation. Deep learning is about finding small blocks that represent features at various levels and gradually combining them into higher-level features.

So, how many features should each layer have? The more features, the more reference information is provided, and accuracy will improve. However, more features mean increased computational complexity and a larger exploration space, leading to sparse data on each feature, which can cause various problems. More features are not necessarily better. Also, how many layers are appropriate? What architecture should be used for modelling? How to conduct unsupervised training? These require an overall design.

16.2 The Basic Idea of Deep Learning

Assume there is a system *S* with *n* layers (*S*1, …, *Sn*), its input is *I*, and output is *O*, represented as: *I* => *S*1 => *S*2 => … => *Sn* => O. If the output *O* equals the input *I*, meaning that after passing through this system, there is no information loss, and it remains unchanged, this implies that the input *I* has undergone no information loss through each layer *Si*, i.e., at any layer *Si*, it is another representation of the original information (i.e., input I). Deep learning requires automatic feature learning. Suppose there is a set of inputs *I* (such as a set of images or text), and a system *S* (with n layers) is designed. By adjusting the parameters in the system so that its output remains the input I, a series of hierarchical features of the input I can be automatically obtained, i.e., *S*1, …, *Sn*. For deep learning, the idea is to design multiple layers, where the output of each layer is the input to the next layer, thereby achieving a hierarchical representation of input information.

The above assumes that the output is strictly equal to the input, which is actually impossible, as information processing does not increase information, and most processing will lose information. This restriction can be somewhat relaxed, for instance, by minimizing the difference between input and output, which leads to another class of different deep learning methods.

16.3 Shallow Learning and Deep Learning

(1) Shallow learning is the first wave of machine learning

The previously introduced BP neural network, invented in the late 1980s, sparked a wave of machine learning that continues to this day. It was discovered that using the BP algorithm, an artificial neural network model could learn statistical patterns from a large number of training samples, thereby predicting unknown events. This statistical-based machine learning method showed superiority in many aspects compared to past systems based on manual rules. At this time, artificial neural networks, although also called multi-layer perceptrons, were actually shallow models containing only one hidden layer node, hence also referred to as shallow learning.

In the 1990s, various shallow machine learning models were proposed one after another, such as support vector machines (SVM), Boosting, and maximum entropy methods (logistic regression, LR). The structure of these models can basically be seen as having one hidden layer node (such as SVM, Boosting) or no hidden layer node (such as LR). These models achieved great success both in theoretical analysis and application. In contrast, due to the difficulty of theoretical analysis and the need for much experience and skill in training methods, shallow artificial neural networks were relatively quiet during this period.

(2) Deep learning is the second wave of machine learning

In 2006, Geoffrey Hinton, a professor at the University of Toronto in Canada and a leading figure in the field of machine learning, along with his student Ruslan Salakhutdinov, published an article in "Science" that initiated the wave of deep learning in academia and industry. This article had two main points: ① Multi-hidden layer artificial neural networks have excellent feature learning capabilities, and the features learned provide a more essential depiction of the data, which is beneficial for visualisation or classification; ② The difficulty in training deep neural networks can be effectively overcome through "layer-wise pre-training," which in this article was achieved through unsupervised learning.

Most current classification, regression, and other learning methods are shallow structure algorithms, whose limitation lies in their limited ability to represent complex functions with finite samples and computational units, and their generalisation ability is somewhat restricted for complex classification problems. Deep learning can achieve complex function approximation by learning a deep non-linear network structure, representing input data in a distributed manner, and demonstrating a strong ability to learn the essential features of a dataset from a small number of samples. In other words, the advantage of multiple layers is that complex functions can be represented with fewer parameters.

The essence of deep learning is to learn more useful features by constructing machine learning models with many hidden layers and massive training data, thereby ultimately improving the accuracy of classification or prediction. Therefore, "deep models" are the means, and "feature learning" is the goal. Unlike traditional shallow learning, deep learning differs in that: ① It emphasises the depth of the model structure, usually having dozens of layers or more. ② It clearly highlights the importance of feature learning, meaning that through layer-by-layer feature transformation, the feature representation of samples in the original space is transformed into a new feature space, making classification or prediction easier. Compared to the method of constructing features with manual rules, using big data to learn features can better depict the rich intrinsic information of the data.

16.4 Deep Learning and Neural Networks

Deep learning is a new field in machine learning research, motivated by the desire to establish and simulate neural networks that analyse and learn like the human brain. It mimics the mechanism of the human brain to interpret data, such as images, sounds, and text. Deep learning can be trained in supervised, unsupervised, or semi supervised ways. Early work often focused on unsupervised feature learning, but modern deep models are predominantly trained in a supervised manner.

The concept of deep learning originates from the study of artificial neural networks. A multi-layer perceptron with multiple hidden layers is a type of deep learning structure. Deep learning forms more abstract high-level representations of attribute categories or features by combining low-level features, to discover distributed feature representations of data. Deep learning is a branch of machine learning, which can be simply understood as the development of neural networks. Approximately two to three decades ago, neural networks were a particularly popular research direction in the ML field; however, they gradually fell out of favor due to the following reasons:

① They are relatively prone to overfitting, parameters are difficult to tune, and they require a lot of training tricks.
② The training speed is relatively slow, and in cases with fewer layers (≤ 3), the performance is not superior to other methods.

Therefore, for more than 20 years, neural networks received little attention, and this period was basically dominated by SVM and boosting algorithms. However, a dedicated gentleman, Hinton, persisted and eventually, along with others (Bengio, Yann LeCun, etc.), proposed a practically feasible deep learning framework.

Deep learning and traditional neural networks have both similarities and many differences. The similarity lies in the fact that deep learning adopts a layered structure similar to neural networks, consisting of a multi-layer network with an input layer, hidden layers (multiple), and an output layer. Connections exist only between nodes of adjacent layers, with no connections within the same layer or across layers. Each layer can be regarded as a logistic regression model; this layered structure closely resembles the structure of the human brain.

To overcome the issues in neural network training, DL employs a training mechanism quite different from neural networks. In traditional neural networks, the back-propagation method is used, which, simply put, involves using an iterative algorithm to train the entire network. Initial values are set randomly, the current network output is calculated, and then the parameters of the preceding layers are adjusted based on the difference between the current output and the label until convergence (overall, it is a gradient descent method). In contrast, deep learning generally uses a layer-wise training mechanism. The reason for this is that if the backpropagation mechanism is used, for a deep network (more than 7 layers), the residuals propagated to the foremost layers become too small, leading to the so-called gradient diffusion.

16.5 Deep Learning Training Process

If all layers are trained simultaneously, the complexity will be very high. If one layer is trained at a time, the bias will be transmitted layer by layer. This will face the opposite problem to supervised learning mentioned above, as the neurons and parameters of a deep network are too many, leading to severe underfitting.

In 2006, Hinton proposed an effective method for building multi-layer neural networks on unsupervised data, which can be simply divided into two steps:

① First, construct single-layer neurons layer by layer, so that each time a single-layer network is trained.
② After all layers are trained, use the wake-sleep algorithm for fine-tuning.

The weights between layers, except for the topmost layer, are made bidirectional, so the topmost layer remains a single-layer neural network, while other layers become graphical models. The upward weights are used for "recognition," and the downward weights are used for "generation." Then, the wake-sleep algorithm is used to adjust all the weights, ensuring that recognition and generation are consistent, meaning that the topmost representation generated can accurately reconstruct the nodes of the lower layers as much as possible. For example, if a node at the top layer represents a human face, all images of human faces should activate this node, and the image generated downward should roughly represent a human face. The wake-sleep algorithm is divided into two parts: wake and sleep.

① Wake phase: The recognition process generates an abstract representation (node state) of each layer through external features and upward weights (recognition weights), and modifies the downward weights (generation weights) between layers using gradient descent. This means "if reality is different from what I imagine, change my weights so that what I imagine is like this."
② Sleep phase: The generation process generates the state of the lower layers through the top layer representation (concepts learned during wake) and downward weights, while modifying the upward weights between layers. This means "if the scene in my dream does not match the corresponding concept in my mind, change my recognition weights so that this scene appears to me as this concept."

The specific training process of deep learning is as follows:

① Use bottom-up unsupervised learning.

Start from the bottom layer and train layer by layer up to the top layer. Use unlabelled data (labelled data can also be used) for layer-wise training of each layer's parameters. This step can be regarded as an unsupervised training process, which is the biggest difference from traditional neural networks. This process can be seen as a feature learning process. First, train the first layer with unlabelled data, learning the parameters of the first layer (this layer can be seen as obtaining a hidden layer of a three-layer neural network that minimises the difference between output and input). Due to the model's capacity limitations and sparsity constraints, the resulting model can learn the structure of the data itself, thus obtaining features with greater representational power than the input; after learning the n − 1 layer, use the output of the n − 1 layer as the input for the n layer, and train the n layer, thereby obtaining the parameters of each layer.

② Top-down supervised learning.

Train with labelled data, transmitting errors from top to bottom to fine-tune the network. Further adjust the parameters of the entire multi-layer model based on the parameters obtained in the first step; this step is a supervised training process. The first step is similar to the random initialisation process of neural networks. Since the first step of deep learning is not random initialisation but is obtained by learning the structure of the input data, this initial value is closer to the global optimum, thus achieving better results. Therefore, the effectiveness of deep learning largely depends on the feature learning process of the first step.

16.6 Common Methods of Deep Learning

16.6.1 Autoencoder

One of the simplest methods of deep learning is to utilise the characteristics of artificial neural networks. An artificial neural network (ANN) is inherently a hierarchical system. If a neural network is given, we assume its output is the same as its input, and then train to adjust its parameters to obtain the weights in each layer. Naturally, this results in several different representations of the input I (each layer represents a different representation), and these representations constitute the features. An autoencoder is a neural network that attempts to reproduce the input signal as closely as possible. To achieve this reproduction, the autoencoder must capture the most important factors that can represent the input data and find the main components that can represent the original information.

The specific process is outlined as follows.

(1) Given unlabelled data, use unsupervised learning features

In previous neural networks, as shown in Fig. 16.1a, the input samples are labelled, i.e., (input, target), so the parameters of the previous layers are changed based on the difference between the current output and the target (label) until convergence. But now there is only unlabelled data, as shown in the figure on the right. So how is this error obtained?

As illustrated in Fig. 16.2, upon passing through the encoder, the input is transformed into an encoding, which represents the input. So how do we know that this encoding represents the input? By adding a decoder, the decoder will output information, and if this output information is very similar to the initial input signal (ideally the same), then it is reasonable to believe that this encoding is reliable. Therefore, by

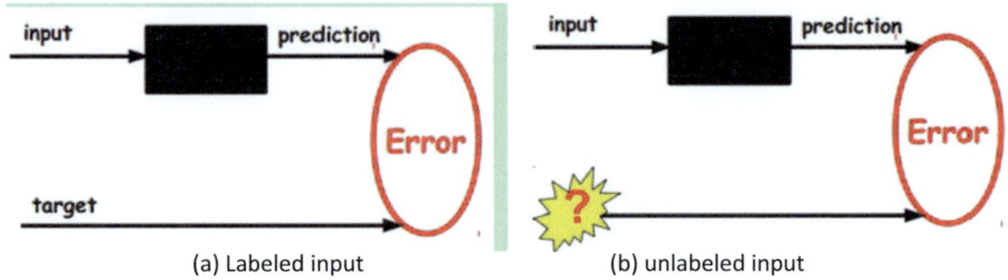

Fig. 16.1 Neural network input

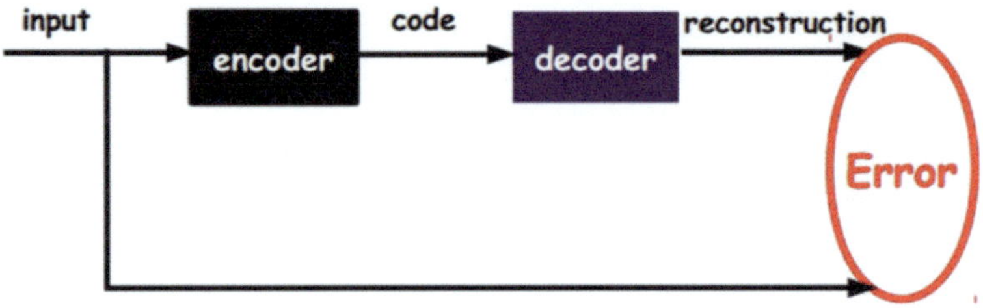

Fig. 16.2 Encoder and decoder

adjusting the parameters of the encoder and decoder to minimise the reconstruction error, the first representation of the input signal is obtained, which is the encoding. Since the data is unlabelled, the source of the error is directly obtained by comparing the reconstruction with the original input.

(2) Generate features through the encoder, then train the next layer step by step

The first layer's encoding is obtained above, and minimising the reconstruction error indicates that this encoding is a good representation of the original input signal, or it is identical to the original signal (different expression, reflecting the same thing). The second layer is trained in the same way as the first layer, using the first layer's output encoding as the input signal for the second layer, and similarly minimising the reconstruction error to obtain the second layer's parameters and the encoding of the second layer's input, which is the second representation of the original input information. Other layers can be trained in the same manner (when training this layer, the parameters of the previous layers are fixed, and their decoders are no longer needed). Figure 16.3 shows the layer-by-layer training model.

(3) Supervised fine-tuning

Using the above method, many layers can be obtained. As for how many layers are needed (or how deep it needs to be, there is currently no scientific evaluation method), it requires experimentation. Each layer will obtain a different representation of the original input. Of course, the more abstract, the better, just like the human visual system.

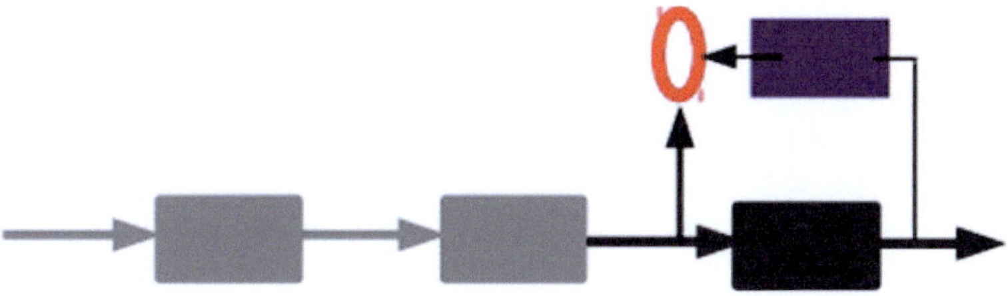

Fig. 16.3 Training model layer by layer

At this point, the autoencoder cannot yet be used to classify data because it has not learned how to associate an input with a class. It has only learned how to reconstruct or reproduce its input. In other words, it has only learned to obtain a feature that can well represent the input, which can represent the original input signal to the greatest extent. To achieve classification, a classifier (e.g., logistic regression, SVM, etc.) can be added to the top encoding layer of the autoencoder, and then trained using the standard supervised training method of multilayer neural networks (gradient descent method). That is to say, at this point, the feature encoding of the last layer needs to be input into the final classifier, and fine-tuned through supervised learning with labelled samples, which can be done in two ways: one is to adjust only the classifier, as shown in the black part of Fig. 16.4.

The other way, as shown in Fig. 16.5, is to fine-tune the entire system with labelled samples. If there is enough data, this method is best, allowing for end-to-end learning.

Once supervised training is complete, this network can be used for classification. The top layer of the neural network can serve as a linear classifier, which can then be replaced with a better-performing classifier. Research has shown that adding these automatically learned features to the original features can greatly improve accuracy. There are some variants of the autoencoder, and two are briefly introduced here.

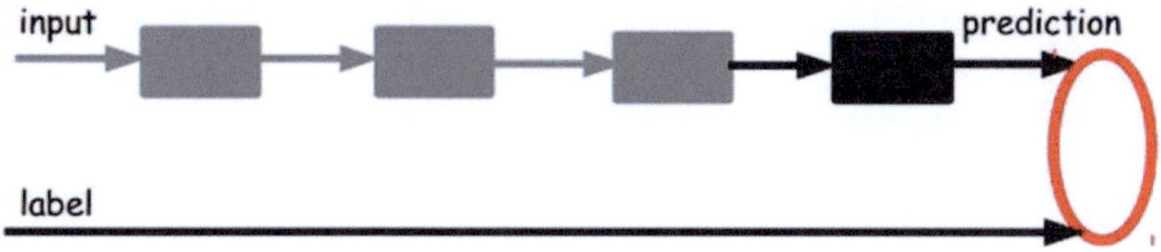

Fig. 16.4 Schematic diagram of adjusting the classifier

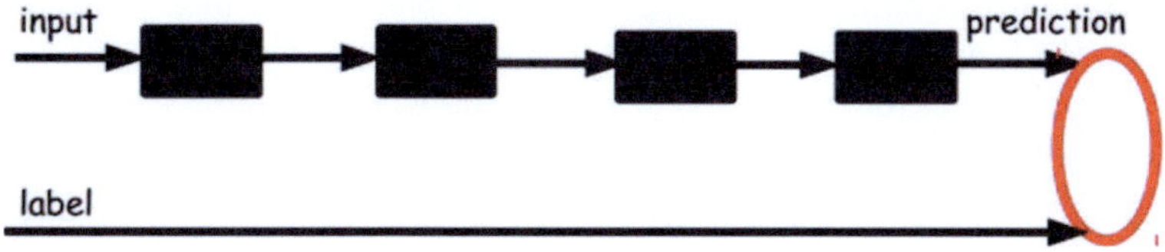

Fig. 16.5 Schematic diagram of fine-tuning the entire system

(1) Sparse Autoencoder

New deep learning methods can be obtained by adding some constraints, for example, if L1 regularity constraints are added to the autoencoder (L1 mainly constrains most nodes in each layer to be zero, with only a few non-zero, which is the source of the name sparse), a sparse autoencoder method can be obtained. It essentially limits the expression encoding obtained each time to be as sparse as possible. Sparse expressions are often more effective than other expressions. The human brain seems to work this way, where a certain input only stimulates certain neurons, while most other neurons are inhibited.

(2) Denoising Autoencoder

Denoising autoencoder (denoising auto... encoders, DA) are based on autoencoders, with noise added to the training data, so the autoencoder must learn to remove this noise to obtain the true input that has not been polluted by noise. This forces the encoder to learn a more robust representation of the input signal, which is why its generalisation ability is stronger than that of a typical encoder. DA can be trained using the gradient descent algorithm.

16.6.2 Sparse Coding

If the restriction that the output must be equal to the input is relaxed, while utilising the concept of basis in linear algebra, namely:

$$O = a_1\Phi_1 + a_2\Phi_2 + \cdots + a_n\Phi_n \tag{16.1}$$

where Φ_i is the basis, a_i is the coefficient.

This leads to an optimisation problem: min$|I - O|$, where I represents the input, O represents the output. By solving this optimisation equation, the coefficients a_i and the basis Φ_i can be obtained.

If an L1 regularity constraint is added to the above equation, it becomes:

$$\mathrm{Min}|I - O| + u(|a_1| + |a_2| + \cdots + |a_n|) \tag{16.2}$$

This method is called sparse coding. In simple terms, it means representing a signal as a linear combination of a set of bases, requiring only a few bases to represent the signal. "Sparsity" is defined as having only a few non-zero elements or only a few elements significantly greater than zero. The requirement for the coefficients a_i to be sparse means that for a set of input vectors, only as few coefficients as possible are significantly greater than zero. The choice to use sparse components to represent input data is because the vast majority of sensory data, such as natural images, can be represented as a superposition of a small number of basic elements, which in images can be surfaces or lines. The human brain has a large number of neurons, but for certain images or edges, only a few neurons are excited, while others remain inhibited.

The sparse coding algorithm is an unsupervised learning method used to find a set of "overcomplete" basis vectors to represent sample data more efficiently. Although techniques like Principal Component Analysis (PCA) can easily find a set of "complete" basis vectors, the aim here is to find a set of "overcomplete" basis vectors to represent input vectors (meaning the number of basis vectors is greater than the dimensionality of the input vectors). The advantage of overcomplete bases is that they can more effectively uncover the structures and patterns hidden within the input data. However, for overcomplete bases, the coefficients a_i are no longer uniquely determined by the input vectors. Therefore, in the sparse coding algorithm, an additional criterion of "sparsity" is added to address the degeneracy problem caused by overcompleteness. For example, at the lowest level of feature extraction in images, edge detectors need to be generated. The task here is to randomly select some small patches from the original image, generate the "basis" that can describe them, and then provide a test patch. The reason for generating edge detectors is that edges in different directions can describe the entire image, so edges in different directions naturally become the basis of the image. Sparse coding is divided into two parts:

① Training phase: Given a series of sample images $[x_1, x_2, \ldots]$, a set of bases $[\Phi_1, \Phi_2, \ldots]$, which is the dictionary.

Sparse coding is a variant of the clustering algorithm (k-means), and its training process is quite similar, being a repetitive iterative process. The basic idea is as follows: if the objective function to be optimised contains two variables, such as $L(W, B)$, then one can first fix W, adjust B to minimise L, and then fix B, adjust W to minimise L. By iterating alternately in this manner, L is continuously pushed towards its minimum value. According to the above method, alternately change a and Φ to minimise the following objective function.

$$\min_{\alpha,\emptyset} \sum_{i=1}^{m} \left\| x_i - \sum_{j=1}^{k} \alpha_{i,j\emptyset_j} \right\|^2 + \lambda \sum_{i=1}^{m} \sum_{j=1}^{k} \left|\alpha_{i,j}\right| \tag{16.3}$$

Each iteration is divided into two steps:

a. Fix the dictionary $\Phi\,[k]$, then adjust $a\,[k]$ to minimise the above expression, i.e., the objective function, which is to solve the LASSO (least absolute shrinkage and selection operator) problem.
b. Then fix $a\,[k]$, adjust $\Phi\,[k]$ to minimise the above expression, i.e., the objective function, which is to solve the convex QP (quadratic programming, convex quadratic programming) problem.

Iterate continuously until convergence. In this way, a set of bases that can well represent this series of x, which is the dictionary, can be obtained.

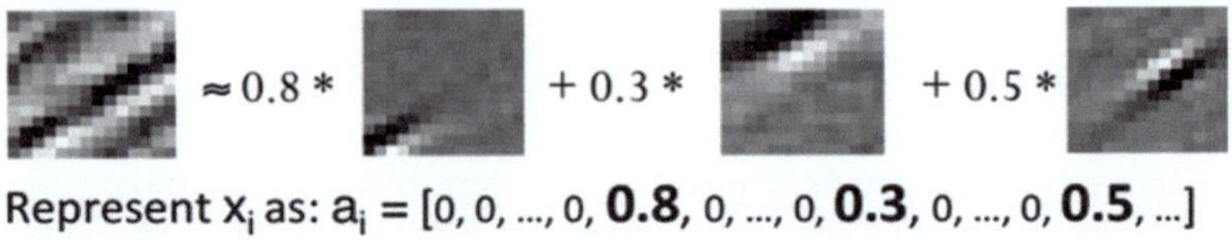

Fig. 16.6 Coding example

② Coding stage: Given a new image x, using the dictionary obtained above, a sparse vector a is obtained by solving a LASSO problem. This sparse vector is a sparse representation of the input vector x, as shown in Eq. (16.4).

$$\min_{\alpha} \sum_{i=1}^{m} \left\| x_i - \sum_{j=1}^{k} \alpha_{i,j} \varnothing_j \right\|^2 + \lambda \sum_{i=1}^{m} \sum_{j=1}^{k} \left| \alpha_{i,j} \right| \tag{16.4}$$

As shown in Fig. 16.6.

16.6.3 Restricted Boltzmann Machine

Assume there is a two-layer graph, as shown in Fig. 16.7, where there are no links between nodes in each layer. One layer is the visible layer, i.e., the input data layer (v), and the other layer is the hidden layer (h). If it is assumed that all nodes are random binary variable nodes (can only take values of 0 or 1), and it is also assumed that the full probability distribution $p(v, h)$ satisfies the Boltzmann distribution, we call this model a restricted Boltzmann machine (RBM).

Since this model is a two-layer graph, all hidden nodes are conditionally independent when v is known (because there are no connections between nodes), i.e., $p(h \mid v) = p(h_1 \mid v) \ldots p(h_n \mid v)$. Similarly, when the hidden layer h is known, all visible nodes

Fig. 16.7 Layer 2 diagram

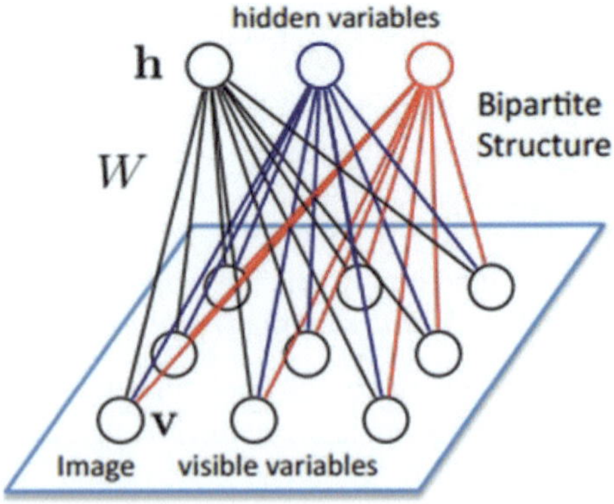

are conditionally independent. Furthermore, since all v and h satisfy the Boltzmann distribution, when input v is given, the hidden layer h can be obtained through $p(h \mid v)$, and after obtaining the hidden layer h, through $p(v \mid h)$ can also obtain the visible layer. If by adjusting the parameters, the visible layer obtained from the hidden layer, v_1, can be made identical to the original visible layer v, then the resulting hidden layer is another representation of the visible layer. Therefore, the hidden layer can serve as a feature of the input data of the visible layer, making it a method of deep learning.

How to train, that is, how to determine the weights between the visible layer nodes and the hidden nodes? Some mathematical analysis is required, which means establishing a model.

The energy of the joint configuration can be expressed as Eq. (16.5).

$$E(v, h; \theta) = -\sum_{ij} W_{ij} v_i h_j - \sum_i b_i v_i - \sum_j a_j h_j \tag{16.5}$$

$$\theta = \{W, a, b\} \text{ model parameters}$$

The joint probability distribution of a certain configuration can be determined through the Boltzmann distribution (and the energy of this configuration), as shown in Eq. (16.6).

$$P_\theta(v, h) = \frac{1}{Z(\theta)} \exp(-E(v, h; \theta)) = \frac{1}{Z(\theta)} \prod_{ij} e^{W_{ij} v_i h_j} \prod_i e^{b_i v_i} \prod_j e^{a_j h_j} \tag{16.6}$$

$$Z(\theta) = \sum_{h,v} \exp(-E(v, h; \theta))$$

Because the hidden nodes are conditionally independent (since there are no connections between the nodes), that is:

$$p(h|v) = \prod_j P\left(h_j|v\right) \tag{16.7}$$

It is relatively easy to [factorize the above equation] to obtain the probability of the jth node in the hidden layer being 1 or 0, given the visible layer v:

$$P\left(h_j = 1|v\right) = \frac{1}{1 + \exp\left(-\sum_i W_{ij} v_i - a_j\right)} \tag{16.8}$$

Similarly, given the hidden layer h, the probability of the ith node in the visible layer being 1 or 0 can also be easily obtained:

$$P(v_i = 1|h) = \frac{1}{1 + \exp\left(-\sum_j W_{ij} h_j - b_i\right)} \tag{16.9}$$

Given a sample set that satisfies independent and identically distributed conditions: D = {$\mathbf{v}^{(1)}$, $\mathbf{v}^{(2)}$, …, $\mathbf{v}^{(N)}$}, it is necessary to learn the parameters $\theta = \{W, a, b\}$.

Maximise the following log-likelihood function (Maximum Likelihood Estimation: for a certain probability model, it is necessary to choose a parameter that maximises the probability of the current observed samples):

$$L(\theta) = \frac{1}{N}\sum_{n=1}^{N} \log P_\theta\left(V^{(n)}\right) - \frac{\lambda}{N}\|W\|_F^2 \tag{16.10}$$

That is, by differentiating the maximum log-likelihood function, the parameter W corresponding to the maximum L can be obtained, as shown in Eq. (16.11).

$$\frac{\partial L(\theta)}{\partial W_{ij}} = Ep_{data}[v_i h_i] - Ep_\theta[v_i h_i] - \frac{2\lambda}{N} W_{ij} \tag{16.11}$$

If the number of layers in the hidden layer is increased, a deep Boltzmann machine (DBM) can be obtained; if a Bayesian belief network (i.e., a directed graphical model, where there are still no connections between nodes within a layer) is used in the part close to the visible layer, and a restricted Boltzmann machine is used in the part farthest from the visible layer, a deep belief net (DBN) can be obtained, as shown in Fig. 16.8.

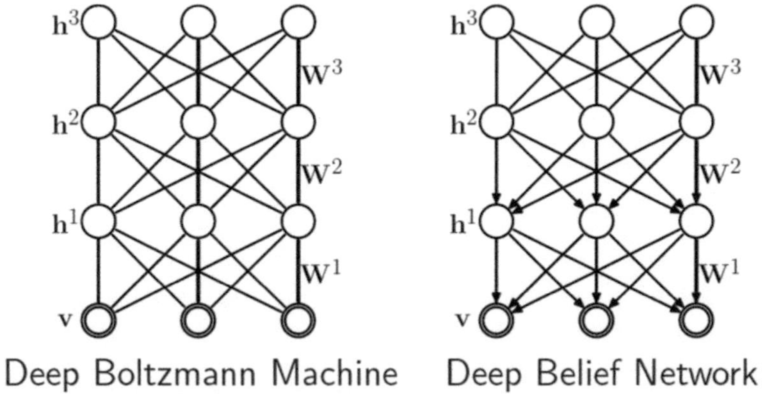

Fig. 16.8 DBM and DBN

16.6.4 Deep Belief Networks

Deep Belief Networks (DBNs), as shown in Fig. 16.9, are probabilistic generative models. In contrast to traditional discriminative models' neural networks, generative models establish a joint distribution between observed data and labels, evaluating both P(Observation | Label) and P(Label | Observation), whereas discriminative models only evaluate the latter, i.e., P(Label | Observation). When applying the traditional BP algorithm in deep neural networks, DBNs encounter the following issues:

① A labelled sample set is required for training;
② The learning process is relatively slow;
③ Inappropriate parameter selection can lead to learning converging to a local optimum.

DBNs are composed of multiple layers of restricted Boltzmann machines, a typical type of neural network as shown in Fig. 16.10. These networks are "restricted" to a visible layer and a hidden layer, with connections existing between layers but not within the units of a layer. The hidden layer units are trained to capture the higher-order data correlations exhibited in the visible layer.

Initially, without considering the top two layers that form an associative memory, the connections of a DBN are determined by top-down generative weights, with RBMs acting like building blocks. Compared to traditional and deep layered sigmoid belief networks, it facilitates the learning of connection weights.

The process begins with an unsupervised greedy layer-wise method to pre-train and obtain the weights of the generative model. This unsupervised greedy layer-wise method was proven effective by Hinton and is referred to as contrastive divergence. During this training phase, a vector v is generated in the visible layer, through which values are passed to the hidden layer. Conversely, the input to the visible layer is randomly selected to attempt to reconstruct the original input signal. Finally, these newly visible neuron activation units will forward propagate to reconstruct the hidden

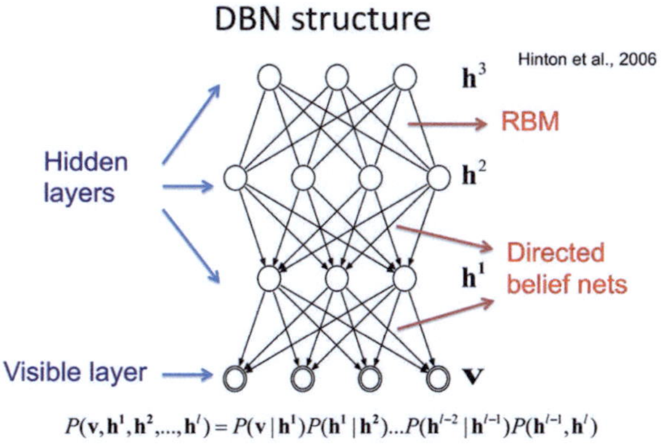

Fig. 16.9 DBNs model

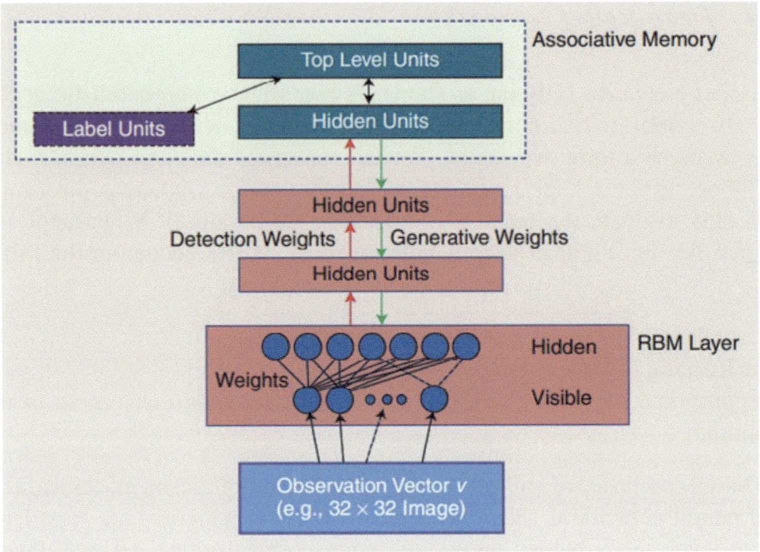

Fig. 16.10 DBN framework diagram

layer activation units, obtaining h. These backward and forward steps are commonly known as Gibbs sampling, and the difference in correlation between the hidden layer activation units and the visible layer input serves as the main basis for weight updates.

This significantly reduces training time, as only a single step is needed to approach maximum likelihood learning. Adding each layer to the network improves the log probability of the training data, which can be understood as getting closer to the true expression of energy. This meaningful extension and the use of unlabelled data are decisive factors in any deep learning application.

In the top two layers, weights are connected together so that the output of the lower layers will provide a reference clue or association to the top layer, thus linking it to its memory content. Ultimately, what is desired is discriminative performance.

After pre-training, a DBN can be adjusted for discriminative performance using the BP algorithm with labelled data. Here, a set of labels is attached to the top layer (extending associative memory), and a network classification surface is obtained through a bottom-up, learned recognition weight. This performance is better than a network trained solely with the BP algorithm. This can be intuitively explained as the BP algorithm of DBNs only needs to perform a local search in the weight parameter space, which is faster and requires less convergence time compared to feedforward neural networks.

The flexibility of DBNs makes them relatively easy to extend. One extension is convolutional DBNs (convolutional deep belief networks, CDBNs). DBNs do not consider the two-dimensional structural information of images, as the input is simply a one-dimensional vectorization of an image matrix. CDBNs address this issue by utilizing the spatial relationships of neighbouring pixels, achieving transformation invariance of the generative model through a model called convolutional RBMs, and can easily transform to high-dimensional images. DBNs do not explicitly handle learning the temporal connections of observed variables, although there is current research in this area, such as stacked temporal RBMs, which, as an extension, have temporal convolution machines for sequence learning. This application of sequence learning brings an exciting future research direction to speech signal processing problems.

Current research related to DBNs includes stacked autoencoders, which replace the RBMs in traditional DBNs with stacked autoencoders. This allows for the training of deep multi-layer neural network architectures using the same rules, but it lacks the strict parameterization requirements of layers. Unlike DBNs, autoencoders use discriminative models, making it difficult for this structure to sample the input sampling space, which makes it harder for the network to capture its internal representation. However, denoising autoencoders can effectively avoid this problem and perform better than traditional DBNs. They achieve this by adding random noise during the training process and stacking to produce field generalization performance. The process of training a single denoising autoencoder is similar to the process of training a generative model with RBMs.

16.6.5 Convolutional Neural Networks

Convolutional neural networks (CNNs) are a type of artificial neural network that have become a research hotspot in the fields of speech analysis and image recognition. Their weight-sharing network structure makes them more similar to biological neural networks, reducing the complexity of the network model and the number of weights. This advantage is more evident when the network's input is a multi-dimensional image, allowing the image to be directly used as the network's input, avoiding the complex feature extraction and data reconstruction processes of traditional recognition algorithms. Convolutional networks are a multi-layer perceptron specifically designed for recognising two-dimensional shapes, and this network structure has a high degree of invariance to translation, scale, tilt, or other forms of deformation.

CNNs are influenced by early Time-Delay Neural Networks (TDNN). Time-Delay Neural Networks reduce learning complexity by sharing weights across the time dimension, making them suitable for processing speech and time-series signals.

CNNs are the first learning algorithm to successfully train a multi-layer network structure. They utilise spatial relationships to reduce the number of parameters that need to be learned, thereby improving the training performance of the general forward BP algorithm. CNNs, as a deep learning architecture, were proposed to minimise data preprocessing requirements. In a CNN, a small part of an image (local receptive field) serves as the input at the lowest level of the hierarchical structure, and information is then transmitted to different layers. Each layer uses a digital filter to obtain the most significant features of the observed data. This method can capture significant features of observed data that are invariant to translation, scaling, and rotation, because the local receptive field of the image allows neurons or processing units to access the most basic features, such as oriented edges or corners.

(1) History of Convolutional Neural Networks

In 1962, Hubel and Wiesel proposed the concept of the receptive field through their research on the visual cortex cells of cats. In 1984, Japanese scholar Fukushima proposed the neocognitron based on the concept of the receptive field, which can be regarded as the first implementation of a convolutional neural network and the first application of the receptive field concept in the field of artificial neural networks. The neocognitron decomposes a visual pattern into many sub-patterns (features), which are then processed in a hierarchically connected feature plane. It attempts to model the visual system so that it can recognise objects even when they are displaced or slightly deformed.

Typically, the neocognitron contains two types of neurons, namely S-cells responsible for feature extraction and C-cells for deformation resistance. Two important parameters are involved in S-cells: the receptive field and the threshold parameter. The former determines the number of input connections, while the latter controls the response level to feature sub-patterns. Many scholars have been committed to improving the performance of the neocognitron. In traditional neocognitrons, the amount of visual blur brought by C-cells in each S-cell's receptive field is normally distributed. If the blur effect at the edge of the receptive field is greater than that at the centre, the S-cell will accept the greater deformation tolerance caused by this non-normal blur. Generally, it is hoped that the difference between the effects produced by the training pattern and the deformed stimulus pattern at the edge and centre of the receptive field becomes increasingly significant. To effectively form this non-normal blur, Fukushima proposed an improved neocognitron with a double C-cell layer.

Van Ooyen and Niehuis introduced a new parameter to improve the discrimination ability of the neocognitron. In fact, this parameter acts as an inhibitory signal, suppressing the excitation of neurons to repetitive excitation features. Most neural networks memorise training information in weights. According to Hebb's learning rule, the more a certain feature is trained, the easier it is to be detected in future recognition processes. Some scholars have combined evolutionary computation theory with the neocognitron, weakening the training of repetitive excitation features, thereby enabling the network to focus on different features to help improve discrimination ability. The above describes the development process of the neocognitron, and convolutional neural networks can be regarded as a generalised form of the neocognitron, with the neocognitron being a special case of convolutional neural networks.

(2) Network Structure of Convolutional Neural Networks

As shown in Fig. 16.11, a convolutional neural network is a multi-layer neural network, with each layer composed of multiple two-dimensional planes, and each plane consisting of multiple independent neurons.

The input image is convolved with three trainable filters and an additive bias, as shown in Fig. 16.11. After convolution, three feature maps are generated in the C1

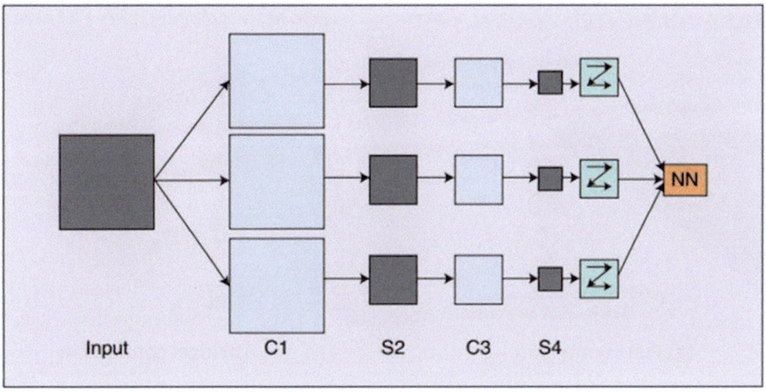

Fig. 16.11 Conceptual diagram of convolutional neural networks

layer. Then, four pixels in each group of the feature maps are summed, weighted, and biased, and a Sigmoid function is applied to obtain three feature maps in the S2 layer. These maps are then filtered to obtain the C3 layer. This hierarchical structure produces S4 in the same way as S2. Finally, these pixel values are rasterised and connected into a vector input to a traditional neural network to obtain the output.

Generally, the C layer is the feature extraction layer, where each neuron's input is connected to the local receptive field of the previous layer, extracting the local features. Once these local features are extracted, their positional relationships with other features are also determined. The S layer is the feature mapping layer, with each computational layer of the network composed of multiple feature maps. Each feature map is a plane, and all neurons on the plane have equal weights. The feature mapping structure uses a Sigmoid function with a small influence function kernel as the activation function of the convolutional network, making the feature mapping have displacement invariance.

Additionally, because neurons on a mapping plane share weights, the number of free parameters in the network is reduced, lowering the complexity of network parameter selection. Each feature extraction layer (C layer) in a convolutional neural network is followed by a computational layer (S layer) for local averaging and secondary extraction. This unique two-stage feature extraction structure gives the network a high tolerance for distortion when recognising input samples.

(3) On Parameter Reduction and Weight Sharing

As mentioned above, one of the key features of CNNs is the reduction in the number of parameters that need to be trained in a neural network through receptive fields and weight sharing. As shown in Fig. 16.12a, if there is an image with 1000×1000 pixels and 1 million hidden layer neurons, then if they are fully connected (each hidden layer neuron is connected to every pixel of the image), there would be $1000 \times 1000 \times 1{,}000{,}000 = 10^{12}$ connections, which means 10^{12} weight parameters. However, the spatial relationships in images are local, just as humans perceive external images

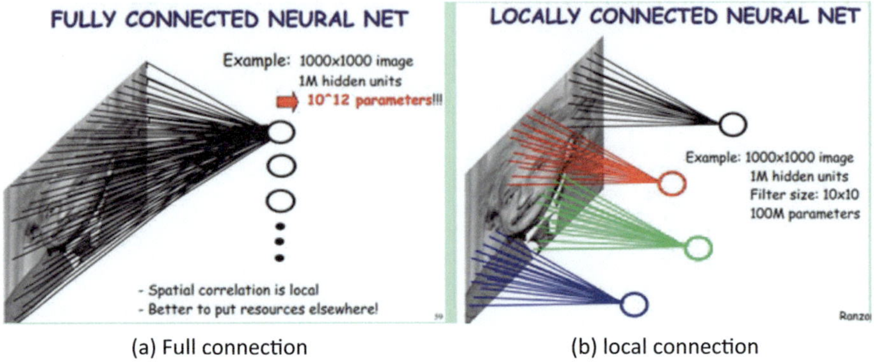

Fig. 16.12 Hidden layer neuron connections

through a local receptive field. Each neuron does not need to perceive the entire image globally; each neuron only perceives a local image area, and then at a higher level, these neurons that perceive different local areas are integrated to obtain global information. This reduces the number of connections, thereby reducing the number of weight parameters that need to be trained in the neural network. As shown in Fig. 16.12b, if the local receptive field is 10×10, each hidden layer neuron only needs to connect to this 10×10 local image, so 1 million hidden layer neurons would have only 100 million connections, i.e., 10^8 parameters. This reduces the original number by four orders of magnitude, making training less laborious, though it still feels substantial.

Each neuron in the hidden layer connects to a 10×10 image area, meaning each neuron has $10 \times 10 = 100$ connection weight parameters. If these 100 parameters for each neuron are the same, meaning each neuron uses the same convolution kernel to convolve the image, then there are only 100 parameters. Regardless of the number of neurons in the hidden layer, there are only 100 parameters between the two layers, which is weight sharing, another important feature of convolutional neural networks.

If a filter, or a convolution kernel, extracts a feature from an image, and multiple features need to be extracted, multiple filters are added, each with different parameters to represent different features extracted from the input image, such as different edges. Thus, each filter convolves the image to obtain different feature representations of the image, known as feature maps. Therefore, 100 convolution kernels result in 100 feature maps, which together form a layer of neurons.

The number of neurons in the hidden layer is related to the size of the input image, the size of the filter, and the stride of the filter across the image. For example, if the input image is 1000×1000 pixels and the filter size is 10×10, assuming the filter does not overlap, i.e., the stride is 10, then the number of neurons in the hidden layer is $(1000 \times 1000)/(10 \times 10) = 100 \times 100$ neurons. If the stride is 8, meaning the convolution kernel overlaps by two pixels, the number of neurons would be different. This is for one filter, or the number of neurons for one feature map; if there are 100 feature maps, it is 100 times more. Thus, the larger the image, the greater the disparity in the number of neurons and the number of weight parameters that need to be trained.

In summary, the core idea of convolutional networks is to combine the concepts of local receptive fields, weight sharing (or weight replication), and temporal or spatial subsampling to achieve a certain degree of invariance to displacement, scale, and deformation.

16.7 Handwritten Character Recognition Based on Convolutional Neural Networks

The previous section introduced the basic principles of CNNs, and this section introduces the application of CNNs in handwritten character recognition. The weight-sharing network structure of CNNs makes them more similar to biological neural networks, reducing the complexity of the network model and the number of weights. When the network input is a multidimensional image, the advantages of NN become more apparent, allowing the image to be directly used as network input, avoiding the complex feature extraction and data reconstruction processes of traditional recognition algorithms. CNNs offer numerous advantages in two-dimensional image processing, including their ability to autonomously extract image features, including colour, texture, shape, and the topological structure of the image, especially in applications requiring invariance to displacement, scaling, and other forms of distortion, providing robustness and computational efficiency. The generalisation ability of CNNs is significantly superior to other methods, and convolutional neural networks have been applied in pattern classification, object detection, and object recognition.

16.7.1 Convolutional Neural Network Structure for Handwritten Character Recognition

As previously introduced, CNNs typically have at least two non-linear trainable convolutional layers, two non-linear fixed convolutional layers (pooling layers or subsampling layers), and a fully connected layer, with at least five hidden layers in total, as shown in Fig. 16.11.

The following introduces the classic convolutional neural network LeNet-5 structure used for handwritten digit recognition based on the MNIST database, as shown in Fig. 16.13. The network structure of LeNet-5 for handwritten digit recognition consists of 8 layers: the input layer, the first convolution (C1 layer), the first subsampling (S2 layer), the second convolution (C3 layer), the second subsampling (S4 layer), the third convolution (C5 layer), the fully connected (F6 layer), and the output layer. The operational procedures for each layer are as follows.

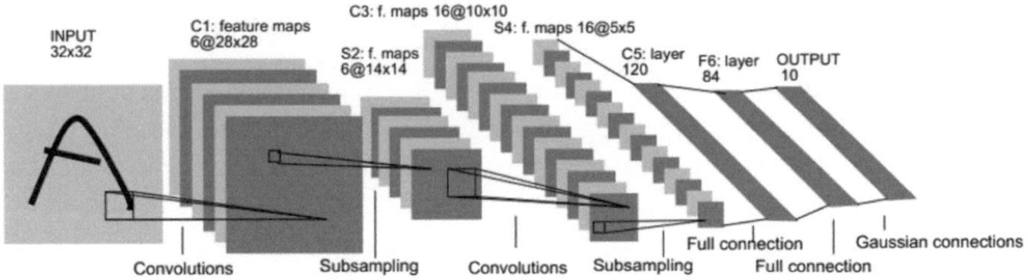

Fig. 16.13 Structure of LeNet-5 convolutional neural network

(1) Input Layer

As shown in Fig. 16.14, the input layer uses the Minis database. This database has a training set of 60,000 handwritten digit images and a test set of 10,000 images. The input images are processed to a size of 32 × 32, so that potential prominent features, such as stroke breakpoints or corner points, can appear at the centre of the receptive field of the highest layer feature monitoring sub-receptive field.

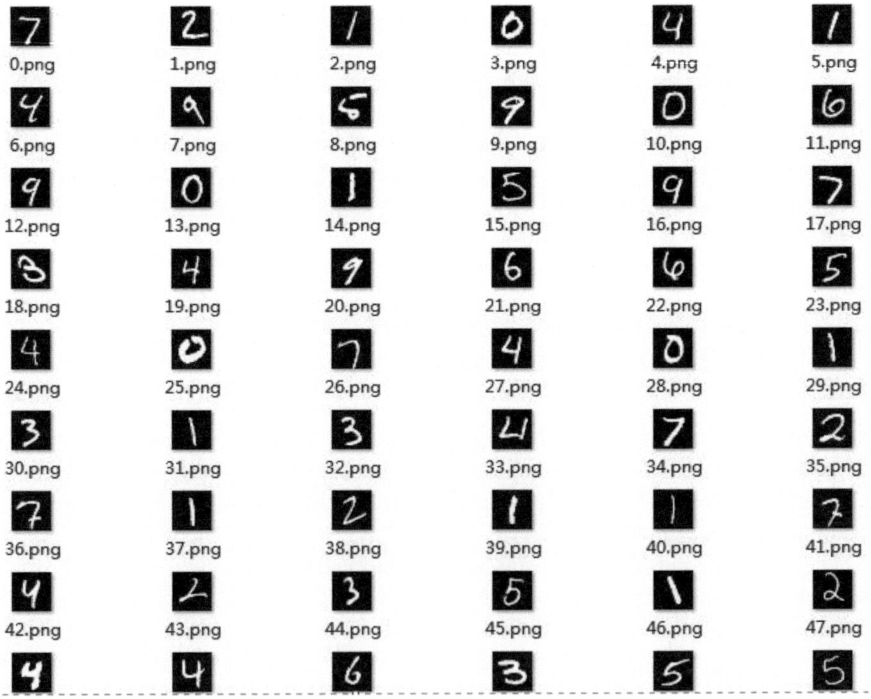

Fig. 16.14 Input image sample

(2) C1 Layer

The C1 layer is a convolutional layer composed of 6 feature maps. Convolution is performed using 6 filters of size 5×5, resulting in 6 feature maps in the C1 convolutional layer. Each neuron in the feature map is connected to a 5×5 neighbourhood in the input image, i.e., a 5×5 convolution kernel is used to convolve the input layer. The size of the feature map output from the C1 layer is $(32 - 5 + 1) \times (32 - 5 + 1) = 28 \times 28$. Each filter has $5 \times 5 = 25$ parameters and 1 bias parameter, with a total of 6 filters, so this layer has $(5 \times 5 + 1) \times 6 = 156$ trainable parameters, and the number of neurons is $(28 \times 28) \times 6 = 4707$, with a total of $156 \times (28 \times 28) = 122{,}304$ connections.

(3) S2 Layer

The S2 layer is a subsampling layer. Subsampling utilises the principle of local correlation in images to reduce the amount of data processing while retaining useful information. The input to this layer is $(28 \times 28) \times 6$, and it outputs 6 feature maps of size 14×14. Each unit in the feature map is connected to a 2×2 neighbourhood in the corresponding feature map in C1. The 4 inputs of each unit in the S2 layer are summed, multiplied by a trainable parameter, and then added to a trainable bias. The result is computed through a Sigmoid function. The trainable coefficient and bias control the non-linearity of the Sigmoid function. If the coefficient is relatively small, the operation approximates a linear operation, and subsampling is equivalent to blurring the image. If the coefficient is relatively large, depending on the size of the bias, subsampling can be seen as a noisy "or" operation or a noisy "and" operation. The 2×2 receptive fields of each unit do not overlap, so the size of each feature map in S2 is 1/4 of the size of C1 (1/2 for both rows and columns). The S2 layer has $1 \times 6 + 6 = 12$ trainable parameters, with a neuron count of $(14 \times 14) \times 6 = 1176$, and a total of $2 \times 2 \times 14 \times 14 \times 6 + 14 \times 14 \times 6 = 5880$ connections.

(4) C3 Layer

The C3 layer is also a convolutional layer, which convolves the S2 layer using a 5×5 convolution kernel, resulting in feature maps with an output size of 10×10. There are 16 convolution kernels, so there are 16 feature maps. The number of neurons is $(10 \times 10) \times 16 = 1600$; the number of trainable parameters is $6 \times (3 \times 25 + 1) + 6 \times (4 \times 25 + 1) + 3 \times (4 \times 25 + 1) + 1 \times (6 \times 25 + 1) = 1516$; and the number of connections is $1516 \times 10 \times 10 = 151{,}600$.

(5) S4 Layer

The S4 layer is a subsampling layer, with an input of 16 images of size 10×10, and an output of 16 feature maps of size 5×5. Each unit in the feature map is connected to a 2×2 neighbourhood in the corresponding feature map in C3, similar to the connection between C1 and S2. Each feature map in the S4 layer has 1 factor and 1 bias, so there are $(1 + 1) \times 16 = 32$ trainable parameters, with a neuron count of $(5 \times 5) \times 16 = 400$, and a total of $(2 \times 2 \times 5 \times 5 \times 16) + (5 \times 5 \times 16) = 2000$ connections.

Fig. 16.15 Entering the number to be recognized

(6) C5 Layer

The C5 layer is a convolutional layer with 120 feature maps. Each unit is connected to the entire 5 × 5 neighbourhood of all 16 units in the S4 layer. Since the size of the feature maps in the S4 layer is also 5 × 5, the size of the C5 feature maps is 1 × 1, forming a fully connected layer between S4 and C5. The C5 layer has 48,120 trainable connections.

(7) F6 Layer

The F6 layer inputs 120 units of size 1 × 1, with 84 neurons, fully connected to the C5 layer. Like a classic neural network, the F6 layer calculates the dot product between the input vector and the weight vector, plus a bias, with a total of $120 \times 84 + 84 = 10{,}164$ connections and trainable parameters.

(8) Output Layer

This is the final layer of the entire network, with one unit per class, each with 84 inputs, and outputs a feature vector of size 1 × 10.

16.7.2 Implementation of Convolutional Neural Network for Text Recognition

The ImageSys general image processing system of Beijing Hyundai Fubo Technology Co., Ltd. integrates the function of handwritten character recognition using CNN, as shown in Fig. 16.15. First, open the "File" menu and input the handwritten digits to be recognised, which will be displayed in the image processing window.

Then click the "Digit Recognition Menu", and a window as shown in Fig. 16.16 will pop up, which implements the phased function of recognising handwritten digits using CNN. Click the "Train" button to perform training according to the process described earlier. After the training is completed, click the "Save Parameters" button to save the training results on the local computer. When you need to use the training results to recognise digits, you can directly click the "Load Parameters" button after opening the image to load the trained parameters, and then click the "Recognise" button to obtain the recognition result.

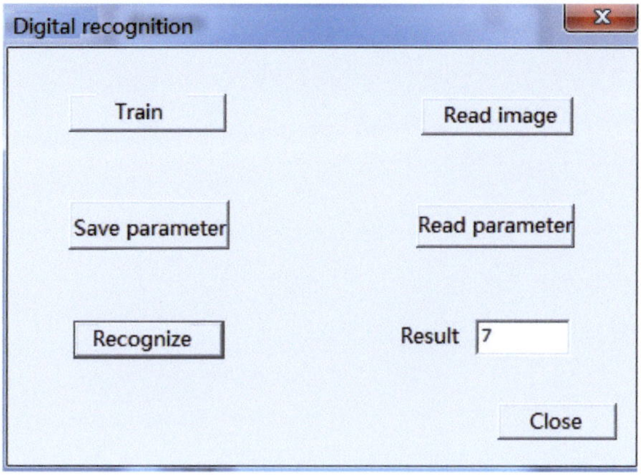

Fig. 16.16 Interface of recognizing numbers with CNN

Reference

1. Ye Yun. Deep Learning and Computer Vision [M]. Beijing: Mechanical Industry Press, 2017.

Chapter 17
Genetic Algorithms

17.1 Overview of Genetic Algorithms

The genetic algorithm (GA) was proposed by Holland of the University of Michigan in 1969 and subsequently summarized by De Jong (1975) and Goldberg (1989), forming a class of evolutionary algorithms [1, 2]. GA represents an idealized random search method based on Darwin's theory of evolution, encompassing genetic variation and natural selection. It attains the optimal solution through genetic operations including selection, crossover, and mutation among individuals. GA starts with a set of solutions called a population, which consists of a certain number of individuals representing chromosomes encoded by genes. A population is utilized to create a new population, with the expectation that the new one will be superior to the old. The parents selected to form offspring (i.e., new solutions) are chosen based on their fitness, indicating that the higher their adaptability, the greater the likelihood of their selection. This process persists across generations until specific conditions (e.g., the number of evolutionary generations, maximum fitness, or average fitness) are satisfied.

Problem-solving is often expressed as finding the extremum of a function. Observe the example of GA implementation results shown in Fig. 17.1, where GA attempts to find the minimum value of a function. This curve represents a search space, with vertical lines representing some solutions (points in the search space), where the bold line represents the optimal solution, and the thin lines represent some other solutions.

According to the standard form of GA, it uses binary genetic coding, i.e., alleles $\Gamma = \{0, 1\}$, individual space $H_L = \{0, 1\}^L$, and reproduction is divided into two independent steps: crossover and mutation. As illustrated in Fig. 17.2, the operational process of GA is outlined as follows:

① Initialisation. Determine the population size N, crossover probability (Pc), mutation probability (Pm); establish the termination criteria for evolution; randomly generate N individuals (appropriate solutions to the problem) as the initial population $P(0)$; set the generation $t = 0$.

© Chemical Industry Press 2026

B. Chen and S. Chen, *Machine Vision Technology*, Advanced and Intelligent Manufacturing in China, https://doi.org/10.1007/978-981-95-1184-6_17

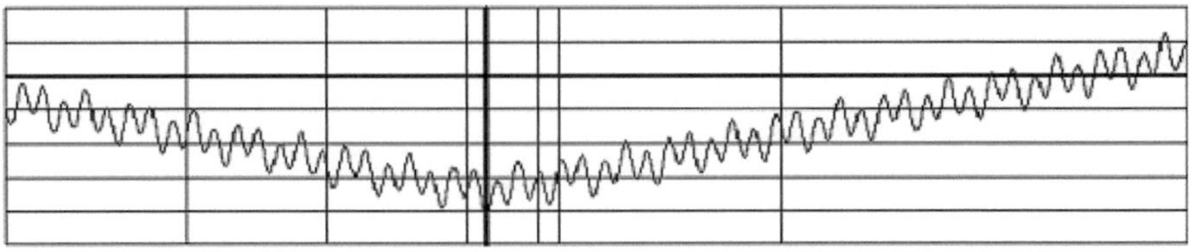

Fig. 17.1 Results of GA implementation

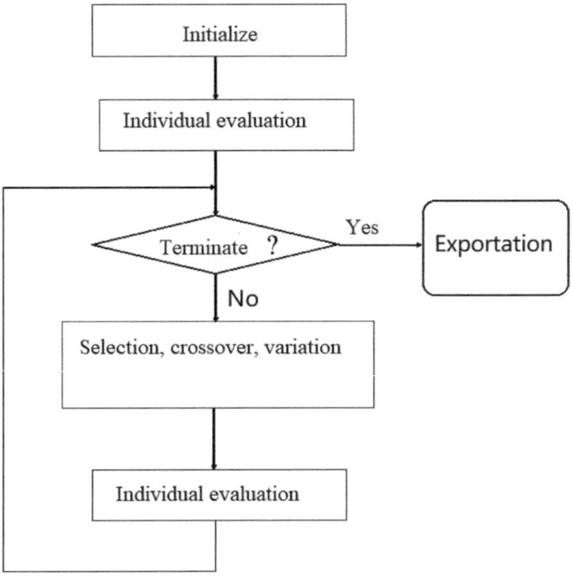

Fig. 17.2 Genetic algorithm process

② Individual evaluation. Calculate or estimate the fitness of each individual in the tth generation population $P(t)$.

③ Population evolution. Perform the following genetic operation steps through generational alternation until a new population is generated.

a. Selection. Use the selection operator to select M/2 pairs of parents from P(t) based on fitness (the better the fitness, the greater the chance of selection).
b. Crossover. Perform crossover on the selected M/2 pairs of parents with probability P_c to form M intermediate individuals. If P_c is 0, the intermediate individuals are merely exact copies of the parents.
c. Mutation. Independently perform mutation on each gene locus (position on the individual's chromosome) of the M intermediate individuals with probability P_m to form M candidate individuals. Select N individuals (offspring) from the M candidate individuals and/or the old generation population based on fitness to reconstitute the new generation population P(t + 1).

④ Termination test. If the termination criteria are met, output as the optimal solution maximum fitness in P(t + 1), terminate the calculation; otherwise set $t = t + 1$ and return to step ②.

In addition, GA has various extensions and variations. GA replaces the single-point search method by utilising the population of individuals, thus the search is conducted in a parallel manner.

As can be seen from the above, the basic operational process of GA is quite ordinary, and different execution methods need to be adopted according to different problems. The key is to solve the following two issues.

The first issue is how to create individuals (chromosomes) and what kind of genetic representation form to choose? Related to this are the two basic genetic operators of crossover and mutation.

The second issue is how to select the parents for crossover. There are many methods to consider, but the main idea is to select better parents in the hope that better parents can produce better offspring. You might think that merely taking the new offspring as the new generation population might cause the optimal individual (chromosome) to be lost from the previous population, which is entirely possible. Therefore, a so-called elitism method is often adopted. This means that at least the optimal solution is copied unchanged into the new population, so that the optimal solution found can be maintained until the end of the operation.

Some of the issues of concern will be introduced later, and you may still be puzzled as to why GA works. This can be partially explained by the schema theorem, although this theory still has its imperfections.

17.2 Simple Genetic Algorithm

Holland's original GA is the simple genetic algorithm (SGA). The simple GA is the most basic and important GA. The difference from other GAs lies in the genetic representation, selection mechanism, crossover, and mutation. The technical characteristics of the simple GA are summarised in Table 17.1. They will be introduced separately below.

Table 17.1 Simple GA technical characteristics

Genetic expression	Binary string
Recombination	N-point crossing, uniform crossing
Mutation	Reverse bits with a fixed probability
Parental selection	Proportional to fitness
Survival selection	All progeny replace parents
Feature	Emphatic crossing

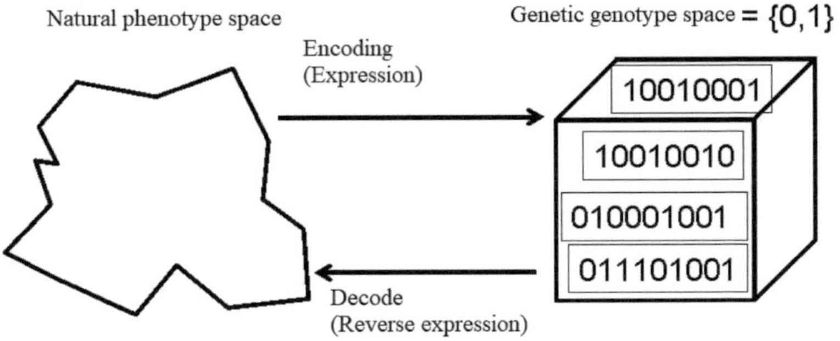

Fig. 17.3 Relationship between phenotype and genotype

17.2.1 Genetic Expression

When starting to solve a problem with GA, the issue of genetic expression needs to be considered. Chromosomes should contain the information of the solution they represent in some way. Genetic expression is also known as genetic coding or chromosome encoding. The most commonly used genetic expression method is to use binary strings, so chromosomes can be expressed in the following form:

Chromosome 1: 1101100100110110
Chromosome 2: 1101111000011110

As shown in Fig. 17.3, each chromosome has a binary string, and each bit in this string can represent a certain feature of the solution. Alternatively, the entire string represents a number, i.e., each phenotype can be expressed by a genotype. Of course, there are many other encoding methods, which mainly depend on the problem being solved itself, for example, integer or real number encoding can be used, and sometimes permutation encoding is useful.

17.2.2 Genetic Operators

As known from Sect. 17.1, selection, crossover, and mutation are the most important parts of GA, and the performance of GA is mainly affected by these three genetic operations. The following will explain the three genetic operators of GA: crossover, mutation, and selection.

(1) Crossover

After determining the encoding used, we can proceed with the recombination step. In biology, the most general form of recombination is crossover. Crossover selects genes from the chromosomes of the parents to create an intermediate individual. The simplest form of crossover is to randomly select a crossover point, copy the part

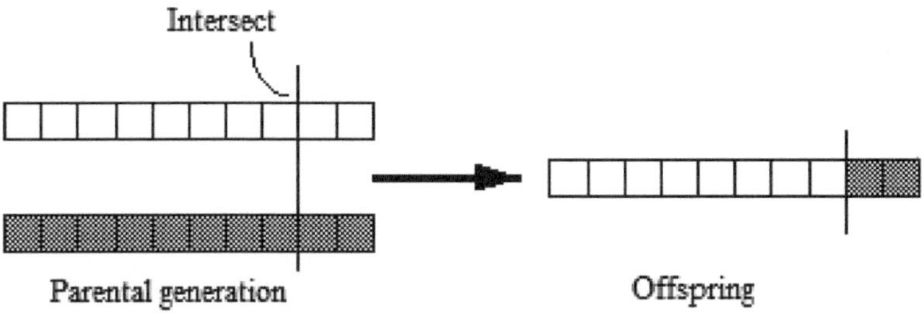

Fig. 17.4 Single point crossing (example: **1101100101** + 1011101111 = **11011001111**)

before this point from the first parent, and copy the part after this point from the second parent.

Crossover can be represented in the following form:

Chromosome of Parent 1: *11011* | *00100110110*
Chromosome of Parent 2: 11011 | 11000011110
Chromosome of Intermediate Individual 1: *11011* | 11000011110
Chromosome of Intermediate Individual 2: 11011 | *00100110110*

Here, | is the crossover point.

This crossover method is called single-point crossover. There are other ways to perform crossover, such as selecting more crossover points. Crossover can be quite complex, mainly depending on the encoding of the chromosomes. Special crossovers for specific problems can improve the performance of GA.

Below is an introduction to how crossover is implemented in binary encoding.

① Single-point crossover: This is the crossover method introduced above, as shown in Fig. 17.4. However, for single-point crossover, the head and tail of a chromosome cannot be passed together to the offspring. If both the head and tail of a chromosome contain good genetic information, the offspring obtained through single-point crossover will not be able to share these two good features.

② Two-point crossover: Two crossover points are selected. The binary string from the start of the chromosome to the first crossover point is copied from the first parent, the segment from the first crossover point to the second crossover point is copied from the second parent, and the remaining part is copied again from the first parent, as shown in Fig. 17.5. It is evident that using two-point crossover can avoid the aforementioned defect, and therefore, it is generally considered that two-point crossover is superior to single-point crossover.

③ N-point crossover: In fact, the problem can be generalized through each gene position on the chromosome. Adjacent genes on the chromosome have more opportunities to be passed together to the offspring obtained by N-point crossover, which will bring about unwanted correlation issues between adjacent genes. Therefore, the effectiveness of N-point crossover will depend on the position of the genes on the chromosome. Genes of solutions with related characteristics should be encoded together.

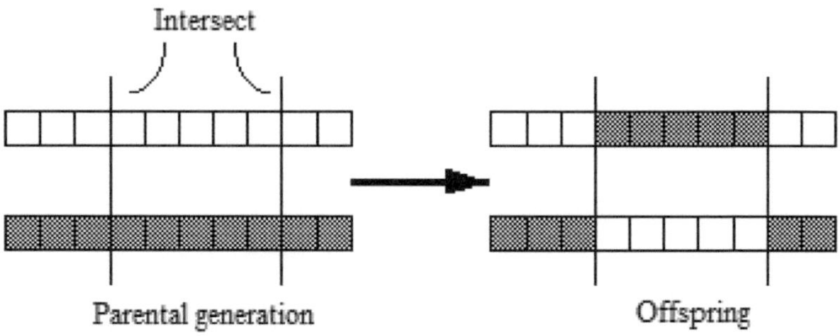

Fig. 17.5 Intersection of two points (example: **0110001011** + 1111011110 = **0111011111**)

④ Uniform crossover: Each bit (gene) is randomly copied from either the first or the second parent, as shown in Fig. 17.6. A crossover mask is used in the figure, where a mask value of 1 means copying the gene from the first parent, and a mask value of 0 means copying the gene from the second parent. Therefore, this recombination operator can avoid the locus problem.

⑤ Arithmetic crossover: Intermediate individuals are produced through some arithmetic operation, for example, 11001011 + 11011101 (AND) = 11001001.

(2) Mutation

After crossover, mutation occurs. This is to prevent all solutions in the population from falling into a local optimum of the problem being solved. Mutation randomly changes the intermediate individuals. For binary encoding, we can randomly change the selected bits from 1 to 0 or from 0 to 1, as shown in Fig. 17.7.

Below is an example of mutating the chromosomes of two intermediate individuals:

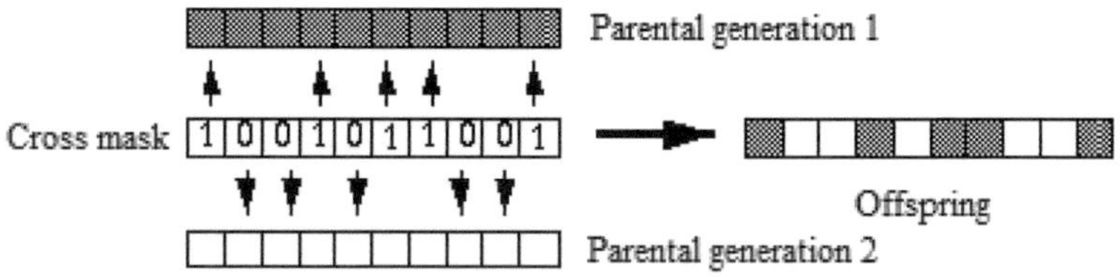

Fig. 17.6 Uniform crossing (example: **1100101110** + 1101110101 = **1101111100**)

Fig. 17.7 Variation

Chromosome of intermediate individual 1: 110 ***1*** 111000011110
Chromosome of intermediate individual 2: 110110 ***0*** 1001101 ***1*** 0
Chromosome of candidate individual 1: 110 ***0*** 111000011110
Chromosome of candidate individual 2: 110110 ***1*** 1001101 ***0*** 0

(3) Selection

As understood from the introduction above, chromosomes are selected from the population of parents used for reproduction. The question is how to select these chromosomes. According to Darwin's theory of evolution, the best individuals should survive and create new offspring. There are many methods to select (reproduce) the best chromosomes, such as roulette wheel selection, local selection, tournament selection, truncation selection, and steady-state selection. Here, we only introduce a commonly used selection method—roulette wheel selection.

Roulette wheel selection chooses parents according to fitness. The better the chromosome, the more chances it has to be selected. Imagine placing all the chromosomes in the population on a roulette wheel, where each individual's chromosome has its position according to the fitness function, as shown in Fig. 17.8. Then, when a ball is thrown to select a chromosome, those with higher fitness will be selected more frequently.

The following algorithm can be used for simulation.

① Sum. Calculate the sum of the fitness of all chromosomes in the population, denoted as S;
② Selection. Generate a random number between the interval (0, S), denoted as r;
③ Loop. Accumulate the fitness from 0 in the population, denoted as s. When s is greater than r, stop and return the chromosome at that position.

Of course, for each population, step ① is performed only once.

Additionally, the new offspring generated through crossover and mutation completely replace the old parents as the new generation population, entering the generational alternation process of the next round of genetic operations until the termination criteria are met. In a simple genetic algorithm, crossover plays a major role.

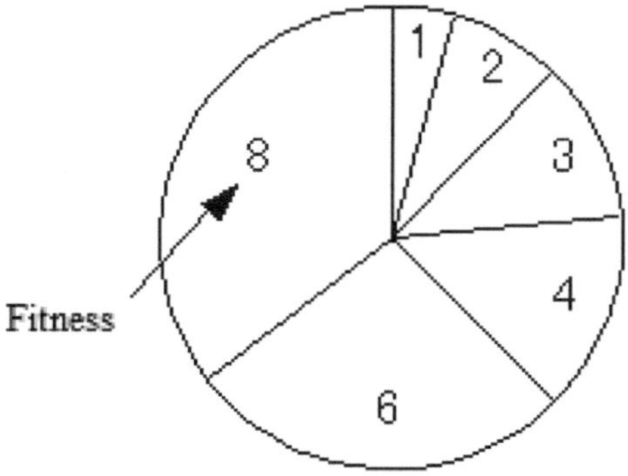

Fig. 17.8 Roulette selection method

17.3 Genetic Parameters

17.3.1 Crossover Rate and Mutation Rate

In GA, there are two basic parameters—crossover rate and mutation rate. The crossover rate indicates the frequency at which crossover occurs. Without crossover, the intermediate individuals will be exact copies of the parents. If crossover occurs, the intermediate individuals are generated from various parts of the parents' chromosomes. If the crossover rate is 100%, then all intermediate individuals are formed by crossover. If the crossover rate is 0, the entire new generation population is formed by exact copies of the chromosomes of the old generation population, but this does not mean that the new generation population is identical. Crossover is intended to allow new chromosomes to have the good parts of the old chromosomes, and perhaps the new chromosomes will be better. It is advantageous to retain some parts of the population for the next generation.

The mutation rate indicates the frequency at which parts of the chromosome are mutated. Without mutation, candidate individuals are produced without any changes after crossover (or copying). If mutation occurs, parts of the chromosome are altered. If the mutation rate is 100%, all chromosomes will change; if it is 0, there will be no change.

Mutation is intended to prevent GA from converging to a local optimum, serving to restore the diversity of individuals. However, mutation should not occur too frequently, otherwise GA will effectively become a random search.

17.3.2 Other Parameters

There are other GA parameters, such as population size pop_size (population size), which is also an important parameter.

Population size indicates how many individuals' chromosomes are in a generation population. If there are too few chromosomes, GA has little opportunity for crossover, and only a small search space is explored. Conversely, if there are too many chromosomes, GA execution will be very slow. Research shows that beyond a certain limit (which mainly depends on the encoding and the problem itself), increasing the population size is useless, as it does not speed up problem-solving.

17.3.3 Determining Genetic Parameters

Here, reference suggestions for determining genetic parameters will be provided. However, these are only general suggestions, and if you are considering implementing GA, you may need to experiment with GA for specific problems, as there is no general theory applicable to all problems to describe GA parameters so far. Therefore, these suggestions are merely summaries of experimental research results, and these experiments are often conducted using binary encoding.

① Crossover rate P_c: The crossover probability should generally be high, between 80 and 90%. However, experimental results also show that for some problems, a crossover rate of around 60% is optimal.
② Mutation rate P_m: The mutation rate should be low, with the best mutation rate between 0.5 and 1%. It can also be selected according to population size pop_size and chromosome length 1/chromosome_length, with a typical mutation rate between 1/pop_size and 1/chromosome_length.
③ Population size: It may be hard to believe, but a very large population size usually does not improve GA performance (i.e., the speed of finding the optimal solution). An excellent population size is about 20–30. However, there have been reports of using 50–100 as the best instances. Some studies also show that the optimal population size depends on the encoding and the size of the encoded string, which means if you have a 32-bit chromosome, then the population size should also be 32. However, this is undoubtedly twice the optimal population size for a 17-bit chromosome.

17.4 Fitness Function

In evolutionary search, GA basically does not use external information, relying only on the objective function, i.e., the fitness function, to conduct searches using the fitness value of each individual (chromosome) in the population. The objective function of GA is not constrained by continuous differentiability, and the domain can be any set. The only requirement for the objective function is that it can compute a comparable non-negative result for the input.

In specific applications, the design of the fitness function must be determined in conjunction with the requirements of the problem being solved. It should be noted that the evaluation of the fitness function is the basis for the selection operation, and the design of the fitness function directly affects the performance of the Genetic Algorithm (GA). Here, we only introduce the basic principles and key points of fitness function design, focusing on discussing the impact of the fitness function on GA performance and providing corresponding countermeasures.

17.4.1 Mapping the Objective Function to the Fitness

Function In many problem-solving scenarios, the goal is to find the minimum value of the function g(x). Since the fitness function in GA needs to be ranked and the selection probability calculated on this basis, the value of the fitness function must be positive. Therefore, in many cases, it is necessary to map the objective function to a fitness function in the form of finding the maximum value with non-negative function values.

Under conventional search methods, to convert a minimisation problem into a maximisation problem, one simply needs to multiply the function by − 1. However, for GA, this method is not sufficient to ensure non-negative values in all situations. To address this, the following method can be used for conversion:

$$f(x) = \begin{cases} C_{\max} - g(x) & g(x) < C_{\max} \\ 0 & others \end{cases} \quad (17.1)$$

Clearly, there are multiple ways to choose the coefficient C_{max}. C_{max} can be the maximum value of g(x) during the evolutionary process or the maximum value of g(x) in the current population. It can also be the maximum value of g(x) in the previous K generations. Ideally, C_{max} should be an appropriate input value independent of the population. If the objective function is a maximisation problem, to ensure its non-negativity, the following transformation can be used:

$$f(x) = \begin{cases} u(x) + C_{\min} & u(x) + C_{\min} > 0 \\ 0 & others \end{cases} \quad (17.2)$$

Here, C_{max} can be the minimum value of g(x) in the current generation or the previous K generations, or it can be a function of the population variance.

The method of mapping the objective function to the fitness function described above is known as the boundary construction method. However, this method sometimes faces difficulties in pre-estimating boundary values and cannot be precise.

17.4.2 Fitness Function Scaling

When applying GA, especially when dealing with small-scale populations, the following issues often arise:

① In the early stages of genetic evolution, some exceptional individuals are usually produced. If the proportional selection method is used, these exceptional individuals, due to their outstanding competitiveness, will dominate the selection process, thereby affecting the global optimisation performance of the algorithm.

② During the genetic evolution process, although the diversity of individuals in the population still exists, the average fitness of the population often approaches the fitness of the best individual, weakening the competitiveness among individuals. This causes the targeted optimisation process to tend towards a random wandering process without a target.

We usually refer to the above issues as the deception problem of GA. To overcome the first type of deception problem mentioned above, efforts should be made to reduce the competitiveness of certain exceptional individuals, which can be achieved by reducing the corresponding fitness function values. For the second type of deception problem, efforts should be made to enhance the competitiveness among individuals, which can be achieved by amplifying the corresponding fitness function values. This adjustment of fitness scaling is an important technique for maintaining the level of competition during the evolutionary process.

Common methods of fitness scaling include:

(1) Linear Scaling

Assuming the original fitness function is f, and the transformed fitness function is f', linear scaling can be expressed by the following formula:

$$f' = af + b \tag{17.3}$$

The coefficients in the formula a and b can be determined in various ways, but they must satisfy the following two conditions:

① The average value of the original fitness f_{avg} must equal the average value of the transformed fitness f'_{avg}, to ensure that individuals with average fitness have an expected reproduction number of 1 in the next generation, i.e.

$$f'_{avg} = f_{avg} \tag{17.4}$$

② The maximum value of the transformed fitness f'_{max} must equal a specified multiple of the average value of the original fitness f_{avg}, to control the reproduction number of the fittest individuals in the next generation, i.e.

$$f'_{\text{max}} = C_{\text{mult}} f_{\text{avg}} \tag{17.5}$$

where C_{mult} is to achieve the desired reproduction number of the optimal individuals in the population. Experiments show that for a typical population (population size 50 ~ 100), C_{mult} can range from 1.2 to 2.0.

It must be noted that using linear transformation may result in negative fitness values. The reason is that in the later stages of the algorithm, excessive scaling of the original fitness function may occur, causing some individuals in the population to transform into negative values due to their original fitness being far below the average. There are many solutions to this; when the C_{mult} coefficient cannot be adjusted, the simplest method is to map the minimum value of the original fitness f_{min} to the minimum value of the transformed fitness f'_{min}, ensuring that $f'_{\text{min}} = 0$. However, it is still necessary to maintain $f'_{\text{avg}} = f_{\text{avg}}$.

(2) σ Truncation

σ Truncation is a preprocessing method used before the above linear transformation, primarily utilising the population standard deviation σ information. The aim is to more effectively ensure that the transformed fitness does not become negative. The corresponding expression is as follows:

$$f' = f - (f_{avg} - c\sigma) \tag{17.6}$$

where c is a constant.

(3) Power Function Transformation

The transformation formula is as follows:

$$f' = f^k \tag{17.7}$$

In the formula, the exponent k is related to the problem being solved and can be adjusted as needed. In an experimental instance of machine vision, k is set to 1.005.

(4) Exponential Transformation

The transformation formula is as follows:

$$f' = e^{-af} \tag{17.8}$$

The basic idea of this transformation method is derived from the simulated annealing method, where the coefficient a determines the compulsion of reproduction; the smaller its value, the more reproduction tends to those individuals with higher fitness.

17.4.3 The Impact of Fitness Function Design on GA

Besides the above scale transformations of the fitness function that can overcome the deception problem of GA, the design of the fitness function is directly related to the selection operation of GA, so its impact on GA is also reflected in other aspects.

(1) The fitness function influences the stopping criteria of GA iterations

Strictly speaking, there is currently no consensus on the stopping criteria for GA iterations. When the maximum value of the fitness function is known or the lower bound of the fitness of a near-optimal solution can be determined, it is generally used as the stopping criterion for GA iterations upon discovering a solution that meets the maximum value or near-optimal solution. However, in many combinatorial optimisation problems, the maximum fitness value is unclear and is itself the object of the search, making it difficult to determine the lower bound of fitness. Therefore, in many application cases, if it is found that the evolution of individuals in the population has stabilised, in other words, if a certain proportion of individuals in the population are found to be completely identical, the iteration process is terminated.

(2) The fitness function and problem constraints

GA relies solely on fitness to evaluate and guide the search, and the inherent constraints of the problem cannot be explicitly represented. Therefore, we can check whether new individuals violate the constraints in each iteration during the evolutionary process. If a violation is detected, the individual is removed as invalid. This method is effective for solving weakly constrained problems but performs poorly for strongly constrained problems. This is because, in such cases, finding an invalid individual is as difficult as finding an optimal individual.

As a countermeasure, a penalty method can be used. The basic idea of this method is to penalise individuals for violating constraints and incorporate this penalty into the design of the fitness function. Thus, a constrained optimisation problem is transformed into an unconstrained optimisation problem with an additional cost or penalty.

For example, a constrained minimisation problem:

$$\begin{aligned} &\text{Minimise: } g(x) \\ &\text{Subject to: } b_i(x) \geq 0 \quad i = 1, 2, \ldots, n \end{aligned}$$

Through the penalty method, the above problem can be transformed into the following unconstrained problem.

$$\text{Minimise: } g(x) + r \sum_{i=1}^{M} \Phi[b_i(x)] \tag{17.9}$$

where Φ is the penalty function, and r is the penalty coefficient.

There are many ways to determine the penalty function, and the following setting is made for all individuals that violate constraints.

$$\Phi[b_i(x)] = b_i^2(x) \tag{17.10}$$

Under certain conditions, when the value of the penalty coefficient r approaches infinity, the unconstrained solution can converge to the constrained solution. In practical applications, in GA, r is usually assigned different values for various constraints, so that the penalty component for constraint violations is appropriate. The idea of adding a penalty to the fitness function is simple and intuitive, but the penalty function value can change drastically at the constraint boundary, often causing problems, which should be noted. Additionally, when using GA to solve constrained problems, measures can also be taken in the design of coding and genetic operations.

17.5 Schema Theorem

The execution process of GA involves a large number of random operations, so it is necessary to analyse its mathematical mechanism. For this purpose, the concept of schema is first introduced.

A schema, also known as a similarity template, is a string using the character set {0, 1, *}. Examples of schemas include: 010*1, *110*, *****, 10101, …

The symbol * is a wildcard character. For binary strings, inserting the wildcard character * into the {0, 1} string can generate all possible schemas. Therefore, any schema can be constructed using {0, 1, *}. We say a schema matches a specific string if: the 1s in the schema match the 1s in the string, the 0s in the schema match the 0s in the string, and the * in the schema can be either 0 or 1 in the string. Thus, a schema can represent several strings, such as: *10*1 represents 01001, 01011, 11001, 11011. It can be seen that the advantage of defining schemas is that it allows us to easily describe the similarity of strings.

We introduce two properties of schemas: order and defining length.

The number of non-* characters is called the order O of the schema, i.e., the number of determined positions (positions of 0 or 1). As shown in Table 17.2.

An order of O represents a pattern of length N with 2^{N-O} different strings.

The distance between the two furthest non-* characters is called the defining length δ of the pattern, which is the distance between the first and last determined positions, as shown in Table 17.3. For example, in the pattern H = 1*1*, the first determined position is 1, and the last determined position is 3, so $\delta(H) = 3 - 1 = 2$.

A string represented by a pattern (such as a chromosome) is said to contain that pattern, as shown in Table 17.4.

A string of length N contains 2^N patterns.

A string of length N contains 3^N different patterns.

A population of P strings of length N contains between 2^N and $\min(P \times 2^N, 3^N)$ patterns, so relative to the population size, GA has a greater effect on the number of patterns, as shown in Table 17.5.

Table 17.2 Order of schemas

Schema	Order O	Representative string
***	0	000 001 010 011 100 101 110 111
1	1	010 011 110 111
*10	2	010 110
1*1	2	101 111
101	3	101

Table 17.3 Defining length of patterns

Schema	Defining length δ
**** *1**	0
10 10**	1
1*1*	2
1*11 0**1 1001	3

Table 17.4 Contained patterns

Schema	Contained schemas
1	1 *
00	00 0* *0 **
110	110 11* 1*0 1** *10 *1* **0 ***
1011	1011 101* 10*1 10** 1*11 1*1* 1**1 1***
	*011 *01* *0*1 *0** **11 **1* ***1 ****

Table 17.5 Effect of GA on pattern quantity

N	P	Schema quantity
6	20	64 ~ 729
20	50	1,048,576 ~ 52,428,800
40	100	1.099511×10^{12} ~ 1.099511×10^{14}
100	300	1.267650×10^{30} ~ 3.802951×10^{32}

17.5.1 *Geometric Interpretation of Patterns*

A chromosome (string) of length N can be viewed as a point (vertex of a hypercube) in a discrete N-dimensional search space, as shown in Fig. 17.9.

Patterns can be viewed as hyperplanes (i.e., hyper-edges or hyper-surfaces of a hypercube) in the search space, as shown in Fig. 17.10.

Lower-order hyperplanes include more vertices (2^{N-O}).

(1) Fitness of Patterns/Hyperplanes

Define the fitness f of a pattern (hyperplane) as the average fitness of the chromosomes (vertices) containing a pattern (hyperplane). It is possible to estimate the

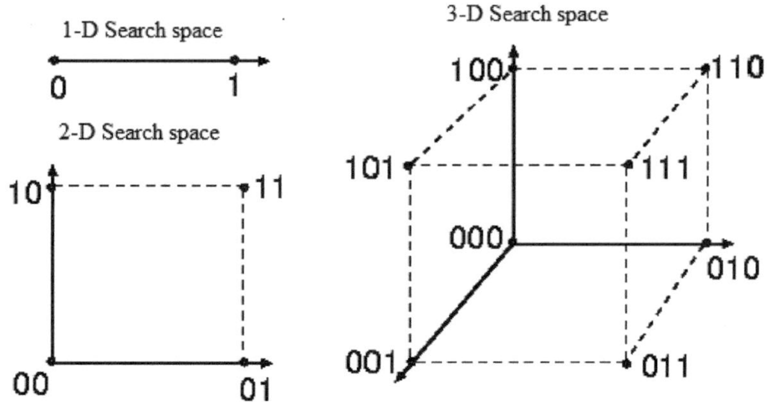

Fig. 17.9 Points of a string

Fig. 17.10 Hyperplane

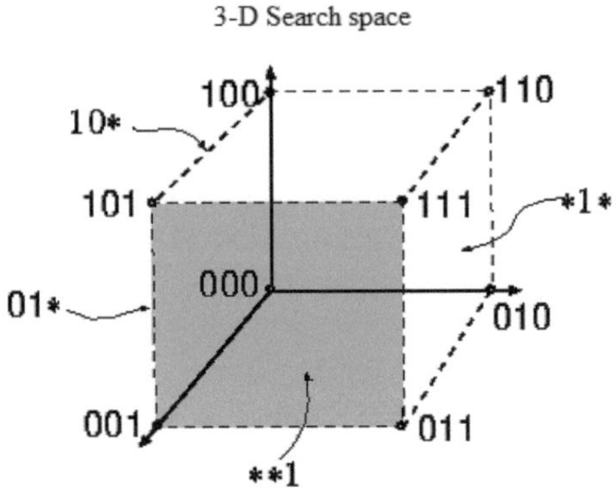

fitness of patterns contained in a population based on the fitness of the chromosomes including the population *J*, as shown in Table 17.6.

When a pattern contains more chromosomes, the estimate of fitness is more accurate for lower-order patterns on average, such as the 0th order pattern "**…*" which contains every string.

(2) Observations on Patterns

If only the method of proportional fitness allocation is applied (without crossover or mutation), by increasing (or decreasing) the number of chromosomes, sampling patterns with above (or below) average fitness generation after generation, then:

a. Patterns with longer defining lengths δ have a higher probability of being disrupted by crossover;
b. Higher-order patterns have a higher probability of being disrupted by mutation;
c. Patterns with low order and short defining lengths are called building blocks, which undergo GA operations with minimal disruption, allowing GA to use building blocks with relatively high fitness to achieve a global optimal solution.

Table 17.6 Fitness of patterns

Population	f	Schema	f
101	5	***	(5 + 1 + 2 + 3)/4 = 2.75
100	1	**0	(1 + 2 + 3)/3 = 2
010	2	**1	5/1 = 5
110	3	*0*	(5 + 1)/2 = 3
		*00	1/1 = 1
		*01	5/1 = 5
		1	(2 + 3)/2 = 2.5
		…	…

17.5.2 Pattern Theorem

Next, we analyse the impact of several important GA operations on patterns.

(1) Impact of Replication on Patterns

Assume that at a given time t, the population $A(t)$ contains m specific patterns H, denoted as:

$$m = m(H, t) \tag{17.11}$$

During the replication process, any string A_i (i = 1, 2, ..., n) is selected for replication with a probability of $f_i / \sum f_i$. Therefore, it can be expected that after replication is completed, at time $t + 1$, the number of a specific pattern H will become:

$$m(H, t+1) = m(H, t)nf(H)/\sum f_i = m(H, t)f(H)/\overline{f} \tag{17.12}$$

or written as:

$$\frac{m(H, t+1)}{m(H, t)} = \frac{f(H)}{\overline{f}} \tag{17.13}$$

where n is the population size (total number of individuals), $f(H)$ represents the average fitness of the strings corresponding to pattern H at time t, and $\overline{f} = \sum f_i / n$ is the average fitness of the entire population.

It is evident that after the replication operation, the number of a specific pattern will change proportionally to the ratio of the average fitness of that pattern to the average fitness of the entire population. In other words, patterns with fitness higher than the average fitness of the entire population will increase in number in the next generation, while those with fitness lower than the average will decrease. Additionally, the processing of all patterns H in population A is parallel, meaning that after the replication operation, all patterns simultaneously increase or decrease according to the proportion of their average fitness to the overall average fitness. In summary, the effect of the replication operation on patterns is that the number of patterns with fitness above the average will increase, while the number of patterns with fitness below the average will decrease.

To further analyse the growth in the number of patterns with fitness above the average, let:

$$f(H) = (1 + c)\overline{f} \qquad c > 0 \tag{17.14}$$

Then the above equation can be rewritten as the following difference equation:

$$m(H, t+1) = m(H, t)(1 + c) \tag{17.15}$$

Assuming c is a constant, we obtain:

$$m(H, t) = m(H, 0)(1 + c)^t \tag{17.16}$$

It is evident that the number of schemata with fitness above the average will grow exponentially.

From the analysis of the replication process, it can be seen that although the replication process successfully controls the number of schemata in a parallel manner, increasing or decreasing exponentially, it only replicates certain high-fitness individuals entirely or discards some low-fitness individuals, and never generates new schema structures. Therefore, its improvement on performance is limited.

(2) The Impact of Crossover on Schemata

The crossover process is an organised yet random exchange of information between strings. It creates new structures while minimally disrupting the high-fitness schemata selected by the replication process. To observe the impact of crossover on schemata, consider a string with $N = 7$ and two representative schemata contained within this string.

$$A = 0111000$$
$$H_1 = *1 * * * *0$$
$$H_2 = * * *10 * *$$

First, let's review the single-point crossover process. A matching partner is randomly selected, then a crossover point is randomly chosen, and the corresponding segments are exchanged. Assuming the randomly selected crossover point for the given string is 3, it is easy to see its impact on the two schemata H_1 and H_2. The crossover point is marked with a separator "|".

$$A = 011|1000$$
$$H_1 = *1 * | * * * 0$$
$$H_2 = * * *|10 * *$$

Unless the matching partner of string A has the same fixed positions in the schema as A (we ignore this possibility), schema H_1 will be disrupted because the "1" at position 2 and the "0" at position 7 will be allocated to different offspring individuals (these two fixed positions are separated by the separator representing the crossover point). Similarly, it is evident that schema H_2 will continue to exist because the "1" at position 4 and the "0" at position 5 are transferred intact to the next generation of individuals. Although the crossover point in this example is randomly selected, it is not difficult to see that schema H_1 is more prone to disruption than schema H_2. On average, the crossover point is more likely to fall between the two boundary points. If analysed quantitatively, the defining length of schema H_1 is 5. If the crossover point is always randomly selected from $N - 1 = 7 - 1 = 6$ possible positions, then it is clear that the probability of schema H_1 being disrupted is:

$$p_{\mathrm{d}} = \delta(H_1)/(N-1) = 5/6 \tag{17.17}$$

The probability of its survival is:

$$p_{\mathrm{s}} = 1 - p_{\mathrm{d}} = 1/6 \tag{17.18}$$

Similarly, the defining length of schema H_2 is $\delta(H_2) = 1$, its probability of being disrupted is $p_{\mathrm{d}} = 1/6$, and the probability of survival is $p_{\mathrm{s}} = 1 - p_{\mathrm{d}} = 5/6$. Extending to the general case, the lower bound of the crossover survival probability for any pattern can be calculated as:

$$p_s \geq 1 - \frac{\delta(H)}{N-1} \tag{17.19}$$

where the greater-than sign indicates that there is also a possibility of the pattern not being destroyed when the crossover point falls within the defined length.

In the previous discussion, we assumed the probability of crossover to be 1. In general, if the probability of crossover is set as p_{c}, then the above equation becomes:

$$p_s \geq 1 - p_c \frac{\delta(H)}{N-1} \tag{17.20}$$

If considering the effects of both replication and crossover, the number of a specific pattern H in the next generation can be estimated by the following equation:

$$m(H, t+1) \geq m(H, t) \frac{f(H)}{\overline{f}} \left[1 - p_c \frac{\delta(H)}{N-1} \right] \tag{17.21}$$

It can be seen that patterns with higher than average fitness and shorter defined lengths will appear more in the next generation.

(3) The Impact of Mutation on Patterns

Mutation is a random replacement at a single position in the string with a probability p_{m}, and thus it may destroy a specific pattern. For a pattern H to survive, it means all its defined positions must survive. Therefore, since the probability of the gene value at a single position surviving is $(1 - p_{\mathrm{m}})$, and each mutation occurs statistically independently, a specific pattern survives only when all its $O(H)$ defined positions survive. Here, $O(H)$ is the order of pattern H. Thus, the survival rate of a specific pattern after mutation is:

$$(1 - p_m)^{O(H)} \tag{17.22}$$

Since $p_{\mathrm{m}} \leq 1$, the above equation can also be approximately expressed as:

$$(1 - p_m)^{O(H)} \approx 1 - O(H)p_m \tag{17.23}$$

Considering the above replication, crossover, and mutation operations comprehensively, the change in the number of a specific pattern H is:

$$m(H, t + 1) \geq m(H, t)\frac{f(H)}{\bar{f}}\left[1 - p_c\frac{\delta(H)}{N - 1}\right](1 - O(H)p_m) \tag{17.24}$$

The above expression can also be approximately represented as:

$$m(H, t + 1) \geq m(H, t)\frac{f(H)}{\bar{f}}\left[1 - p_c\frac{\delta(H)}{N - 1} - O(H)p_m\right] \tag{17.25}$$

where a smaller cross-multiplication term is ignored.

The introduction of variation requires slight modifications to the previous analytical conclusions. Thus, the complete conclusion is: patterns with short defining lengths, low order, and above-average fitness will grow exponentially in the offspring. This conclusion is very important and is often referred to as the schema theorem of GA.

According to the schema theorem, as GA progresses generation by generation, those short, low-order, high-fitness patterns will become more prevalent, and the final string obtained is a combination of these patterns. Therefore, it is expected that performance will improve and eventually approach the global optimum.

17.6 Application of Genetic Algorithms in Pattern Recognition

17.6.1 Problem Setting

Pattern recognition refers to the frequent unintentional feature matching of graphics in our daily lives, such as the template matching problem of "please find the shape similar to (a) from (b) ~ (f) in Fig. 17.11", where we can easily determine that shape (c) is the correct answer.

Performing the same task with a computer is not as easy. For example, the shapes shown in Fig. 17.11 are input as binary images (where each point is either black or white, with no intermediate shades), and the shapes are rotated to overlap each other. At this point, it is necessary to check the degree of overlap between them or set some feature quantities such as the number of arcs and angles, and decide their correspondence based on comparing these feature quantities.

When the types of given shapes are numerous and not fixed, this processing becomes more difficult, such as the problem of "determining the shape similar to the template shape given in Fig. 17.12a from Fig. 17.12b". This problem is also

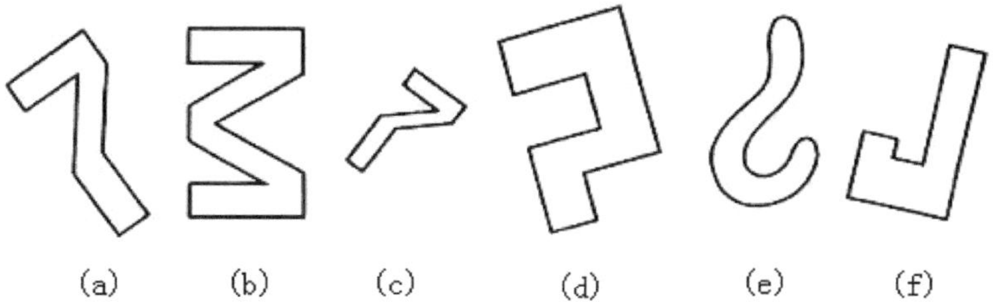

Fig. 17.11 A simple example of pattern recognition

easy for humans but more difficult for computers than the problem in Fig. 17.11. The degrees of freedom of similar shapes in the image, such as position, size, and rotation angle, are many unclear factors, and the more other shapes in the image, the more difficult the processing becomes. For humans, both the local features such as the connection relationships between the endpoints of lines forming the shape and the global features of the overall shape can be instantly grasped. Since computers do not have such concepts, humans must teach computers those solving methods in advance. However, humans still lack the biological or information processing knowledge of how to recognize shapes, so it is also a difficult task for computers to process.

The pattern recognition problem of shapes shown in Fig. 17.11 or Fig. 17.12 is a difficult problem for computers, but from the viewpoint of practical applications of computers, it is an important problem. For example, in the visual design of automatic working machinery in factories, there is the problem of using a robotic arm to pick out only the currently needed parts from the parts coming from the conveyor belt.

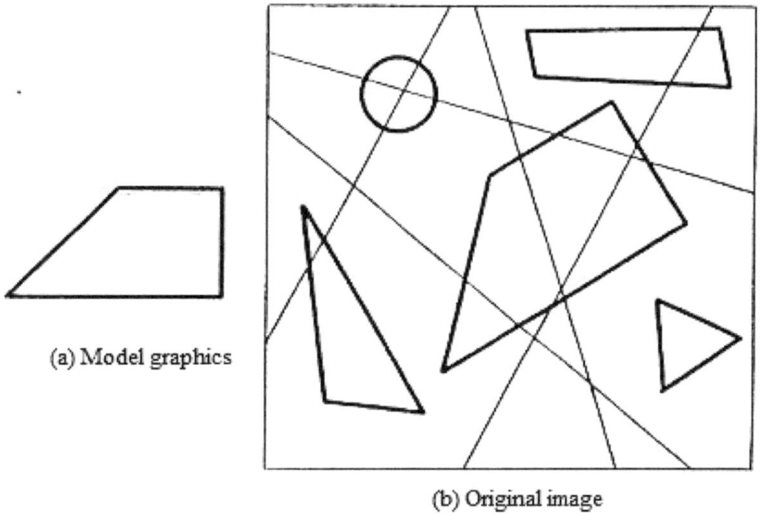

Fig. 17.12 Extraction of similar shapes in binary images

Also, in the increasingly important research on automobile intelligence in recent years, it is necessary to process the extraction of traffic signs and text on road signs from the image seen from the driver's seat. Here, we will discuss how to apply GA to such pattern recognition problems of shapes.

The engineering problem that should be solved here is the position detection problem of binary similar shapes.

For example, suppose a binary template shape like that in Fig. 17.12 is given, and a binary image containing shapes similar to this shape is also given. In this case, the task is to find the position, size, and rotation angle of the shape in the binary image that is similar to the template shape.

For this problem, to apply a simple GA, it is necessary to describe its mathematical nature. Here, as the template shape to be selected, assume a sequence of n points on the xy two-dimensional plane, P, is as follows:

$$\mathrm{P} = \{p_1(x_1, y_1), p_2(x_2, y_2), \ldots, p_n(x_n, y_n)\} \tag{17.26}$$

where, the coordinates of each point p_i ($i = 1, 2, \ldots, n$) are relative coordinates expressed with the centroid p_c of p as the origin. Align the centroid p_c of the template with the absolute coordinates of xy (x_c, y_c), then rotate by θ and scale by M times. The point sequence Q at this time is represented by the following formula.

$$\mathrm{Q} = \left\{q_1\left(x_1^*, y_1^*\right), q_2\left(x_2^*, y_2^*\right), \ldots, q_n\left(x_n^*, y_n^*\right)\right\} \tag{17.27}$$

where, (x_j^*, y_j^*) ($j = 1, 2, \ldots, n$) are the transformed absolute coordinates of each point, expressed by the following formula.

$$\begin{pmatrix} x_j^* \\ y_j^* \end{pmatrix} = M \begin{pmatrix} \cos\theta & -\sin\theta \\ \sin\theta & \cos\theta \end{pmatrix} \begin{pmatrix} x_j \\ y_j \end{pmatrix} + \begin{pmatrix} x_c \\ y_c \end{pmatrix} \tag{17.28}$$

Given a binary image with a white background and black graphics, assume that the number of black points in the point sequence Q of formula (17.27) is n_b, the matching rate R of the template with the graphics in the image can be defined by the following formula.

$$R = \frac{n_b}{n} \tag{17.29}$$

The template P When dealing with point sequences and binary images, the point sequence P is overlapped at various positions on the image with different sizes and rotation angles to calculate the matching rate as shown in Eq. (17.29). The problem of defining the coordinates of the centroid p of the point sequence P when the maximum R is obtained, along with the magnification M and rotation angle θ, is a problem of position detection for binary similar shapes. By setting each coordinate axis as x, y, M, and θ in a 4-dimensional space, it inevitably becomes a problem of searching for the maximum value when using the matching rate shown in Eq. (17.29) as the evaluation value, thus allowing the effective use of the exploration algorithm GA to solve it.

17.6.2 Application Method of GA

(1) Setting of Genotype and Phenotype

When applying genetic algorithms, it is first necessary to set the genotype and phenotype of each individual. Here, it is assumed that each individual I_k ($k = 1, 2, \ldots$) has the following genotype (referred to as chromosome) G_k:

$$G_k = (x_{ck}, y_{ck}, M_k, \theta_k) \tag{17.30}$$

This represents a point in the 4-dimensional space of the exploration object. At this time, the phenotype H_k of each individual I_k, with the template point sequence P expressed by Eq. (17.26), is transformed according to the parameters shown in Eq. (17.30) and represented as a graphic by Eq. (17.31).

$$H_k = \left\{h_{k1}\left(x_{k1}^*, y_{k1}^*\right), h_{k2}\left(x_{k2}^*, y_{k2}^*\right), \ldots, h_{kn}\left(x_{kn}^*, y_{kn}^*\right)\right\} \tag{17.31}$$

where, $\left(x_{kj}^*, y_{kj}^*\right)$ ($j = 1, 2, \ldots, n$) is expressed by the following equation.

$$\begin{pmatrix} x_{kj}^* \\ y_{kj}^* \end{pmatrix} = M_k \begin{pmatrix} \cos\theta_k & -\sin\theta_k \\ \sin\theta_k & \cos\theta_k \end{pmatrix} \begin{pmatrix} x_j \\ y_j \end{pmatrix} + \begin{pmatrix} x_{ck} \\ y_{ck} \end{pmatrix} \tag{17.32}$$

In this case, to simplify the problem, some restrictions have been placed on the degrees of freedom of similar patterns in the original image. Specifically, the size of similar patterns in the original image is the same as the template, meaning that in Eq. (17.30), M is always 1, and the rotation angle θ is in units of 45°, i.e., fixed at one of 0°, 45°, 90°, …, 315°. The parameter ranges in Eq. (17.30) are:

$$\begin{aligned} &x_{ck},\ y_{ck} \in \text{integers in } [0, 63]; \\ &M = 1; \\ &k = 45n,\ n \in \text{integers in } [0, 7]. \end{aligned} \tag{17.33}$$

Thus, the chromosome G_k is represented within the computer as a bit sequence with a total length of 15 bits as follows.

$$G_k = \underbrace{\overbrace{001\cdots\cdots 0}^{x_{ck}}}_{6\text{ bits}}\,\underbrace{\overbrace{101\cdots\cdots 1}^{y_{ck}}}_{6\text{ bits}}\,\underbrace{\overbrace{0\cdots\cdots 1}^{\theta_k}}_{3\text{ bits}} \tag{17.34}$$

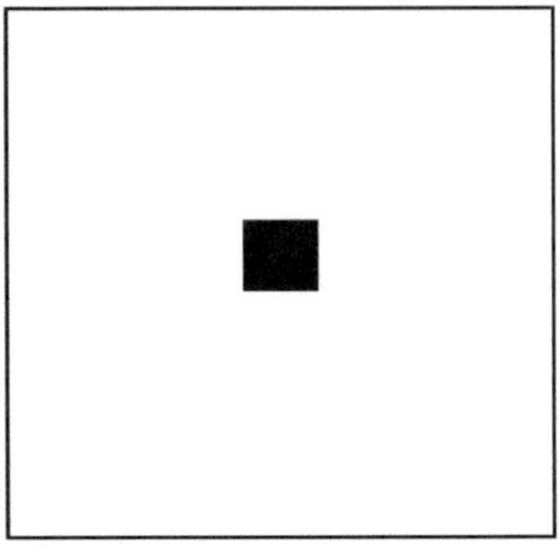

Before transformation
(Background: white; Graphic: Black)

0	0	0	0	0	0	0
0	1	1	1	1	1	0
0	1	2	2	2	1	0
0	1	2	3	2	1	0
0	1	2	2	2	1	0
0	1	1	1	1	1	0
0	0	0	0	0	0	0

After transformation L=3

Fig. 17.13 Blur the original image

Since M_k is a constant, it can be excluded from the chromosome. If the above conditions are not set, all parameters are unknown and can take any value, then the degrees of freedom of similar patterns are too many, in other words, the search space is too large, making the search difficult.

(2) Definition of Fitness

The fitness of each individual to the environment can be measured using the matching rate given by Eq. (17.29). However, even a slight deviation in the pattern can cause a large change in this matching rate. Moreover, if two patterns do not sufficiently overlap, this value will not be large. Here, to avoid this problem, the original image needs to be processed with the following blur. That is, the brightness of black points in the original image is set to L, and the brightness of white points is set to 0, as shown in Fig. 17.13, the image is processed with L-level blur, L being an integer constant.

Next, the fitness $f(I_k)$ of the dye I_k in the form of Eq. (17.33) can be obtained by the following equation.

$$f(I_k) = \frac{\sum_{j=1}^{n} f\left(x_{kj}^*, y_{kj}^*\right)}{L \times n} \tag{17.35}$$

where n is the total number of points constituting the template.

Thus, when the template pattern overlaps with similar patterns in the original image, even with a slight deviation, a certain level of fitness can still be obtained, making the search relatively easier.

(3) Setting of Genetic Operators

Next, the genetic operators used to generate new individuals in generational alternation need to be set. So far, many methods have been described regarding the elimination of individuals from one generation and the selection or generation of individuals for the next generation. Here, the following method is decided upon: rearrange the fitness of each individual in a generation in descending order, unconditionally eliminate a certain proportion of lower-ranked individuals, then randomly select several pairs from the higher-ranked individuals for crossover, generating new individuals pair by pair, keeping the population size (total number of individuals) unchanged.

In the simple genetic algorithm described in Sect. 17.2, the most basic crossover method is single-point crossover, and here it is decided to use two-point crossover.

Additionally, based on a small probability of mutation, each individual's genes undergo random inversion from 0 to 1 or from 1 to 0.

According to the generational alternation conducted by the aforementioned genetic operators, individuals with high fitness may persist for several generations, whereas individuals with low fitness have a high likelihood of being eliminated. Consequently, there is a tendency for biological populations to converge towards superior individuals. When the fitness increases smoothly, the optimal solution can be quickly discovered; however, the possibility of falling into a local optimum also becomes very high, which necessitates attention.

17.6.3 GA-Based Binocular Vision Matching

Binocular vision is a commonly used method for recovering scene depth information, but solving the correspondence problem is the most challenging step in binocular vision. Below, using an apple tree image as an example, the GA-based binocular vision matching method is introduced.

Figure 17.14 illustrates a schematic diagram of measuring the three-dimensional position of an apple using binocular vision. In binocular vision, the optical axes of the left and right cameras are parallel. The three-dimensional position of an apple can be represented by its centroid C, with the projections of centroid C on the left and right image planes being C_l and C_r, respectively. The three-dimensional position $C(x_c, y_c, z_c)$ can be calculated from the displacement between C_l and C_r, and the displacement between C_l and C_r in the world coordinate system xyz formed by C_r and the apple's centroid C is called disparity, i.e.

$$\begin{aligned} x_c &= x_r B/d \\ y_c &= y_r B/d \\ z_c &= FB/d \end{aligned} \tag{17.36}$$

where, x_r, y_r are the coordinates of C_r; $d = x_l - x_r$ is the disparity; x_l, y_l are the projection coordinates of the apple's centroid C on the left image relative to the world coordinate system xyz; F is the focal length of the camera; B is the distance between the left and right cameras, known as the baseline.

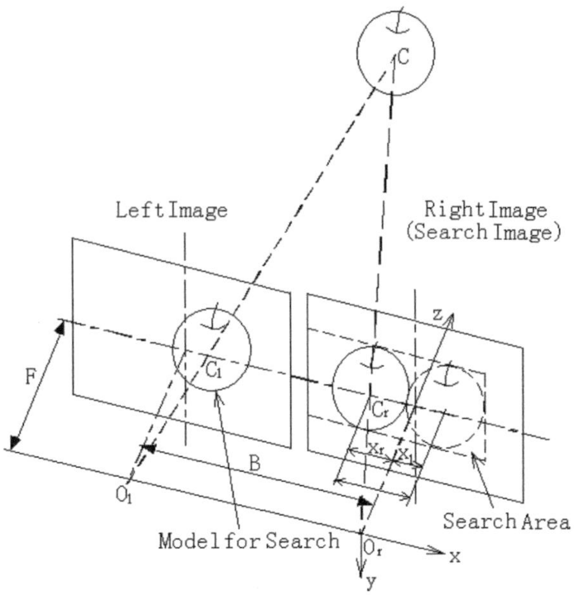

Fig. 17.14 Schematic diagram of apple 3D position measurement based on binocular vision

A set of corresponding points in the apple images (left and right) is determined using a contour-based template matching algorithm. Therefore, each apple contour extracted from the left image is defined as a template; the binary line image containing the apple contour extracted from the right image is the search image. Using the method introduced in Sect. 17.6.2, the contour template matching process involves attempting to find the contour of the same apple in the search image using the selected template from the left image.

As shown in Fig. 17.15, the image acquisition device consists of two sliding rods and cameras mounted on them. The camera can move horizontally left and right on the sliding rods to capture a pair of images (left and right). To obtain the three-dimensional position based on binocular vision, the baseline B in Eq. (17.36) plays a significant role when measuring the disparity d. If a relatively long baseline is set, the estimated three-dimensional position will be more accurate. However, a relatively long baseline will result in a narrow three-dimensional space observable from the left and right cameras. Therefore, there is a balance between measurement accuracy and visible space when choosing the baseline. Given that the actual distance range between the camera and the location of the apple can be considered as 50 ~ 200 cm, the baseline B is determined to be 10 cm, 15 cm, and 20 cm, respectively.

The image processing algorithm is as follows.

Step one involves binarising two colour images (left and right images) of a red apple (Fuji) using the G-Y colour difference method with a threshold of − 5, resulting in two binary images after region segmentation. Noise from small regions is then removed. In step two, the right image undergoes contour tracking to extract the entire contour line as a search image. The left image is segmented into regions, and a

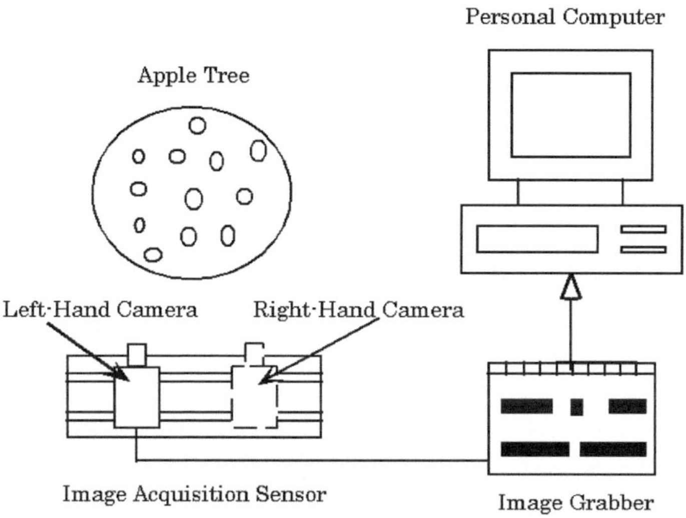

Fig. 17.15 Schematic diagram of image acquisition and processing system

linear discriminant function composed of parameters such as central region moments, circularity, and area is used to classify the apple regions into single apples and composite apples (where multiple apples overlap). For composite apples, distance transformation and dilation separation methods are used to extract individual apples. The apple contours in the left image are extracted as templates. Finally, the GA template matching method is used to find apple contours in the right image that match the templates selected one by one from the left image. The 3D positions of the matched apples in the left and right images can be calculated using centroid coordinates. This search operation is iterated until all apples in the left image, i.e., the templates, are processed.

Figure 17.16a, b are stereo colour images of an apple tree, taken from a distance of 100 cm from the camera to the apples. Figure 17.17a, b show the binary images obtained from colour region segmentation. The eight apples in the left image, marked "1" to "8", correspond to the apples in the right image, marked "a" to "h". Except for the topmost apple, which became too small to serve as a template due to leaf occlusion during separation processing, the remaining eight apples were correctly matched, as shown in Fig. 17.18. In this experiment, 36 sets of images were taken under various lighting conditions with baselines of 10, 15, and 20 cm, and over 95% of the apples were correctly matched.

(a) Left image (b) Right image

Fig. 17.16 Colour binocular image pairs

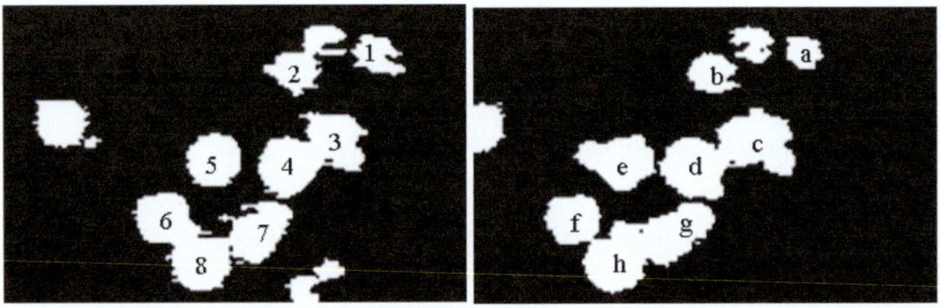

(a) Left image (b) Right image

Fig. 17.17 Binary binocular image pairs

Fig. 17.18 Template matching results based on GA

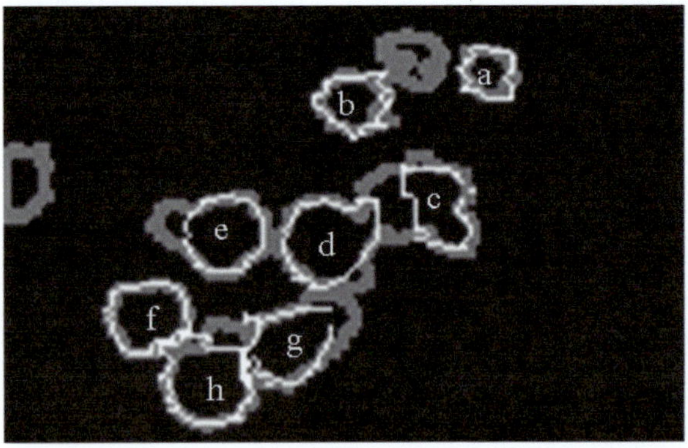

References

1. Sun M, Takahashi T, Zhang S, Bekki E. Matching Binocular Stereo Images of Apples by Genetic Algorithm [J]. Agricultural Engineering Journal, 1999, 8(2): 101–117.
2. Sun M, Apple Fruit recognition and location detection based on image processing [D]. Iwate University, Japan, 1999.

Part II
Machine Vision Application Systems

Chapter 18
General Image Processing System ImageSys

18.1 System Overview

ImageSys is a large-scale image processing system that includes functions such as image/multimedia file processing, image/video capture, image filtering, image transformation, image segmentation, feature measurement and statistics, as well as a development platform [1]. It can process colour, grayscale, static, and dynamic images. The file formats it supports encompass: bitmap files (bmp), TIFF image files (tif, tiff), JPEG image files (jpg, jpeg), text document files (txt), and multimedia video files (avi, dat, mpg, mpeg, mov, vob, flv, mp4, wmv, rm, etc.). Image and video capture utilizes international standard USB and IEEE1394 interfaces, making it compatible with desktop and laptop computers. It supports general consumer CCD digital cameras (via IEEE1394 interface) and PC cameras (via USB interface).

ImageSys, with its extensive and diverse functions, and the numerous available functions provided to users, enables the system to meet the needs of different professions and levels. For educational purposes, it can demonstrate various functions of modern image processing technology to students. In practical applications, it can automatically compute and measure various mathematical data for users; it can use the provided functions to combine various functions for robot vision judgement; Especially for scientific research involving image processing, the system's extensive functionality enables easy conduct of various experiments, rapid discovery of optimal solutions, and the utilization of the provided function library to simplify the creation of custom processing programs.

ImageSys also provides a framework source program that includes functions for image file reading, saving, and capture, basic system settings for window programs, as well as other tedious yet necessary programs unrelated to image processing. Additionally, it includes several image processing programs. Users can easily incorporate their own programs into the framework, saving valuable time and facilitating the referencing of function usage. Figure 18.1 shows the ImageSys operation interface.

© Chemical Industry Press 2026

B. Chen and S. Chen, *Machine Vision Technology*, Advanced and Intelligent Manufacturing in China, https://doi.org/10.1007/978-981-95-1184-6_18

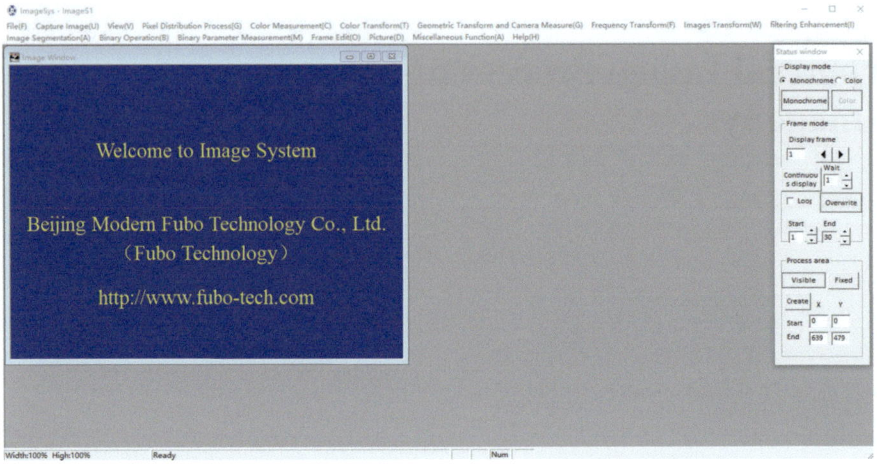

Fig. 18.1 ImageSys operation interface

18.2 Status Window

As shown in Fig. 18.2, the status window is used to display mode, frame mode, and processing area settings.

(1) Display Mode

You can choose grey, colour, pseudo-colour, etc., to represent, and also represent the grey images of R, G, and B components.

(2) Frame Mode

It can display the current frame, or continuously and cyclically display from the start frame to the end frame. For continuous display, it is necessary to set the start frame, end frame, waiting time interval, etc. The image processing results can be shown in the next frame, with the original image retained; or they can be written on the original frame.

(3) Processing Area

The processing area can be set by configuring the start and end coordinates, alternately selecting the pre-set maximum processing area and the middle 1/2 processing area.

The processing area can also be freely set: move the mouse to the starting point of the area to be set, press the "Shift" key and then the left mouse button, move the mouse to the end position of the area to be set, and release the "Shift" key and the left mouse button.

Fig. 18.2 Status window

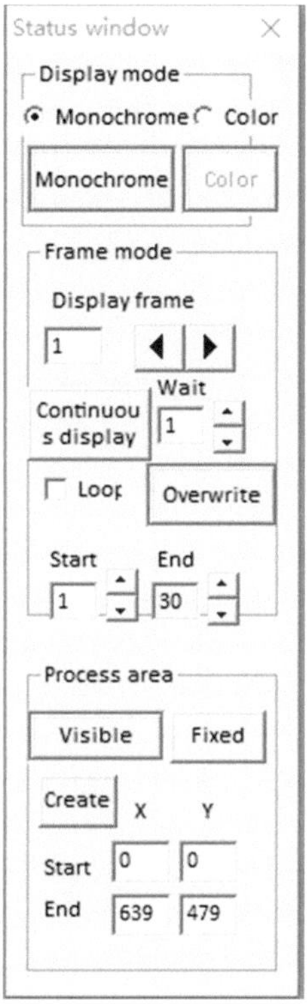

18.3 DirectX Image Capture System

Only the DirectX image capture system is introduced here. Others include A/D image card acquisition, high-speed camera image acquisition, etc., which will not be introduced.

This system supports general consumer CCD digital cameras (IEEE1394 interface) and PC cameras (USB interface).

When using a CCD camera with an IEEE1394 interface, the capture speed to the hard disk is related to the camera's format, usually 25 fps for PAL format and 30 fps for NTSC format; the capture speed to the system frame (memory) is related to the computer's processing speed. When capturing using a USB interface, the capture speed to the hard disk and memory is related to the computer's processing speed. The default capture speed is 15 fps, but it can be set manually. By setting the capture method, images can be captured to the system frame or collected to the hard disk. Figure 18.3 shows the DirectX image acquisition function interface.

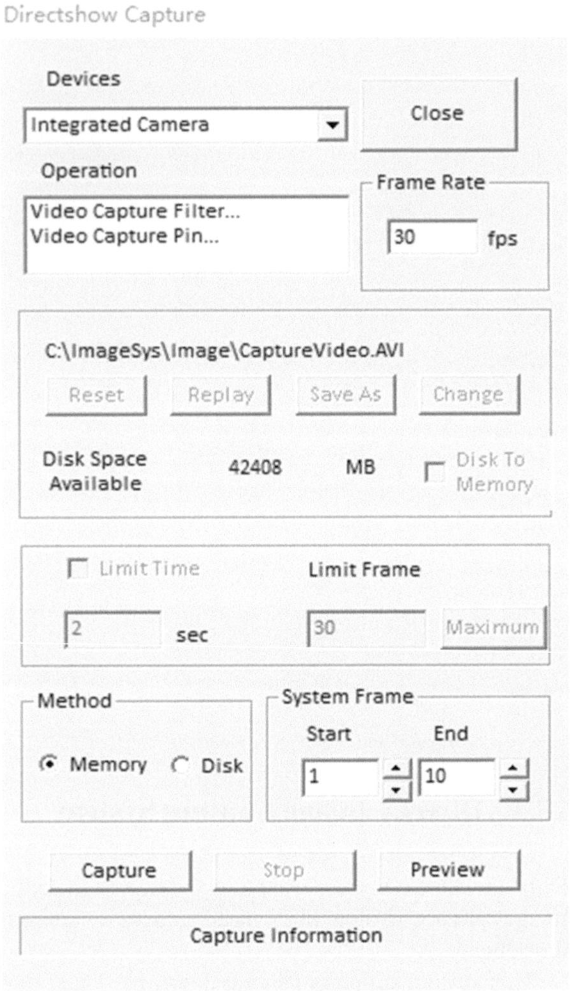

Fig. 18.3 DirectX image capture

18.4 Histogram Processing

18.4.1 Histogram

You can select the type of histogram: grayscale, colour RGB, R component, G component, B component, colour HSI, H component, S component, I component. You can sequentially display the minimum value, maximum value, average value, standard deviation, total pixels, etc., of the pixel area distribution histogram of the selected type. Display the pixel area distribution histogram of the selected type. The histogram can be cut and printed. You can view the distribution of data on the histogram. You

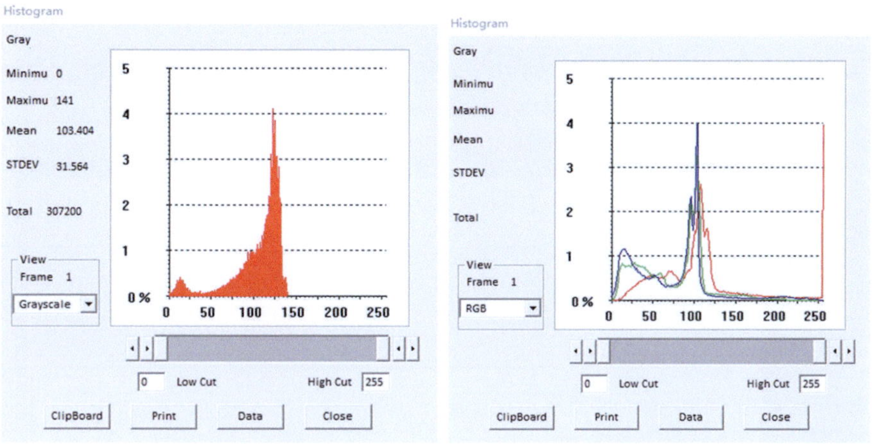

(a) Grayscale mode (b) colour mode

Fig. 18.4 Histogram function interface

can read previously saved data, save current data, and print current data. The saved data can be opened with Microsoft Excel to recreate the distribution chart. Figure 18.4 shows the histogram function interface.

18.4.2 Line Profile

The line profile represents the pixel value distribution along the line drawn by the mouse. The selectable types of line profile distribution graphs include: grayscale, colour RGB, *R* component, *G* component, *B* component, colour HSI, *H* component, *S* component, *I* component, etc.

When a single component is selected, the average value and standard deviation of the line profile information for that component will be displayed on the left side of the window. The line profile can be subjected to moving smoothing and wavelet smoothing. Moving smoothing allows setting the smoothing distance. Wavelet smoothing allows setting the smoothing coefficient, smoothing times, high-frequency removal, and low-frequency removal. High-frequency removal sets high-frequency signals to zero, leaving low-frequency signals, which is the smoothing signal. Low-frequency removal sets low-frequency signals to zero, leaving high-frequency signals, to observe high-frequency signals. Figure 18.5 shows the functional interface of the line profile. The line profile is a very useful image analysis tool.

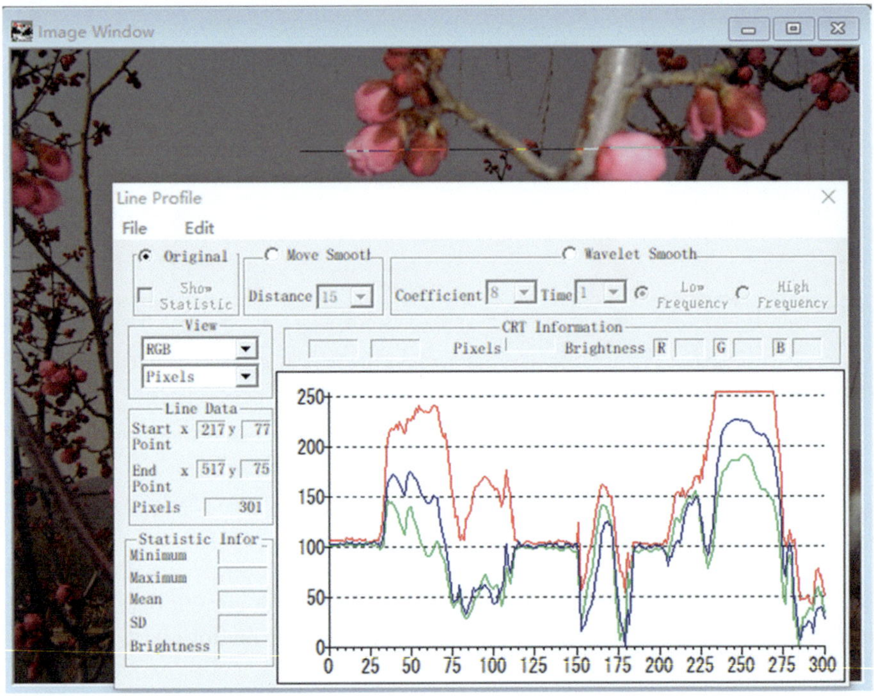

Fig. 18.5 Functional interface of line profile

18.4.3 3D Profile

The *X* axis represents the *x* coordinate of the image, the *Y* axis represents the *y* coordinate of the image, and the *Z* axis represents the pixel's grayscale value. Custom representation and OpenGL 3D display can be used. Sampling space and colour inversion can be set. The *Z* axis height scale, maximum brightness, base brightness, smear colour, background colour, etc., of the distribution graph can be set. Figure 18.6 is an example of a 3D profile.

18.4.4 Cumulative Distribution Graph

A cumulative distribution graph refers to the cumulative curve of pixel values in the vertical or horizontal direction. Opening the function window immediately displays the cumulative distribution of pixels within the processing window. If no processing window is selected, the cumulative distribution of pixels in the entire image is displayed.

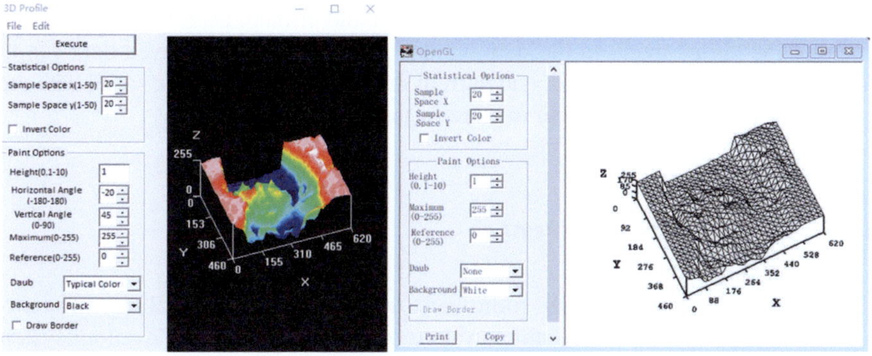

Fig. 18.6 3D profile

The selectable types of cumulative distribution graphs include: grayscale, colour RGB, *R* component, *G* component, *B* component, colour HSI, *H* component, *S* component, *I* component, etc. When a single component is selected, the minimum value, maximum value, average value, standard deviation, total pixels, etc., of the selected type's cumulative distribution graph are displayed. Figure 18.7 shows the vertical cumulative distribution graph of the colour RGB in the dashed window area on the image, with the horizontal axis representing the horizontal coordinate of the processing window and the vertical axis representing the cumulative pixel value.

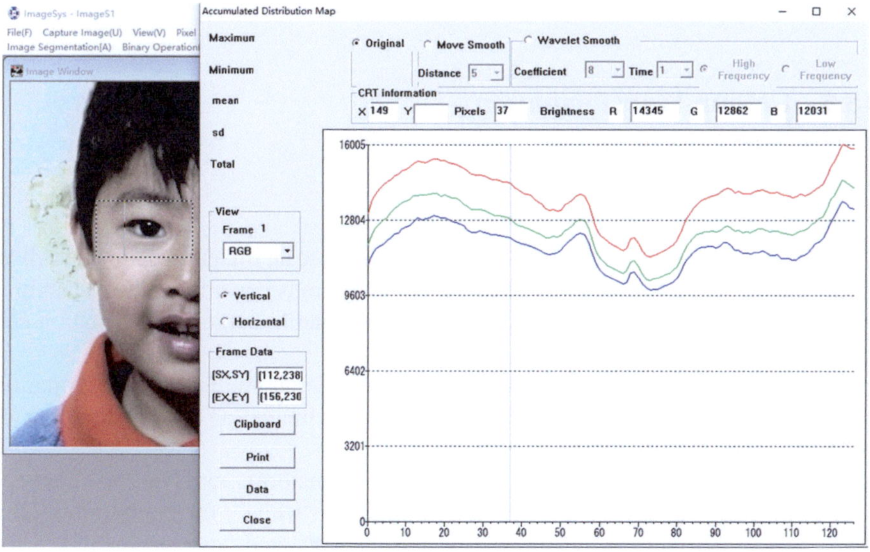

Fig. 18.7 Cumulative vertical distribution of window area

The cumulative distribution graph can be cut and printed. Data can be viewed by opening the "File" menu in the data window, where previously saved data can be read, current data can be saved, and current data can be printed. The saved data can be opened with Microsoft Excel to recreate the distribution graph.

18.5 Colour Measurement

Colour measurement involves coordinate transformation and colour difference measurement based on the brightness values of R, G, B, and the XYZ colour system and HSI colour system advocated by the International Commission on Illumination (CIE).

The reference colour, measured colour, and colour difference can be represented in images and numbers. The content includes: brightness values of R, G, B, values under the HSI colour system, the tristimulus values when transformed to the CIE XYZ colour system, the chromaticity coordinates x, y on the chromaticity diagram of the XYZ colour system, the values in the CIE L^* a^* b^* colour space, and the coordinates u^*, v^* when transformed into the CIE UCS colour space.

The light source during photography can be selected. Source A: tungsten lamp with a correlated colour temperature of about 2856 K; Source B: direct sunlight in the visible wavelength range; Source C: average light in the visible wavelength range; Source D65: average natural light including the ultraviolet range. Figure 18.8 is the functional interface for colour measurement.

18.6 Colour Transformation

18.6.1 Brightness Transformation

Used for brightness transformation of colour or grayscale images. Methods such as linear recovery, pixel extraction, range shift, N-value transformation, L (Logarithmic) transformation, γ (Gamma) transformation, and dynamic range transformation can be selected.

The image can undergo negative processing, reversing the intensity information of the image. By equalising the pixel distribution, the image can become more vivid. Hazy images can be clarified.

Parameters can be set according to the type of transformation. For "pixel extraction", the background can be selected as "black" or "white"; for "range shift", the displacement amount can be set as "displacement Y" and "displacement X"; for "N-value transformation", options include: "2, 4, 8, 16, 32, 64, 128, 256"; for "γ (Gamma) transformation", the γ coefficient can be set between 0 and 1.0, with an initial value of 0.5; the grayscale value can be set by inputting the grayscale value or

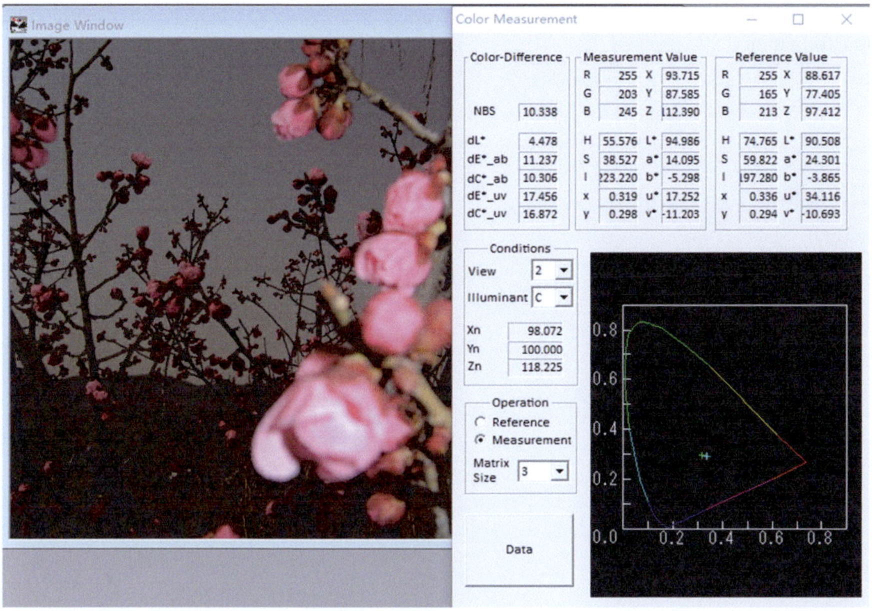

Fig. 18.8 Colour measurement function interface

using the grayscale adjustment handle. Figure 18.9 shows the functional interface of the brightness transformation.

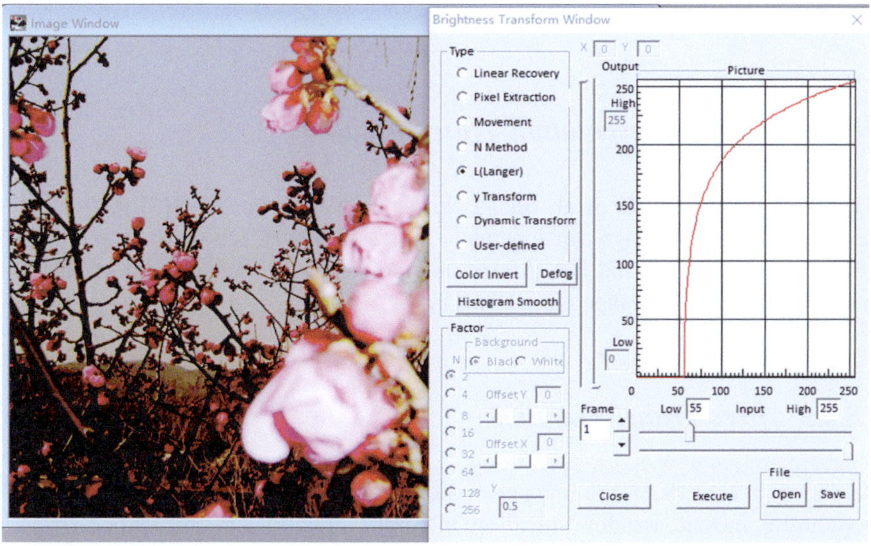

Fig. 18.9 Color brightness change function interface

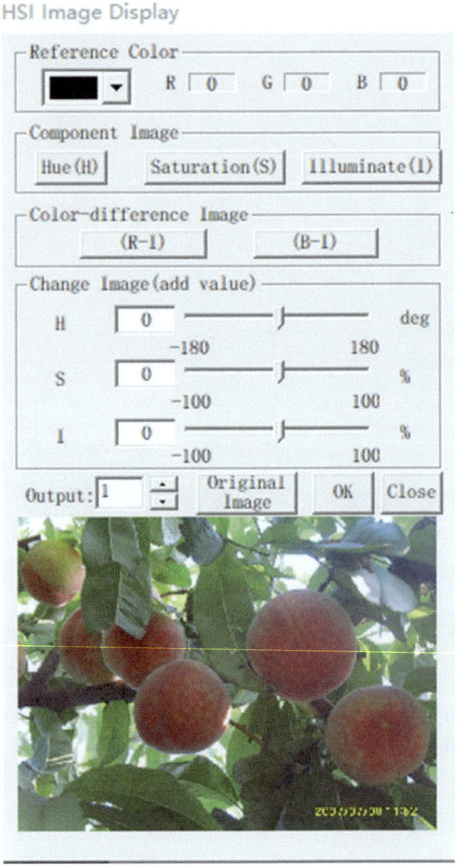

Fig. 18.10 HIS transformation

18.6.2 *HSI Representation Transformation*

The RGB colour values of an image can be converted into an HSI colour value representation. The image can separately represent hue H, saturation S, intensity I, colour difference R-I, and B-I. After freely adjusting each component of HSI, the image colour can be changed, as shown in Fig. 18.10.

18.6.3 *Free Transformation*

As shown in Fig. 18.11, the image can undergo translation, 90° rotation, brightness contouring, mosaic, window smearing, integral averaging, and other processing.

① Translation: Execute scrolling or moving of the image.

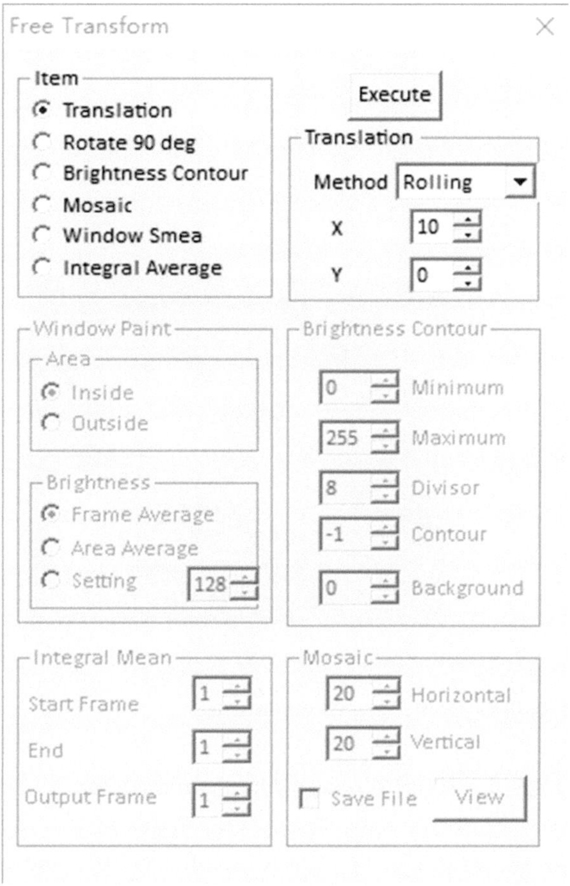

Fig. 18.11 Free transformation

② Brightness Contouring: Draw contour lines for each brightness range. The lower and upper limits of the brightness range can be set, as well as the divisor to divide the brightness range into equal parts, the brightness value of the contour line, and the brightness of the background outside the contour line.

③ Mosaic: Calculate the average brightness of pixels within the set range and draw a mosaic image. The pixel range in the horizontal and vertical directions can be set.

④ Window Smear: Smear inside or outside the processing window with arbitrary brightness.

Set the smearing brightness method: (a) Frame average, the average brightness of pixels around the processing window; (b) Area average, the average brightness of pixels within the processing window; (c) Specify, specify brightness.

⑤ Integral Average: Set multiple frames of images to calculate the average image. Used to remove random noise and improve the image.

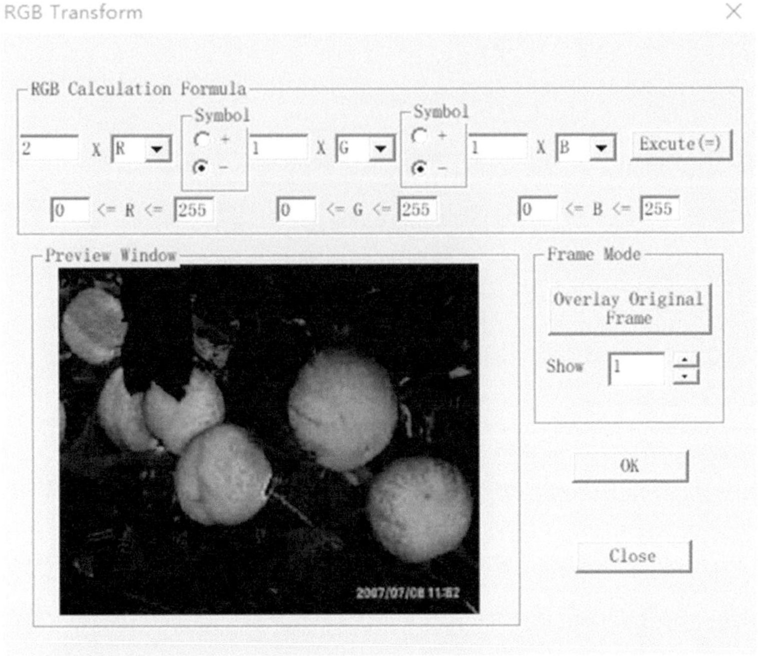

Fig. 18.12 RGB colour transformation

18.6.4 RGB Colour Transformation

As shown in Fig. 18.12, it is used for addition and subtraction operations between the three components *R*, *G*, and *B* of a colour image. It can easily extract component images on *R*, *G*, and *B* in a colour image and enhance certain components.

18.7 Geometric Transformation and Monocular Measurement

18.7.1 Step Transformation

As shown in Fig. 18.13, transformation items such as translation, rotation, and scaling can be selected.

When selecting rotation or scaling, the centre coordinates in the *x* and *y* directions can be set. The default values are the *x* and *y* coordinates of the image centre.

When "rotation" is selected, after setting the rotation angle, the rotated image is automatically displayed in the window.

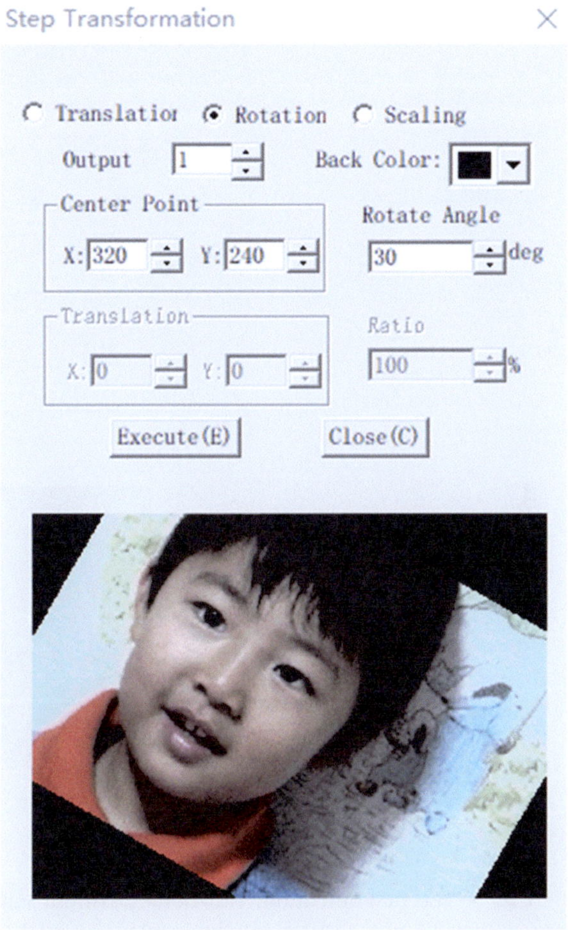

Fig. 18.13 Step transformation

When "translation" is selected, after setting the translation amount, the translated image is automatically displayed in the window.

When "scaling" is selected, according to the set ratio, the resized image is automatically displayed in the window.

18.7.2 Complex Transformations

Figure 18.14 shows the interface and preview of the complex transformation, with optional affine and perspective transformations.

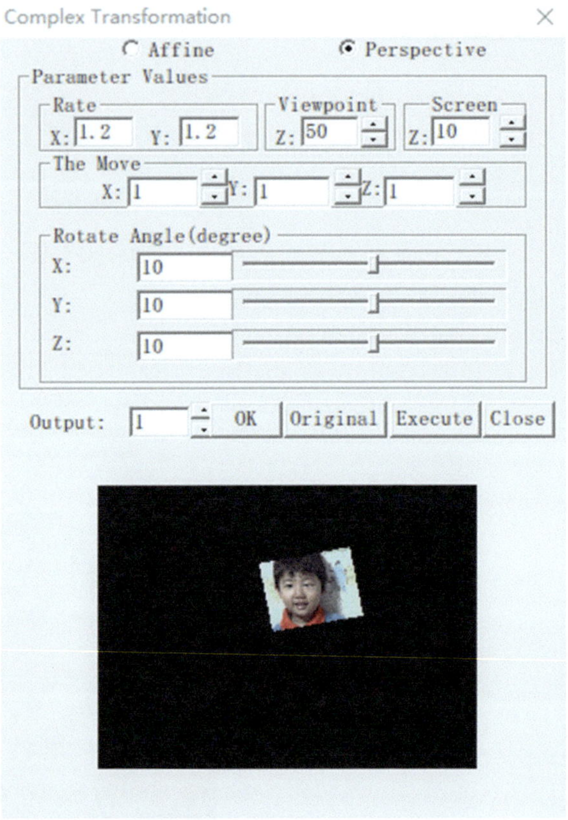

Fig. 18.14 Complex transformation

Can be set: magnification, viewpoint position, screen position, *X*, *Y*, *Z*-direction movement amount, with the rotation angles around the X, Y, and Z axes as the rotation axes.

After clicking "OK", the preview image is displayed on the image interface.

18.7.3 Monocular Measurements

Figure 18.15 shows the calibration interface for monocular measurements.

1. 2D Calibration

Calibration is required before measurement, which can be scaled, multi-point and checkerboard calibration. The calibration function is described in the following.

(1) Load Image: Read the calibration image. Proportional calibration and multi-point calibration can be performed directly on the moving image, without

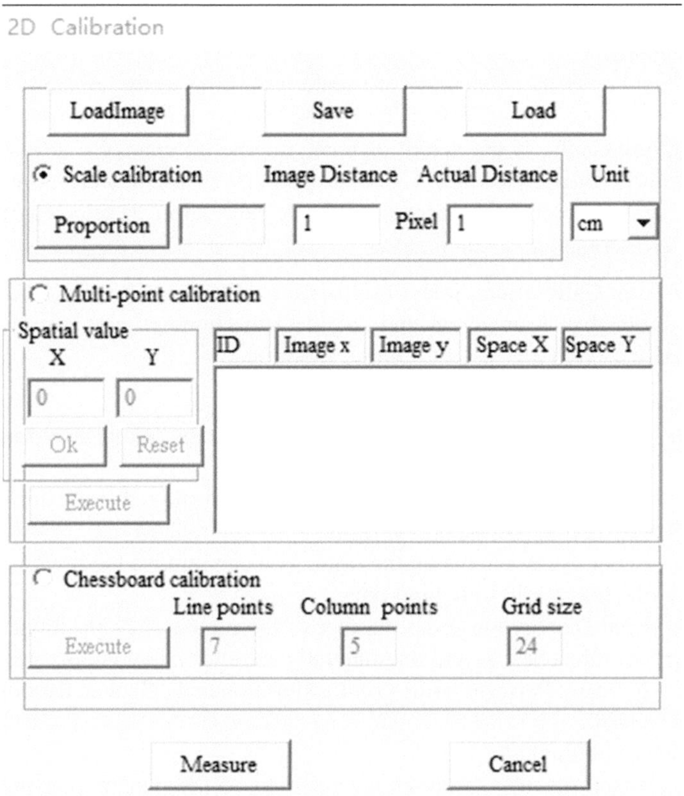

Fig. 18.15 Monocular measurement calibration

reading in the calibration image separately, or reading in the calibration image separately. The calibration of the chessboard needs to read in the calibration image with the chessboard.

(2) Save: Save the current calibration data. The contents include: calibration type, unit, homography matrix data and checkerboard data.

(3) Read: Read in previously saved calibration data. After reading in, the calibration type, unit and chessboard data are displayed on the window, and the homography matrix data is saved in the system.

(4) Unit: Choose the unit of the actual distance, including: pm, nm, um, mm, cm, m, km. The unit is valid for all 3 calibrations.

(5) Proportional Calibration: Select proportional calibration. Scale calibration is generally used in the case of vertical shooting. This includes the following:

   ① Image: The pixel distance of the scale on the image. Using the mouse click scale of the starting point to move to the end point, automatically display the pixel distance.

② Actual: Manually input the actual distance represented by the image distance.

③ Calculate Scale: Calculate the scale according to the image distance and the actual distance. When the mouse moves, the scale will be calculated automatically. If the actual distance is entered manually after the mouse moves, you need to click "Calculate proportion" to calculate the scale. When "Measurement" is performed, if "Scale calibration" is selected, the scale is also automatically calculated.

(6) Multi-point Calibration: Select multi-point calibration. It is necessary to calibrate more than four points and calculate the homography matrix, which is generally used in the case of oblique shooting.

① The interface functions are as follows:

a. Spatial Coordinates X, Y: Manually input the spatial coordinates of the mouse click position.
b. Ok: The entered spatial coordinates are displayed at the corresponding position in the list on the right.
c. Reset: Reset the list on the right. You can remove the bottom item from the list by right-clicking once.
d. List: Display the serial number of the mouse click position, the image coordinates x, y, and the manually entered spatial coordinates X, Y.
e. Execute: Perform multi-point calibration and calculate the homography matrix.

② Operation method:

a. When the mouse clicks the position of the punctuation point on the image, its coordinates are automatically displayed in the list on the right. Click more than 4 position points on the image. You can right click to delete the last item, and click "Clear" to clear them all.
b. The mouse selects an item in the list.
c. Manually enter the spatial X and Y coordinates of the selection. The number 1 item is generally used as the space coordinate point 0, and the space position coordinate here is only used to calculate the homography matrix, and has no relationship with the origin of the coordinate of the image (the default top-left corner).
d. Click "OK" to add the spatial coordinates to the corresponding item in the list.
e. After entering the spatial coordinates of each item in the list, click "Execute", calculate the homography matrix, and display "multi-point calibration is successful!" Hint.

(7) Chessboard Calibration: Select the chessboard calibration. The homography matrix is calculated using the checkerboard calibration, which is generally used in the case of oblique shooting.

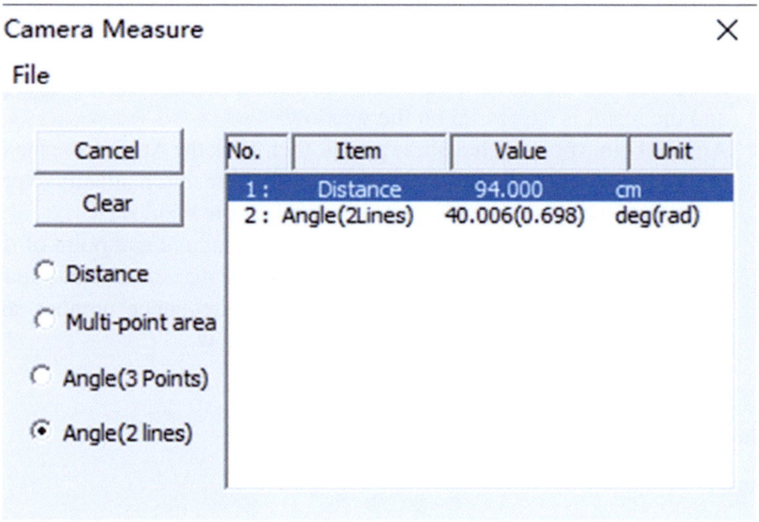

Fig. 18.16 Monocular measurement interface

This is done as follows:

① Read in the calibration image with the chessboard.
② Enter the row corner count, column corner count, and side length of a square. Note: The number of corners and squares on the chessboard are different.
③ Click "Execute" to calculate the homography matrix. Displays "Chessboard calibration success!" Hint.

2. Measure

Click to open the "Plane Measurement" window (Fig. 18.16) and close the "monocular measurement" window.

The interface functions are as follows:

(1) File: Open, save and print the measurement results.
(2) Cancel: Delete the last measurement.
(3) Clear: Clear All measurements.
(4) Result represents window

The measurement result is automatically expressed. When the Angle is measured, the arc is expressed in the bracket.

(5) Measurement items:

① Distance: Click two points in succession on the image, a straight line will be drawn automatically between the two points, and the measurement serial number will be marked at the latter point, and the measurement result will be expressed on the window.

② Multi-point area: click more than 3 points, draw the connection between each point, right click to end the selection point, draw the connection between the last point and the initial point, calculate the area between multiple points, and the result is expressed on the window.

③ Angle of three points: click three points, then click the Angle to be measured, the Angle and measurement sequence number are automatically expressed, and the measurement results are expressed on the window.

④ Angle between two lines: Click the starting point and end point of the two lines respectively, and then click the Angle to be measured to automatically represent the two lines and the measurement sequence number, and the measurement results are expressed on the window.

(6) Save the screen

Save the image that represents the measurement.

3. Cancel

Cancel the measurement and close the window.

4. Steps of operation

(1) Open the "Monocular measurement" dialog box. If you need to read the calibration image, click the "Read image" button to read the calibration image. For proportional calibration and multi-point calibration, the calibration can be performed directly on the measurement image, while for chessboard calibration, the calibration image with the chessboard needs to be read in.
(2) Click "Read calibration" to read the saved calibration file. The contents of the calibration file include: calibration method, unit, scale value for scale calibration, homography matrix data for multi-point calibration and checkerboard calibration, number of corners and individual dimensions of the board. After reading in the calibration, the "measurement" can be performed directly.
(3) Set the unit.
(4) Select "proportional calibration", "multi-point calibration" or "checkerboard calibration", and calibrate accordingly according to the method introduced above.
(5) Click "Save calibration" to save the calibration file.
(6) Click "Measurement" to close the "monocular Measurement" window and open the "Plane Measurement" window to perform the above measurement.
(7) If the measurement is canceled, click the "Cancel" button to close the window.

18.8 Frequency Domain Transformation

18.8.1 Wavelet Transform

Figure 18.17 is the wavelet transform interface, which can perform one-dimensional row transform, one-dimensional column transform, and two-dimensional wavelet transform. During the wavelet transform, any component can be eliminated before performing the inverse transform, and wavelet magnification processing can be applied to the selected area.

(1) One-dimensional Column Transform Example After performing three consecutive column "transforms" on the original image, the "low frequency is set to zero" in the vertical direction, and then three column "restorations" are performed, as shown in Fig. 18.18.
(2) Two-dimensional Transform Example After performing three consecutive two-dimensional wavelet "transforms" on the original image, the "low frequency is set to zero", and then three "restorations" are performed, as shown in Fig. 18.19.

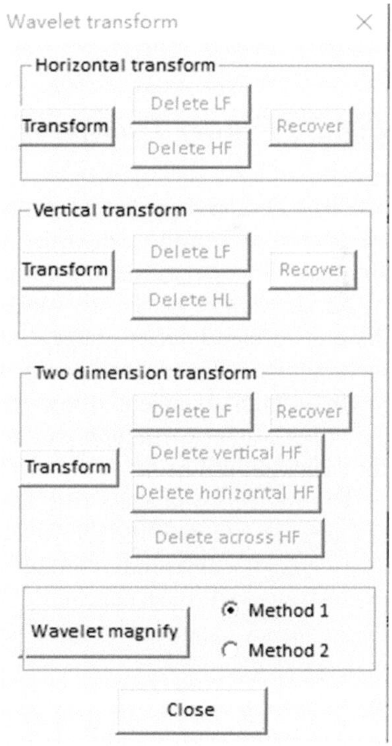

Fig. 18.17 Wavelet transform

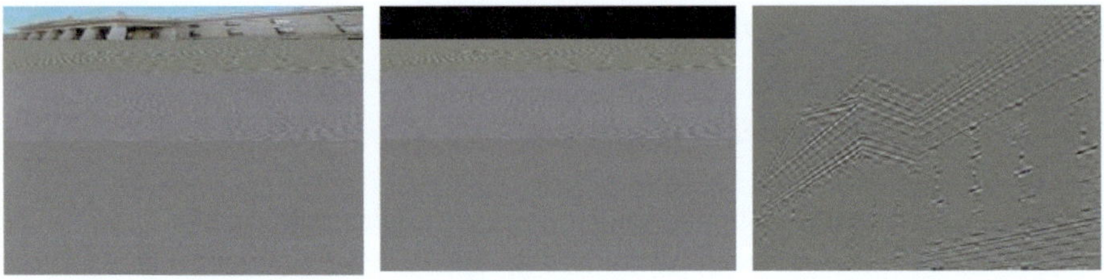

(a) Cubic column transformation (b) De-low frequency (c) Cubic column recovery

Fig. 18.18 Example of one-dimensional column wavelet transform

(a) Cubic 2D transformation (b) De-low frequency (c) Cubic 2D recovery

Fig. 18.19 Example of two-dimensional wavelet transform

18.8.2 Fourier Transform

The Fast Fourier Transform can only be applied to images whose length and width are powers of 2. If the size of the image processing area is not a power of 2, it will automatically be reduced to a power of 2 for processing.

As shown in Fig. 18.20, various types of filters can be selected for filtering the transformed Fourier image, followed by image restoration. The types of filters include: user-defined, ideal low-pass filter, trapezoidal low-pass filter, Butterworth low-pass filter, exponential low-pass filter, ideal high-pass filter, trapezoidal high-pass filter, Butterworth high-pass filter, exponential high-pass filter, etc. Parameters for each filter can be set.

The frequency image's ring features and wedge features can be viewed. Ring features refer to dividing the frequency image into several concentric ring-shaped areas along the polar radius direction in the polar coordinate system, and calculating the total energy in each concentric ring-shaped area. Wedge features refer to dividing the frequency image into several wedge-shaped areas along the polar angle direction in the polar coordinate system, and calculating the total energy in each wedge-shaped area.

An example of processing is shown in Fig. 18.21.

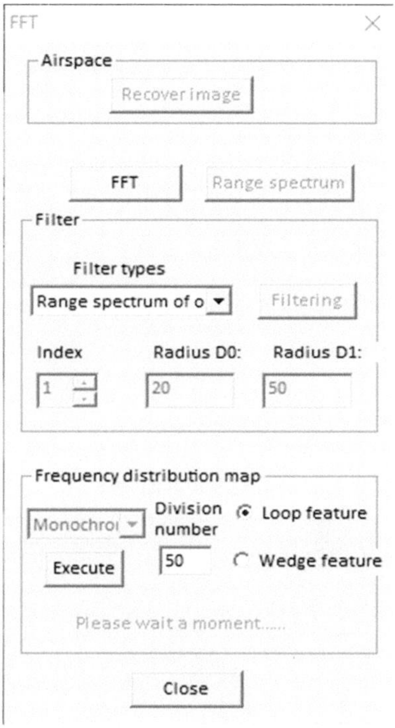

Fig. 18.20 Fourier transform interface

18.9 Image-to-Image Transformation

18.9.1 Image Arithmetic

As shown in Fig. 18.22, perform addition, subtraction, multiplication, division, and logical operations between images. Logical operators include: AND, OR, XOR, XNOR. During arithmetic operations, operation coefficients can be specified arbitrarily. Continuous processing of multiple frames of images can be performed.

18.9.2 Motion Image Correction

(1) Field Transformation: An image captured by a camera device is composed of odd and even scan fields, meaning that a frame image can be synthesized from the pixels of odd and even scan fields. This function creates a frame image from the odd and even fields separately, allowing for continuous processing of multiple frames.

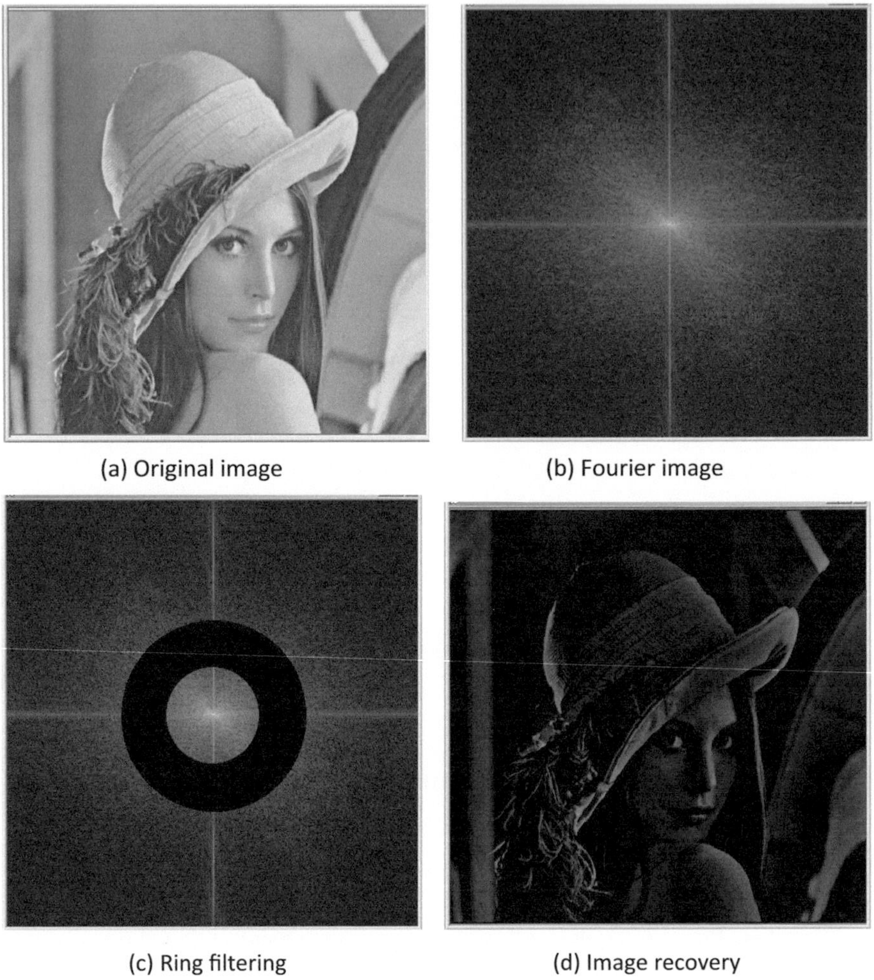

Fig. 18.21 Fourier transform example

(2) Blur Correction: Correction of blurring caused by interlaced scanning during imaging. You can choose odd fields and even fields, using the selected field to create a frame image to replace the original frame image. Figure 18.23 is a schematic of the motion image correction function.

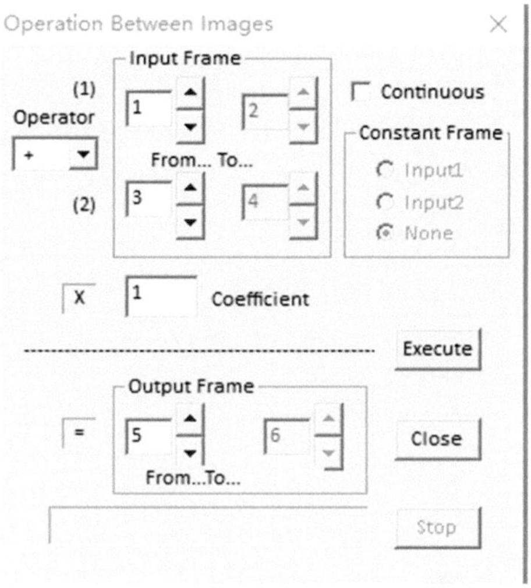

Fig. 18.22 Calculation between images

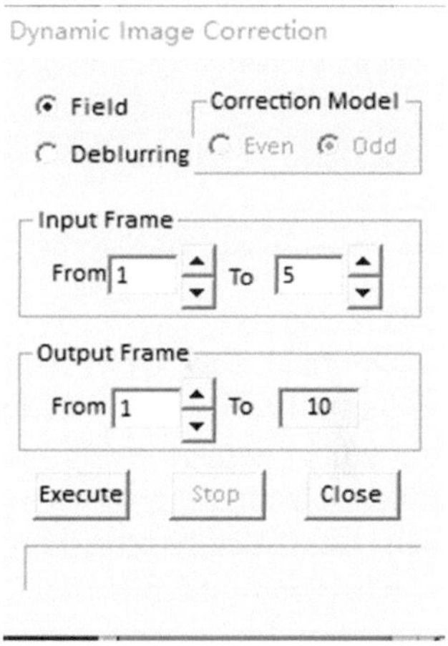

Fig. 18.23 Correction of moving image

18.10 Filter Enhancement

18.10.1 Single Template Filter Enhancement

Filter enhancement involves multiplying each pixel of the image and its surrounding pixels by a series of coefficients (filter operator), dividing the resulting sum by a certain coefficient (divisor), and using the final result as the value of that pixel. Through the above processing, the purpose of enhancing a certain feature of the image or improving the image quality is achieved.

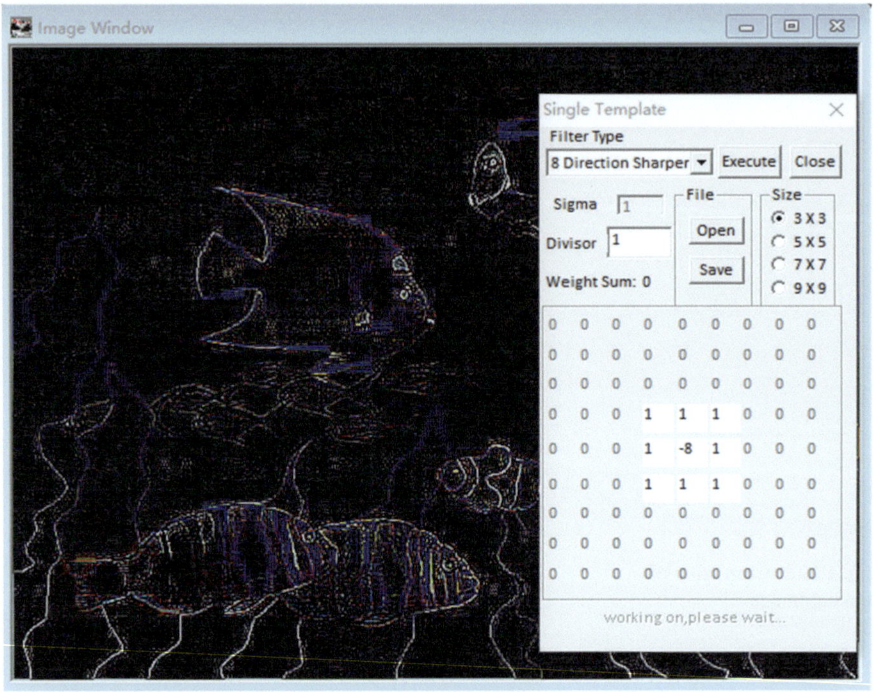

Fig. 18.24 Single template filter enhancement

Optional filter types include: simple mean, weighted mean, 4-direction sharpening, 8-direction sharpening, 4-direction enhancement, 8-direction enhancement, smoothing enhancement, median filtering, sorting, Gaussian filtering, and custom. When selecting the above filter operators, the data of the filter operator and divisor will automatically be displayed in the window.

The size of the filter operator can be chosen: 3×3, 5×5, 7×7, 9×9, etc.

Figure 18.24 shows the function interface and processing example of single template filter enhancement, where an 8-direction sharpening process is applied to a 3×3 area of a colour image.

18.10.2 Multi-template Filter Enhancement

As shown in Fig. 18.25, the optional filter types include: Prewitt operator, Kirsch operator, Robinson operator, general difference, Roberts operator, Sobel operator, Laplacian operation: operator 1, operator 2, operator 3, and user-defined, etc.

Among the above operators, the Prewitt operator, Kirsch operator, and Robinson operator are edge detection and extraction operators based on template matching, each with 9 templates available for user selection. The general difference, Roberts

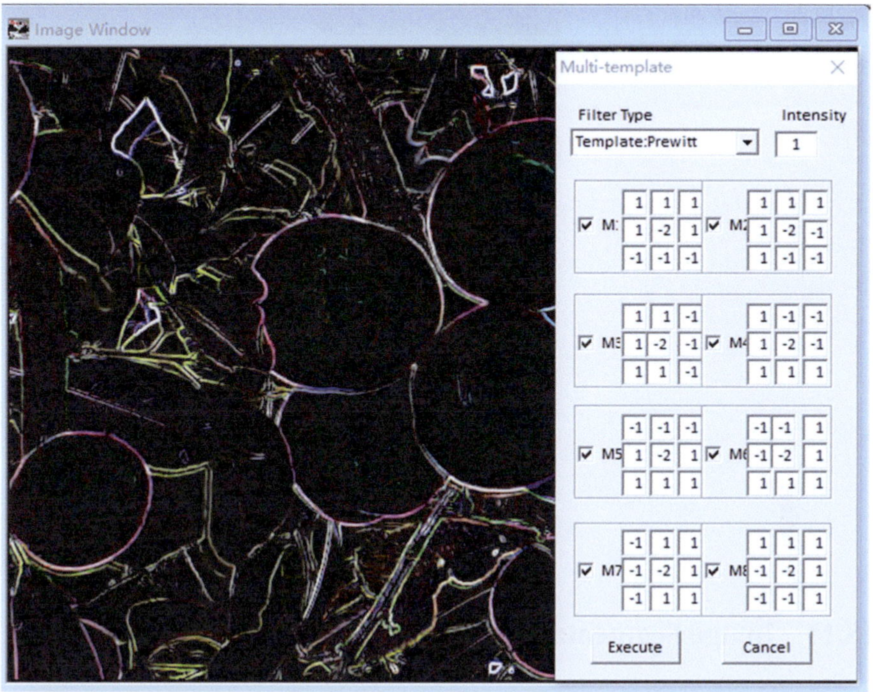

Fig. 18.25 Multi-template filter enhancement

operator, Sobel operator, and the three Laplacian operators are edge detection and extraction operators based on differentiation. After the user selects the type of filter, for operators based on template matching, multiple corresponding templates can be selected simultaneously to achieve the best effect, while for operations based on Laplacian operators, it is a single template.

18.10.3 Canny Edge Detection

As shown in Fig. 18.26, Canny edge detection can be performed step-by-step or with one-click detection. In step-by-step detection, each step is executed in order, displaying the image results of each step. In one-click detection, clicking the "Canny Detection" button only displays the final detection result. After selecting the filter size, the default smoothing scale is automatically adopted, but the smoothing scale can also be set manually. The high threshold (percentage) and low threshold (percentage) can be set according to the detection effect.

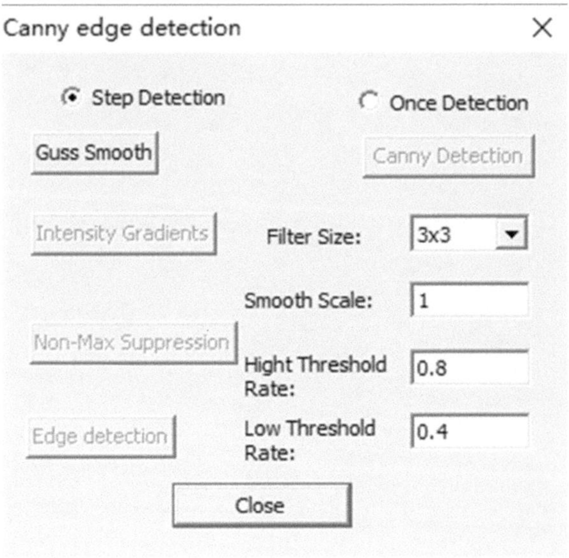

Fig. 18.26 Canny edge detection

18.11 Image Segmentation

The binarization processing of grayscale images or colour images displayed in grayscale mode can have thresholds set manually or automatically calculated by the system to binarize the image. The selectable automatic binarization methods include: modal method, p-parameter method, and Otsu's method.

For the binarization processing of colour images based on the RGB colour system, the effective, ineffective, and threshold ranges of R, G, and B can be set separately; for the binarization processing of colour images based on the HSI colour system, the effective, ineffective, and threshold ranges of H, S, and I can be set separately. The thresholds for both types of colour binarization processing can also be automatically obtained by clicking on the part to be extracted on the image with the mouse, and the size of the mouse click area can be set. Figure 18.27 is the function interface of binarization processing.

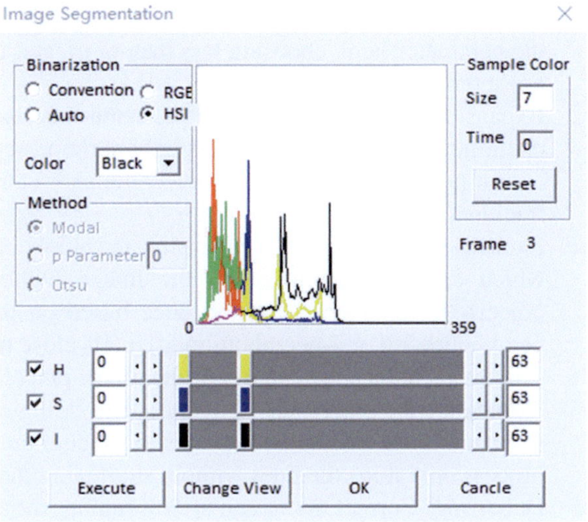

Fig. 18.27 Functional interface of binary processing

18.12 Binary Operations

18.12.1 Basic Operations

As shown in Fig. 18.28, the target object for processing can be selected as "black" or "white", and "8-neighbourhood" or "4-neighbourhood" processing can be selected. The processing items include: noise removal, hole filling, dilation, erosion, exclusive dilation, thinning, deburring, window clearing, contour extraction, etc.

Fig. 18.28 Basic processing of binary image

① Noise removal processing. The number of pixels for noise removal can be set in the parameter item, choosing less than or greater than that number of pixels for noise removal.
② Expansion or corrosion treatment. Execute once, expand or corrode a circle based on the neighbourhood setting (8-neighbourhood or 4-neighbourhood), repeatedly executing expansion and corrosion commands can effectively repair the surface, fractures, holes, etc., of the image.
③ Exclusive expansion. After expansion, the number of objects remains unchanged, which can be used to repair the image without changing the number of objects. Execute once, expand once based on the neighbourhood setting (8-neighbourhood or 4-neighbourhood), parts close to other objects do not expand.
④ Thinning. Reduce the contour of the object pixel by pixel until it is reduced to a "skeleton" one pixel wide (thin line). The "thinning times" can be set, when the set value is "0" (default), it means to execute until it is thinned. This thinning process only thins the lines without shortening them.
⑤ Deburring. Correct the image after thinning, the length of the burr (number of burr pixels) can be set.
⑥ Clear window. Clear unwanted objects on the window. The clearing direction can be set: up, down, left, right.

18.12.2 Special Extraction

It is possible to measure 26 geometric data items of the object, extracting objects based on up to 4 "and" or "or" conditions.

The setting items include: area, perimeter, perimeter/area, area ratio, number of holes, hole area, roundness, equivalent circle diameter, centroid (X), centroid (Y), horizontal projection diameter, vertical projection diameter, projection diameter ratio, maximum diameter, long diameter, short diameter, long diameter/short diameter, projection diameter start X, projection diameter start Y, projection diameter end X, projection diameter end Y, graphic start X (x-coordinate of the initial contact point of the scan), graphic start Y (y-coordinate of the initial contact point of the scan), ellipse major axis, ellipse minor axis, major axis/minor axis.

Effective when selecting two or more items. Indicates the logical relationship between the selected items when extracting objects, "and" or "or" can be selected.

After clicking the target with the mouse, the selected geometric parameters of the target are automatically obtained, and these parameters can be referred to set the extraction threshold. The setting range includes: greater than the threshold, less than the threshold, and between two thresholds.

The set processing conditions can be opened and saved.

Figure 18.29 is the operation interface of special extraction and an extraction example, which is an example of extracting black targets with an area greater than 500 pixels and a perimeter greater than 80 pixels.

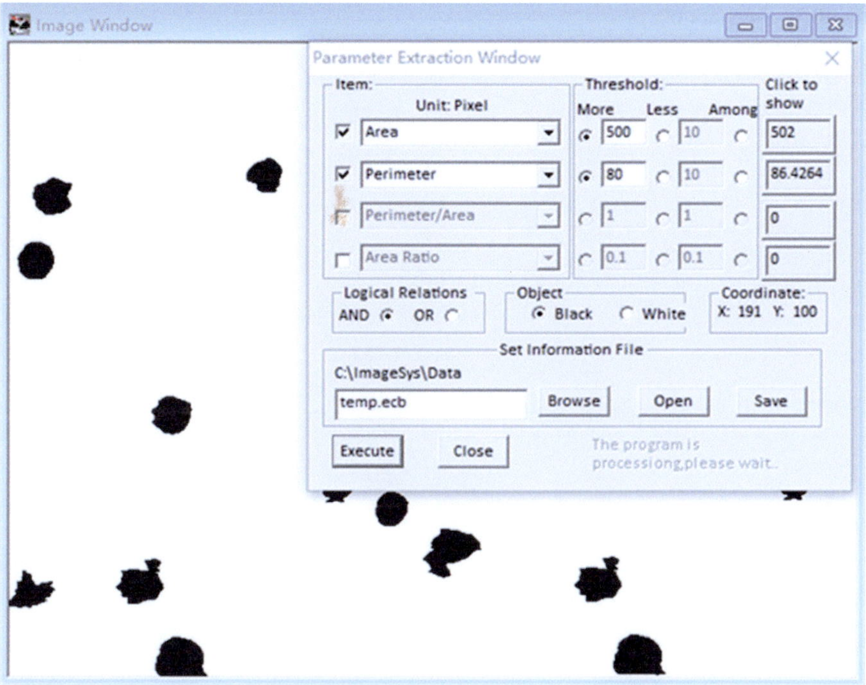

Fig. 18.29 Special extraction function interface and extraction example

18.13 Binary Image Measurement

Includes geometric parameter measurement, straight line parameter measurement, circular separation, and contour measurement, each of which is introduced below.

18.13.1 Geometric Parameter Measurement

General processing and manual processing can be selected. There are a total of 49 items for general automatic parameter measurement; manual measurement can measure the distance between two points, continuous distance, angle between 3 points, angle between two lines, etc.

Before measurement, the scale can be set with the mouse, after setting the scale, the measurement is the actual data, if the scale is not set, the default measurement unit is the number of pixels. The units of the scale include pm, nm, um, mm, cm, m, km, etc. Figure 18.30 is the functional interface of geometric parameter measurement.

Fig. 18.30 Interface of geometric parameter measurement function

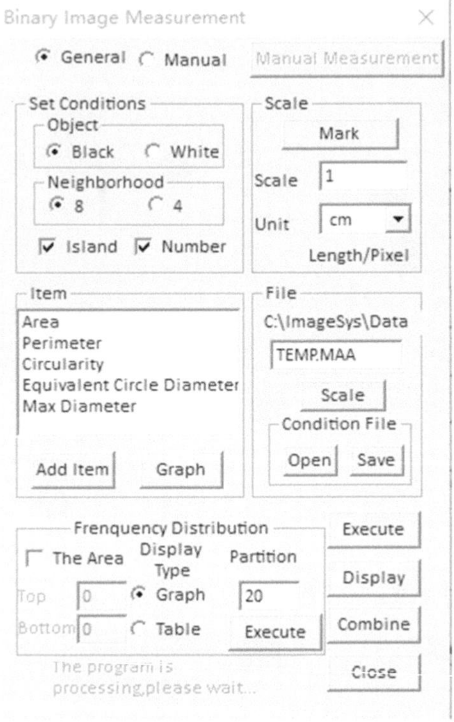

(1) General Automatic Parameter Measurement

The processing object can be selected as "black" or "white", "8-neighbourhood" or "4-neighbourhood" processing can be selected, island processing and non-island processing can be set, and the processing result can be set to label the serial number or not. In island processing, the "island" is treated as a separate object; in non-island processing, the "island" and the object measured outside it are treated as a whole.

① Measurement items. There are the following 39 selectable items (the actual measurement items are 49).

Area, perimeter class:

- Area: Calculated using the number of pixels in the region occupied by the object, excluding the hole area.
- Perimeter: The sum of the distances between adjacent edge pixels in the region occupied by the object.
- Perimeter 2: The sum of the distances between adjacent edge pixels within the region occupied by the object, excluding pixels on the boundary of the processing window.
- Number of Holes: The number of holes within the object's domain.

- Hole Area: The number of pixels of all holes within the region occupied by the object.
- Total Area: The sum of the object's area and the hole area.
- Area Ratio: The object's area (excluding holes) divided by the total area of the processing window.
- Perimeter/Area.
- NCI Ratio: Perimeter ÷ (Total Area)$^{1/2}$.
- Circularity (D): D = 4π × (Total Area) ÷ (Perimeter)^2. D ≤ 1, with a circle's circularity being 1 (maximum).
- Equivalent Circle Diameter: The diameter of a circle with an area equal to that of the object.
- Sphere Volume: The volume of a sphere with the equivalent circle's diameter as its diameter.
- Shape Factor of Circle (C): The reciprocal of circularity, indicating the degree of concavity of the circle; the larger the value, the greater the concavity.

$$C = 1/D = (\text{Perimeter})^2 \div [4\pi \times (\text{Total Area})].$$

- Line Length (Thinned Image): (Perimeter) ÷ 2.

Centroid, Projection Diameter:

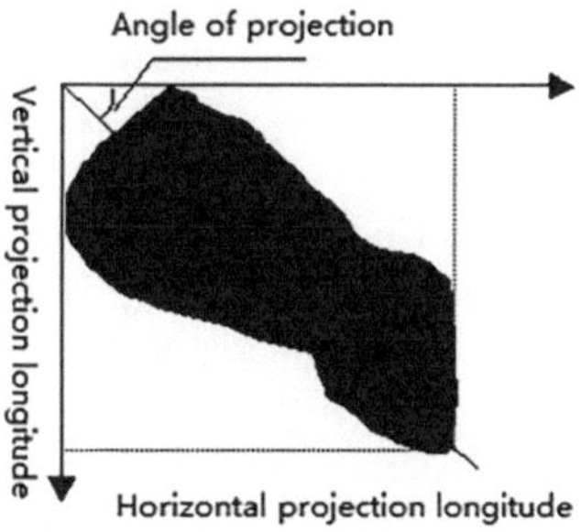

- Centroid: The x-coordinate (X) and y-coordinate (Y) of the centroid.

$$X = (1/n)\sum x; \quad Y = (1/n)\sum y.$$

(n = number of pixels; x = coordinate value of each pixel x; y = coordinate value of each pixel y.)
- Horizontal Projection Diameter: The horizontal diameter projected onto the x-axis.
- Vertical Projection Diameter: The vertical diameter projected onto the y-axis.
- Projection Diameter Angle: The angle between the diagonal of the rectangle formed by the projection diameter (the circumscribed rectangle parallel to the coordinate axis) and the x-axis.

$$\arctan(\text{Vertical Projection Diameter}/\text{Horizontal Projection Diameter}).$$

- Occupancy Rate: The proportion of the object within the rectangle formed by the projection diameter.

$$(\text{Total Area}) \div (\text{Horizontal Projection Diameter} \times \text{Vertical Projection Diameter}).$$

Maximum Diameter:

- Maximum Diameter: The longest straight line within the object. In addition to the length of the maximum diameter, the following four items are automatically measured after selecting the maximum diameter (maximum diameter endpoint x_1, maximum diameter endpoint y_1, maximum diameter endpoint x_2, maximum diameter endpoint y_2).
- Maximum Diameter Angle: The angle between the maximum diameter and the x-axis.
- Diameter Shape Factor: $(\pi/4) \times [(\text{Maximum Diameter})^2 \div \text{Total Area}]$. Minimum is 1 (circle), the larger the value, the further from a circle.
- Long Diameter: The longer side of the smallest rectangle enclosing the object. Equivalent to the long diameter in the case of an ellipse.
- Short Diameter: The shorter side of the smallest rectangle enclosing the object. Equivalent to the short diameter in the case of an ellipse.
- Long Diameter Angle: The angle between the long diameter and the x-axis.

Frame Coordinates:

- Horizontal Projection Diameter Coordinates: Selecting this item will measure the following four items [refer to the diagram (left)]: Horizontal Projection Diameter Start Point X, Vertical Projection Diameter Start Point Y, horizontal projection path endpoint X, vertical projection path endpoint Y.
- Graphic Starting Point Coordinates: After selection, the following two items will be measured [refer to the diagram (right)]: graphic starting point X, graphic starting point Y.

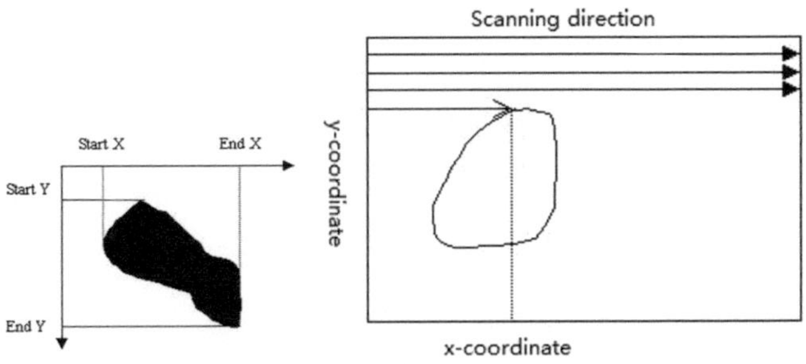

Ellipse Class:

- Ellipse major axis: The major axis of the assumed inertia ellipsoid.

$$m_{\theta\,\max} = \left\{0.5(Mx2 + My2) \pm 0.5\left[(Mx2 - My2)^2 + 4(Mxy)^2\right]^{1/2}\right\}_{\max}$$

$m_{\theta\max}$: moment of inertia for the ellipse major axis; $Mx2$, $My2$, Mxy: second moments about the x axis, y axis, and x, y axes respectively, please refer to the "Region Moments" section below.

$$\text{Ellipse major axis} = (1/m_{\theta\,\max})^{1/2}$$

- Ellipse minor axis: The minor axis of the assumed inertia ellipsoid.

$$m_{\theta\,\min} = \left\{0.5(Mx2 + My2) \pm 0.5\left[(Mx2 - My2)^2 + 4(Mxy)^2\right]^{1/2}\right\}_{\min}$$

$m_{\theta\min}$: moment of inertia for the ellipse minor axis. Ellipse minor axis = $(1/m_{\theta\min})^{1/2}$.
- Ellipse orientation angle: The angle θ between the ellipse major axis and the x axis.

$$\theta = 0.5\arctan\left[2\,Mxy \div (My2 - Mx2)\right].$$

- Ellipse major to minor axis ratio.
- Ellipsoid volume: The volume obtained by rotating the inertia ellipsoid around its major axis.

$$(4/3)\pi \times (\text{major axis}/2) \times (\text{minor axis}/2)^2.$$

- Ellipse shape factor: Indicates the degree of approximation to a circle.

$$\pi \times (\text{major axis} + \text{minor axis}) \div (2 \times \text{circumference})(= a).$$

Circle or ellipse = 1, irregular shape < 1, $0 < a < 1$.

Region Moments Class:

The coordinate system of the image is as follows:

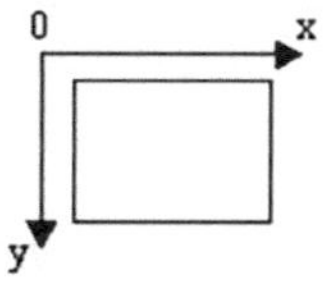

- 0th order moment (M_0): M_0 = area of the object.
- 1st order moment X (Mx_1): First moment about the x axis.

$$Mx_1 = \sum y.$$

- First-order moment Y ($My1$): First-order moment about the y-axis.

$$My1 = \sum x.$$

- Second-order moment X ($Mx2$): Second-order moment about the x-axis.

$$Mx2 = \sum (y - y0)^2; \quad y0\text{: The } y\text{-coordinate of the centroid.}$$

- Second-order moment Y ($My2$): Second-order moment about the y-axis.

$$My2 = \sum (x - x0)^2; \quad x0\text{: The } x\text{-coordinate of the centroid.}$$

- Second-order moment XY (Mxy): Second-order moment about the x and y-axes.

$$Mxy = \sum \sum (x - x0)(y - y0).$$

- Polar moment of inertia (Mo): $M\text{o} = Mx2 + My2$.

Note: The above formulas are only valid for binary images.

Measurement results can be documented, and after opening the document, the measurement results can be saved. The saved data can be read and tabulated by other software. Multiple measurement results can be merged and processed.

② Frequency distribution. Frequency distribution representation can be performed for different measurement items, and distribution graphs or tables can be selected for representation. These charts can be saved, copied, and printed. Figure 18.31 is an example of the frequency distribution graph and table for area measurement results.

(2) Manual Measurement

Measure the distance, angle, etc., specified by the mouse.

- Distance between two points: Click two points on the image in succession, a line will be automatically drawn between the two points, the measurement number will be marked at the latter point, and the measurement result will be displayed in the window.
- Continuous measurement of the distance between two points: Continuously display the distance of the mouse click position.
- Angle between 3 points: After clicking 3 points, click the angle to be measured, the angle and measurement number will be automatically displayed, and the measurement result will be shown in the window.

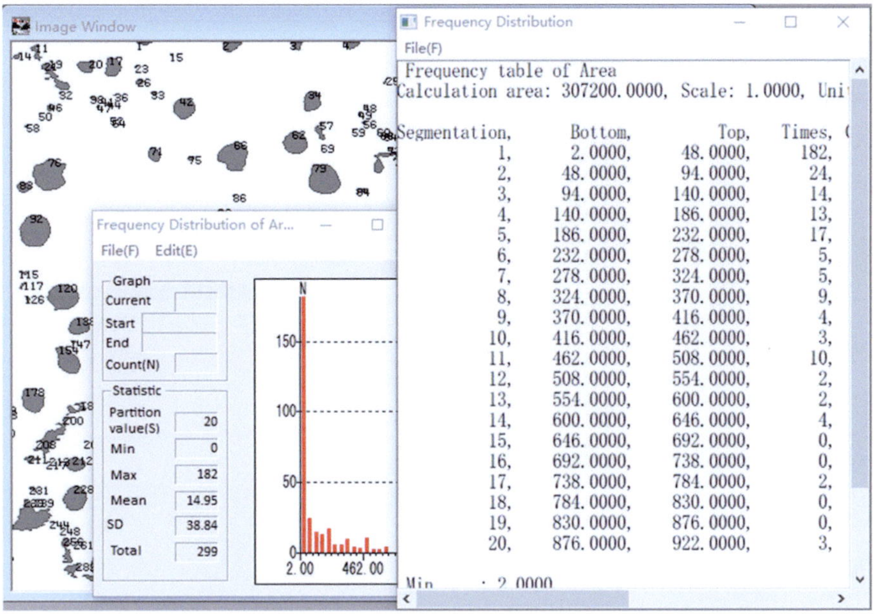

Fig. 18.31 Frequency distribution diagram and illustration of area measurement results

- Angle between two lines: Click the start and end points of two lines respectively, then click the angle to be measured, the two lines and measurement number will be automatically displayed, and the measurement result will be shown in the window.
- Multi-point area: Click more than 3 points, continue to execute the "distance between two points" function, right click the mouse to stop the selection of points, and calculate the surrounding area.

Figure 18.32 is the manual measurement interface and measurement example.

18.13.2 Line Parameter Measurement

As shown in Fig. 18.33, in a binary image, different methods are used to detect the target area's line, and the detection results and parameters are displayed. The following measurement methods can be selected:

① General Hough Transform. Detect line elements in the image using the general Hough Transform.
② Hough Transform Through a Point. Detect line elements passing through a set point.
③ Hough Transform Through a Line. Detect line elements passing through the intersection of a reference line and the target pixel group.

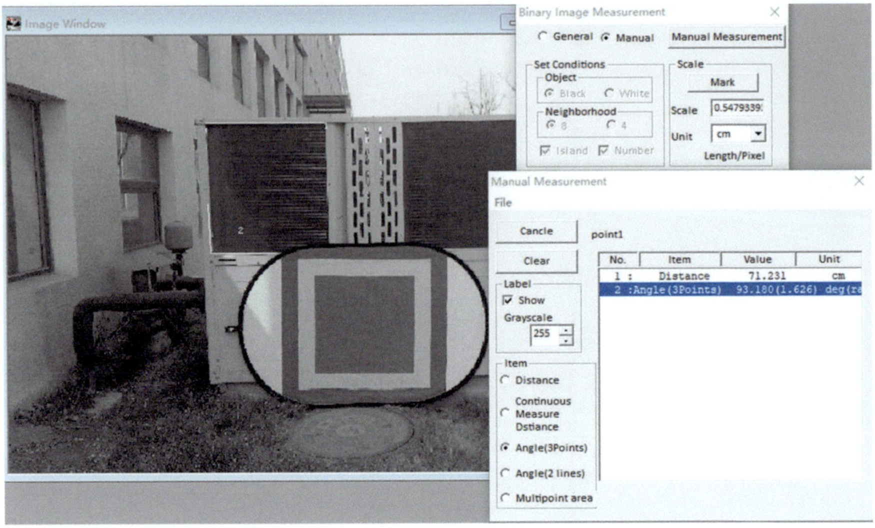

Fig. 18.32 Manual measurement

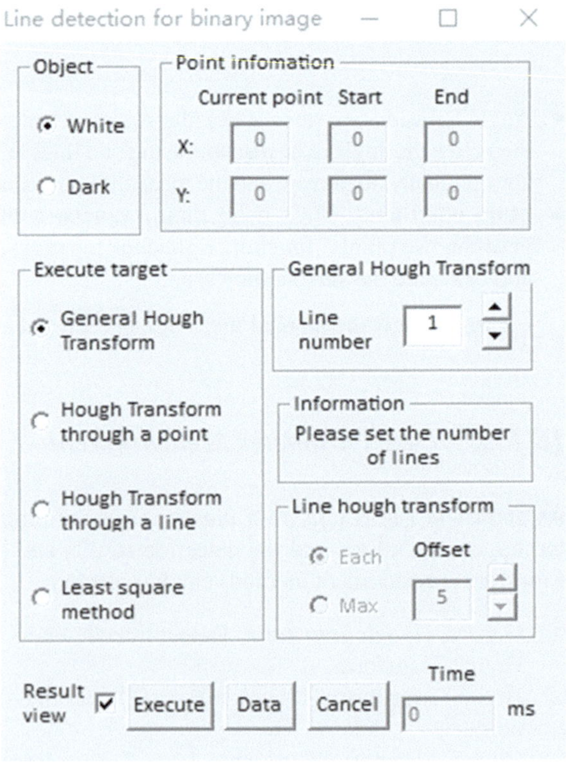

Fig. 18.33 Measurement of linear parameters

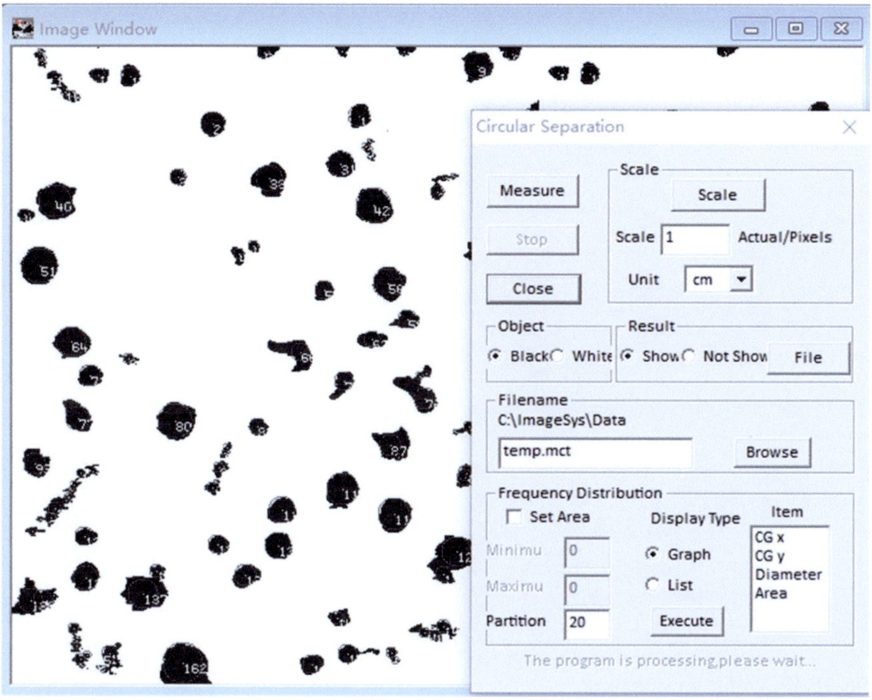

Fig. 18.34 Circular separation

④ Least Squares Method. Detect line elements in the image using the least squares method.

18.13.3 Circular Separation

As shown in Fig. 18.34, circular separation is used to separate circular objects and measure their diameter, area, and centre coordinates. For non-circular objects, measurements and separation are conducted using their inscribed circle. It can also display the frequency distribution of the processing results.

18.13.4 Profile Measurement

As shown in Fig. 18.35, the number of objects to be measured, the contour line length (in pixels) of each object, and the coordinates of each pixel on the contour line are measured. The measurement data can be represented and saved in a document.

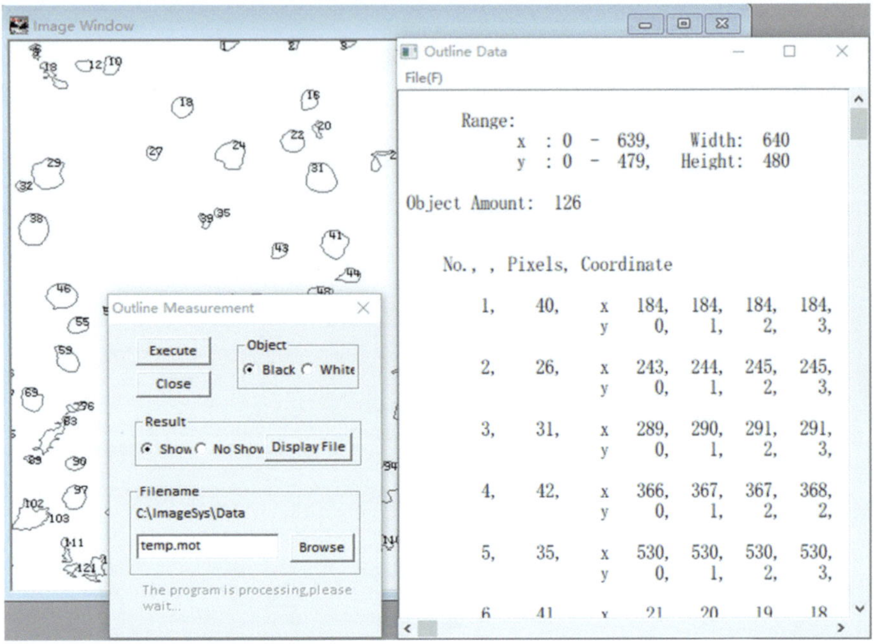

Fig. 18.35 Profile measurement

18.14 Frame Editing

As shown in Fig. 18.36, frame copying can be performed, with copying methods: 1 versus N: copying a single frame image into multiple frame images, N versus N: copying multiple frame images into multiple frame images. Frame clearing can also be performed, setting each pixel value to 0 (black).

18.15 Drawing the Graph

As shown in Fig. 18.37, free lines, polylines, straight lines, rectangles, circles, and smudges (fills) can be directly drawn on the image for correction or free drawing. It has an undo function. Circle drawing is divided into centre/radius circle drawing and 3-point circle drawing. In colour mode, colour and RGB components can be selected.

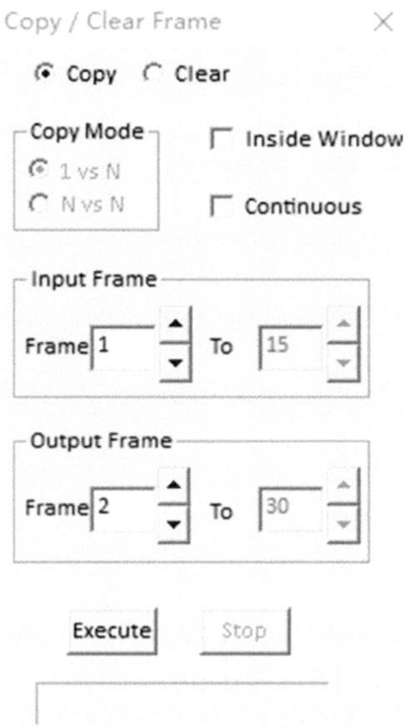

Fig. 18.36 Frame editing function

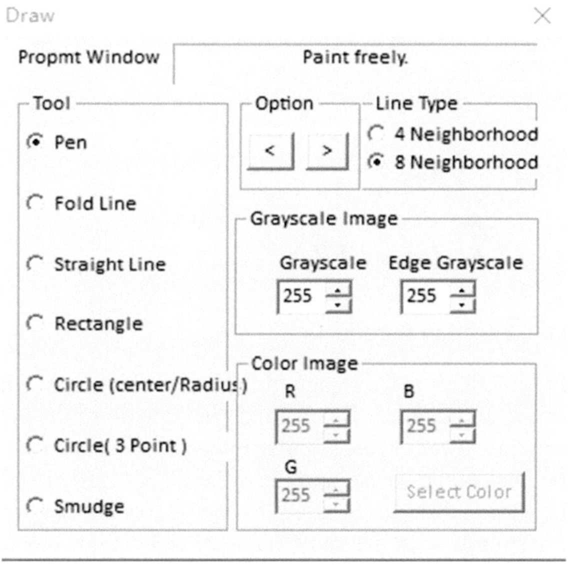

Fig. 18.37 Drawing function window

Fig. 18.38 Pixel value display window

18.16 View

As shown in Fig. 18.38, the colour RGB or greyscale pixel values of a 7 × 7 area around the mouse can be viewed in real-time. The displayed image can be zoomed in or out, and the zoomed image can be saved. The status window can be opened or closed.

18.17 Files

18.17.1 Image Files

(1) Function Introduction

Figure 18.39 is the image file operation interface, where image files can be loaded, saved, and deleted. By clicking "Information," the properties of the file to be loaded can be viewed. Most popular image file formats such as bmp, jpg, jpeg, tif, tiff can be loaded and saved, and the system's specially designated txt image files can also be read and saved. Single files can be loaded, or continuous image files, i.e., image file groups with the same name plus a 4-digit continuous number, can be loaded. When selecting continuous files, file 1 indicates the start file of the continuous files, and file 2 indicates the end file.

① Click "Browse" to open the file browsing window and select the image file to be loaded. The initial setting for the open location is C:\ImageSys\Image. When selecting continuous files, first click the start file with a 4-digit serial number with the mouse, then hold down the "Ctrl" key on the keyboard and click the end file with a 4-digit serial number with the mouse, then open the file.

Fig. 18.39 Image file operation interface

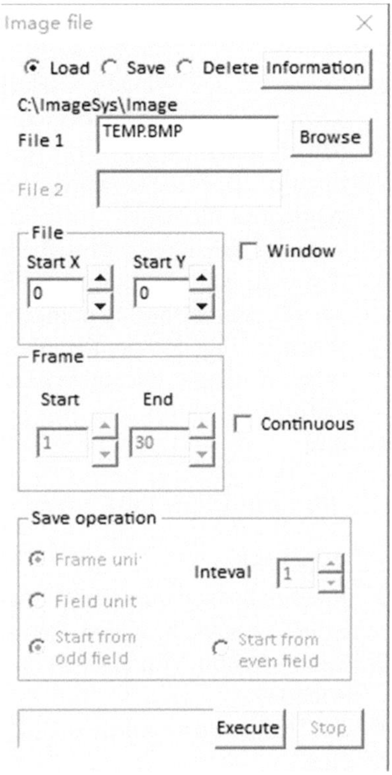

② Within the window, non-selection: operate on the entire frame image; selection: operate on the image within the window. For setting the processing window, please refer to the “Status Window” description.
③ “Start X” and “Start Y” of the file are used to set the starting position of the loaded image file.
④ “Start” and “End” of the frame are effective when setting “Continuous Files,” used to set the start and end frames of continuous images.
⑤ “Storage Operation” is effective when selecting continuous saving (storage, continuous files).

- Frame Unit: Save as a whole frame image unit.
- Field Unit: Save as a scan field unit. After selection, you can further choose “Start from Odd Field” or “Start from Even Field.” An image consists of odd and even scan fields. After saving continuously in field units, the number of images will double.
- Interval: The number of images between the start frame and the end frame interval.

⑥ Run: Start executing bitmap file operation processing.
⑦ Stop: Stop the currently running command.

(2) Method of Use for Loading

① In the "Status Window," select the "Load" mode: greyscale or colour. (The following items are conducted in this window "Image File".)
② Select "Load."
③ Enter or select the file name (browse). When reading continuous files, select the continuous file name.
④ When only reading part of the window, set "Within Window."
⑤ If you do not wish to read the image from the start position (top left corner) of the file, set the starting point for reading (starting *X*, starting *Y*).
⑥ When reading consecutive files, set "Consecutive Files".
⑦ When reading consecutive files, set the start and end frames for reading (start, end).
⑧ Run.

(3) Method to Store the Currently Displayed Frame Image

① Select "Store".
② Enter or browse the name of the file to be saved. When entering the file name, an extension must be added, for example, TEMP, BMP, ABC.JPG, ABC.TIF, etc.
③ Select "Within Window" if you only want to save the image within the processing window.
④ Set "Consecutive Files" to a non-selected state (no tick in the box).
⑤ Run.

(4) Method of Using Continuous Storage

① "Store".
② Enter or browse the file name. When entering the file name, a number or a four-digit sequence number must be added after the file name, and an extension must be added, for example, TEMP1.BMP, ABC0001.JPG, etc. During continuous storage, the system automatically increments the sequence number of the file name, for example, for continuously storing 3 frames of files, if you enter the storage file name TEMP1.BMP, the system stores the results as: TEMP1.BMP, TEMP2.BMP, TEMP3.BMP; if you enter the storage file name ABC0001.JPG, the system stores the results as: ABC0001.JPG, ABC0002.JPG, ABC0003.JPG.
③ Select "Within Window" if you only want to save the image within the processing window.
④ Set "Consecutive Files" to a selected state (tick in the box).
⑤ Set the start frame for continuous saving (start).
⑥ Set the end frame for continuous saving (end).
⑦ Set the storage method (storage operation):

a. Select "Frame Unit" when saving as a whole frame image.
b. Select "Field Unit" when saving as a scan field.

c. After selecting “Field Unit”, further choose “Start from Odd Field” or “Start from Even Field”.
d. Set the interval number of the image (interval).

⑧ Execute “Run” to start saving.
⑨ Execute “Stop” if you want to stop the ongoing saving.

(5) Method to Clear Bitmap Files

① Select “Clear”.
② Enter or browse the file name. When clearing consecutive files, select the consecutive file name.
③ Run.

18.17.2 Multimedia Files

(1) Introduction to the Read-In Function

Figure 18.40 is the interface for reading multimedia files, which can read various video file formats such as avi, mp4, wmv, mkv, flv, rm, dat, mov, vob, mpg, mpeg, etc.

① Load. Select the video image file to read in.
② File. Display the name of the file being read in.
③ Browse. Load the file selection window and choose the multimedia file to read in. When the selected image size differs from the system’s set size, a system setting window will pop up to fill in the image size to be read in. After executing confirm, the system will automatically close and restart according to the set image size and system frame number. If cancel is executed, the original system setting is maintained.
④ Play. Preview the multimedia file to be read in.
⑤ Within Window. Non-selected: read in the entire frame; selected: read in the image within the selected processing window. For setting the processing window, please refer to the “Status Window” description.
⑥ System Frame. Start Frame: Set the start frame for reading into the ImageSys system; End Frame: Set the end frame for reading into the ImageSys system.
⑦ File Frame. Interval: Set the read-in interval for the video image file; Start Point *X*: Set the start point x coordinate on the image of the video file to be read in; Start Point *Y*: Set the start point y coordinate on the image of the video file to be read in; Start Frame: Set the start frame of the video image file to be read in; End Frame: Automatically display the last frame of the loaded video file.
⑧ Run. Start reading in the image.
⑨ Stop. Stop the ongoing read-in operation.
⑩ Close. Close the window.

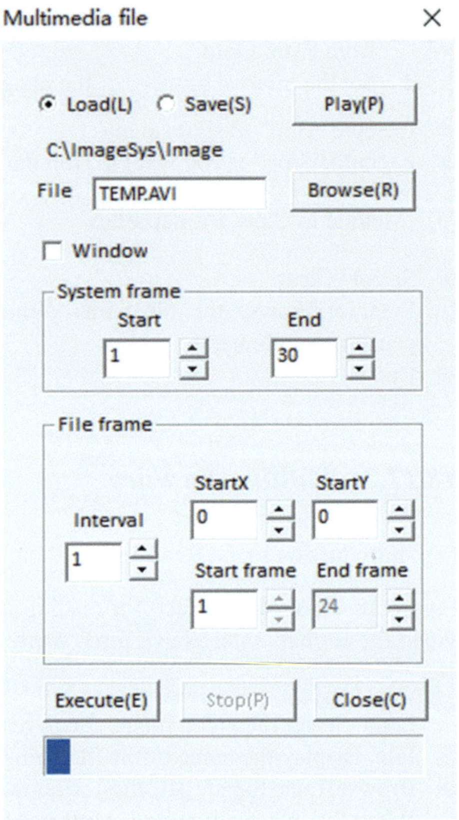

Fig. 18.40 Multimedia file reading interface

(2) Method for Reading Multimedia Files

① Select "Load (L)".
② Select or browse the file to be read in.
③ After selecting the file, if you want to preview the file, you can execute "Play". The playback control panel is shown in Fig. 18.41.
④ If you need to read the image into a specified processing window, select "Within Window".
⑤ Set the start frame and end frame for reading the system frames.
⑥ Set the method for reading the file (video frames), including: image "interval", starting point on the image (starting X, starting Y), and image start frame.
⑦ Run.
⑧ To stop the ongoing reading, execute "Stop".

(3) Introduction to the Save Function

Figure 18.42 is the save interface for multimedia files, which can be saved in avi and mov video file formats, with a choice of various compression modes.

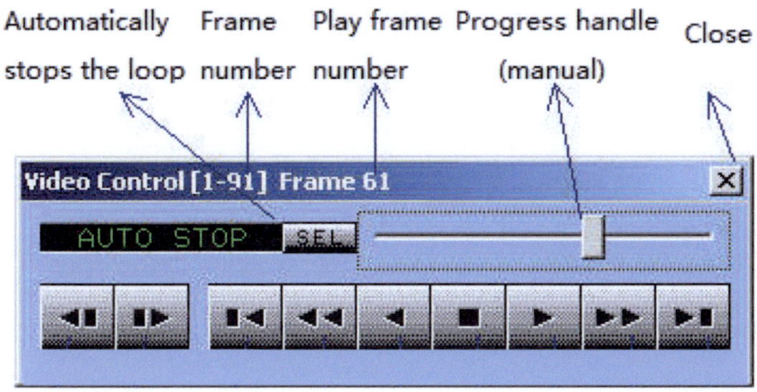

Fig. 18.41 Video playing interface

Fig. 18.42 Multimedia file saving page

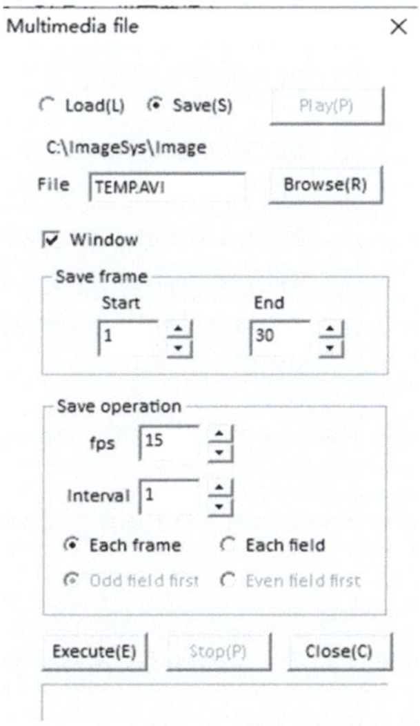

① Save. Choose to save the multimedia file.
② File. You can input or browse the multimedia file name.
③ Browse. Choose the save location and set the save file name.
④ Within the window. Non-selection: save the entire frame image; selection: save the image within the processing window. For the setting of the processing window, please refer to the "Status Window" description.
⑤ Frame storage. Start frame: set the start frame of the system to be saved; end frame: set the end frame of the system to be saved.
⑥ Storage operation. Frame rate: set the playback speed of the video file, generally 30 or 15 fps; interval: the number of intervals from the start frame to the end frame of the system; each frame: save as a whole frame image; each field: save

as a scan field, after selection, you can further choose "odd field first" or "even field first". Odd field first: start saving images from the odd field; even field first: start saving images from the even field.

⑦ Run. Start saving the multimedia file. When saving a colour image, a window for selecting the compression method will appear, allowing you to choose from various compression formats, with the default being uncompressed mode.

⑧ Stop. Stop the ongoing save operation.

⑨ Close. Close the window.

(4) Method for Saving Multimedia Files

① Select "Save".

② Input or browse the name of the multimedia file to be saved.

③ To save only the image within the processing window, select "Within the window".

④ Select the start frame and end frame of the image to be saved.

⑤ Set the save method (storage operation):

a. Set the playback speed of the file (frame rate).
b. Set the interval of the image to be saved (interval).
c. When saving as a whole frame image, select the frame unit (each frame).
d. When saving as a scan field, select the field unit (each field).
e. After selecting the field unit, further choose "odd field first" or "even field first".

⑥ Run save. When the compression selection window appears, choose the compression method.

⑦ To stop the ongoing save, execute "Stop".

18.17.3 Multimedia File Editing

The multimedia file editing function allows for the editing of one or two video (image) files. When two video files are loaded, interleaved editing of the two video files can be performed. A single image file can be inserted into the video, or a single image file can be extracted from the video. The advantage of multimedia file editing is that it is not limited by memory size, allowing for arbitrary settings of the number of video frames to be edited. Editable multimedia file formats include: avi, mp4, wmv, mkv, flv, rm, dat, mov, vob, mpg, mpeg, and others. Figure 18.43 is the multimedia file editing interface.

① Select the number of operation files. Choose to edit one or two files.

② File 1. Select to load the first multimedia file.

③ File 2. Select to load the second multimedia file.

④ Browse. Load the file selection window, and choose the multimedia file to be read.

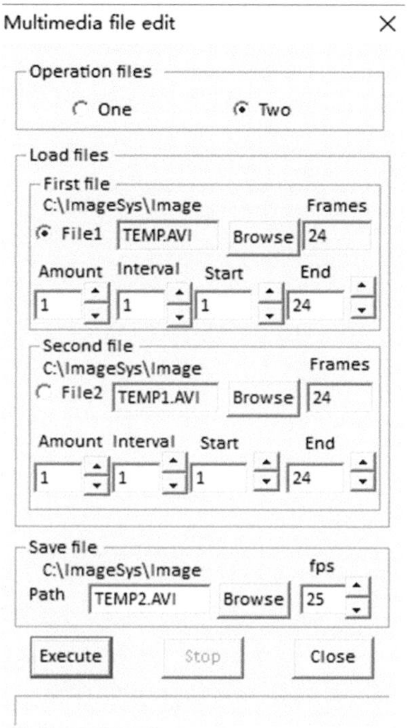

Fig. 18.43 Multimedia file editing interface

⑤ File frame count. Display the frame count of the loaded multimedia file.
⑥ Read frame count. Set the number of frames to be read continuously.
⑦ Interval count. Set the number of intervals to be read.
⑧ Start frame. Set the start frame of the video file to be read.
⑨ End frame. Set the end frame of the video file to be read.
⑩ Save to. Set the file path and file name for saving.
⑪ Run. Start editing the image according to the settings.
⑫ Stop. Stop the ongoing editing operation.
⑬ Close. Close the window.

18.17.4 Adding Watermarks

The main function is to add watermarks to multimedia or image files. You can add a single or multiple watermarks to a single-frame multimedia or image file, and also add watermarks to multi-frame video files. Its main advantage lies in its simplicity and flexibility of operation. Figure 18.44 shows the interface and example of adding watermarks. The interface functions are explained below.

Fig. 18.44 Interface and example of adding watermark function

① Input Text. Enter the watermark text in the text input box.
② Font. Set the properties of the watermark text. Clicking it will bring up the font settings window.
③ Colour. Set the watermark colour. Clicking on colour will bring up the colour selection window.
④ Display. According to the "Font" and "Colour" settings, display the watermark text in the editing window on the screen.
⑤ Confirm. Save the watermark text displayed on the processing screen to the current display frame image.
⑥ Clear Screen. Clear the watermark text displayed on the screen.
⑦ Delete. When adding multiple watermarks, select "Delete" to remove the confirmed watermark text on the screen one by one from the end to the beginning.
⑧ Original File. Display the name of the file read in. You can select the multimedia file to be read in through the "Browse" option that follows. After clicking browse to select a video file, the player will automatically pop up, allowing you to watch the selected video file through the player.
⑨ Save To. Enter or browse the name of the multimedia file to be saved.
⑩ Frame Rate. Set the playback speed of the saved video file. For example, if this number is 15, it indicates a playback speed of 15 fps.

⑪ Save. Execute the operation of adding a watermark.
⑫ Stop. Stop the ongoing operation.
⑬ Close. Close the window.

18.18 System Settings

18.18.1 System Frame Settings

(1) Initial Settings

As shown in Fig. 18.45, the system frame default setting is 640 × 480 pixels, with 4 colour frames. You can close the image window, click "System Frame Settings" in the file menu, open the settings window in Fig. 18.45, set it, confirm, and then restart ImageSys for the settings to take effect.

After starting ImageSys, you can increase or decrease the end frame number as needed, and set the image frame number at any time, referring to the status window in Fig. 18.2. The number of system frames set through the status window becomes invalid after the system is closed.

The setting of the maximum number of frames is related to the size of the computer's memory. Excessive memory usage will affect the computer's running speed.

(2) Restart Settings When Reading Files

When opening an image or multimedia file, if the selected image or multimedia file's image size differs from the current system's image size, a system setting window will pop up to fill in the image size to be read. After selecting "Confirm," the system will automatically close and restart according to the set image size and system frame number, as shown in Fig. 18.46.

Fig. 18.45 System frame setting interface

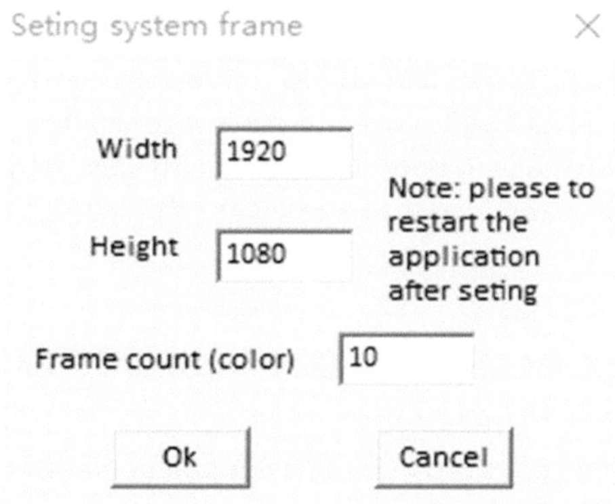

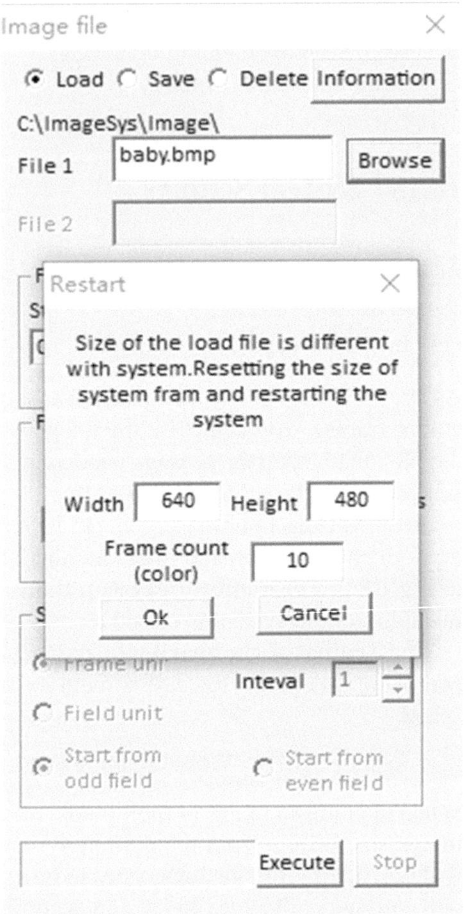

Fig. 18.46 Automatic settings when a file is opened

18.18.2 System Language Settings

The system defaults to a Chinese interface, but an English interface can also be selected. After closing the image window, click "System Language Settings" in the file menu to open the system language settings window in Fig. 18.47. Set it, confirm, and then restart ImageSys for the settings to take effect.

18.19 System Development Platform Sample

ImageSys provides a development platform for a framework source program. Image files, multimedia files, viewing, status window, system frame settings, etc., all fully utilise ImageSys's functional modules, and example programs for greyscale and

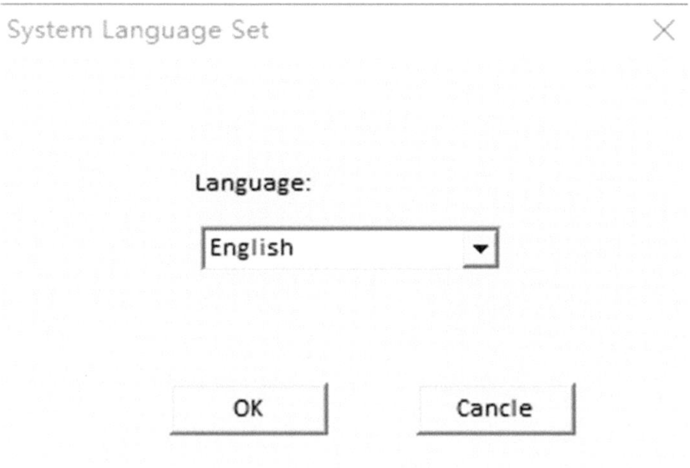

Fig. 18.47 System language setting

colour image processing are provided. On this platform, you can easily add your own menus and dialogue boxes without needing to consider complex auxiliary functions such as image representation and file operations, allowing users to focus on their own image processing algorithm research.

ImageSys provides users with a total of more than 350 functions for image processing, image display, and image access, offering almost all functionalities in the form of functions to users, thereby establishing the system's position as an image processing development platform. Users can use ImageSys to find solutions and write their own programs using the provided functions, significantly improving research and development efficiency.

Reference

1. Chen Bingqi. Detailed Explanation of Machine Vision Technology and Application Examples [M]. Beijing: Chemical Industry Press, 2014.

Chapter 19
2D Motion Image Measurement and Analysis System MIAS

19.1 System Overview [1]

The 2D Motion Image Measurement and Analysis System (MIAS) primarily tracks, measures, and visualizes the motion trajectories of selected targets. Measurement parameters include: coordinate position, speed, acceleration, angle, angular velocity, angular acceleration, displacement, and other datasets. It can perform measurements automatically, manually, or using identification tracking as required. The tracking trajectory can be synchronously represented with the image, and Measurement data can be presented in charts or other intuitive formats.

This system can be applied in various fields, including the analysis of human motion, object motion, and the behavior of animals, insects, microorganisms, etc.; analysis of stress deformation; analysis of vibration and impact of floating objects; analysis of the speed of falling objects; robot vision feedback, etc.

The main functions and features of this system include:

① Multiple measurement and tracking methods. By automatically tracking information such as colour, shape, and brightness, the system measures the movement trajectories of motion points. Tracking methods include: fully automatic, semi-automatic, manual, and identification point tracking.

② Multiple target setting function. Up to 4096 targets can be tracked and measured within the same frame.

③ Rich measurement and representation functions. It can measure over ten parameters, including position, distance, speed, acceleration, angle, angular velocity, angular acceleration, distance between two points, angle between two lines (or three points), angular variables, displacement, and relative coordinate positions, and can represent these measurements in chart or data form. It can also store specified display frames or continuous frames (dynamic) and has powerful dynamic representation functions: dynamically represent trajectory line graphs, vector graphs, and various measurement results, as well as synchronous representation with data.

© Chemical Industry Press 2026

B. Chen and S. Chen, *Machine Vision Technology*, Advanced and Intelligent Manufacturing in China, https://doi.org/10.1007/978-981-95-1184-6_19

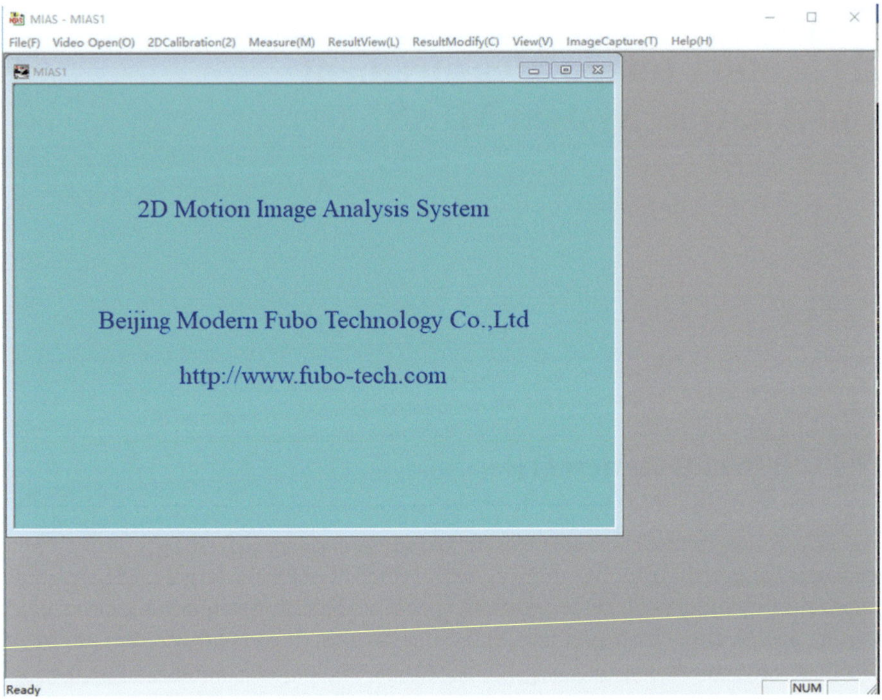

Fig. 19.1 Two-dimensional motion image measurement and analysis system MIAS

④ Convenient and practical correction functions. Modify and correct specified target trajectories. It can perform smoothing to remove angular noise from target motion trajectories, making them more curve-like. It can perform interpolation correction to eliminate the jagged appearance of images (trajectories). It can also merge data, connecting two result files (trajectories). It is also possible to set the reference frame of the object trajectory, add or delete target frames, etc.

Figure 19.1 shows the initial interface of the 2D Motion Image Measurement and Analysis System MIAS.

19.2 Files

The MIAS system can perform multiple operations on 2D result files, including: opening previously saved 2D result trajectory files for subsequent viewing or processing; merging multiple 2D result files, with two methods of merging frames and merging targets; saving the current "result correction" trajectory file; saving the current image as a .BMP type file; printing the currently displayed image, with the option to set the printer and preview the image effect before printing.

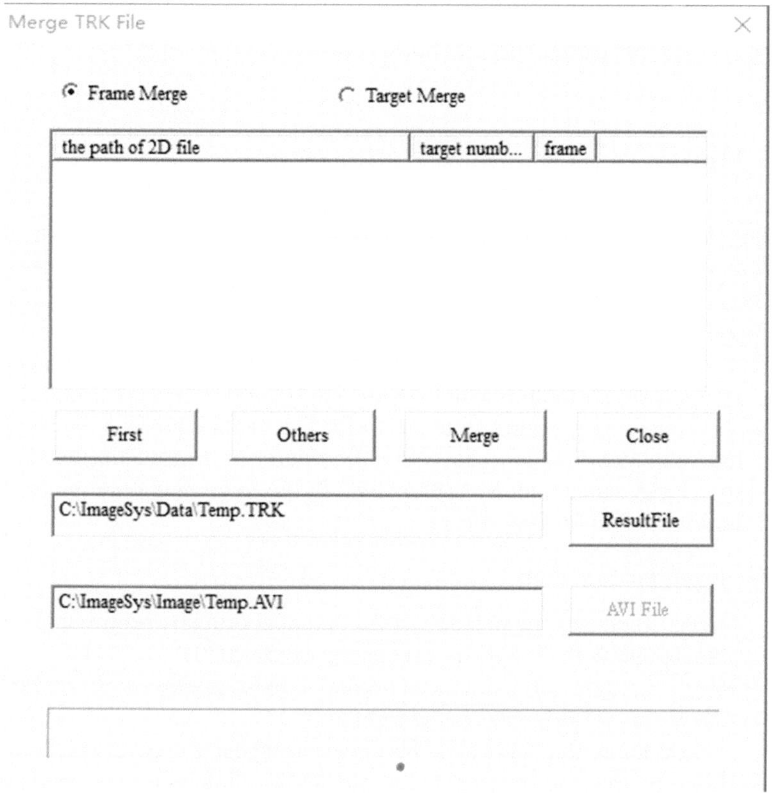

Fig. 19.2 2D result file merging interface

Figure 19.2 is the 2D result file merging interface, and its functions are introduced below.

① Frame merging. On a frame-by-frame basis, merge targets on the same frame number from two or more 2D result files into one sequence image.
② Target merging. On a target-by-target basis, connect two or more 2D result files. The number of targets in each 2D file must be the same. This merging method is mainly used for merging multiple 2D result files in the same scenario.
③ First 2D file. Select a 2D measurement file to use its measurement results as the basis for merging. When "frame merging" is selected, the video image corresponding to the file is played when browsing after merging. When "target merging" is selected, the image and target of this file appear first in the merged file.
④ Other 2D files. Open other 2D measurement files.
⑤ Merge. Execute the merge.
⑥ Close window.
⑦ Merged result file. Set the save path and file name for the merged result file.

⑧ AVI file. Effective when "target merging" is selected. It refers to saving the result image as an AVI format after merging two or more 2D result files.

19.3 Motion Images and 2D Calibration

Click the "Video Open" menu to load continuous image files and video image files. Continuous image files refer to image files with the same name plus consecutive numbers, and file types include: bmp, jpg, tif, etc. Video files include avi, flv, mp4, wmv, mpeg, rm, mov, etc.

Click the "2D Calibration" menu to bring up the 2D calibration interface shown in Fig. 19.3. Scale calibration, multi-point calibration, checkerboard calibration and coordinate transformation can be performed. Among them, the first three calibration functions are the same as "2D calibration" of ImageSys, which can be referred to "18.7.3 Monocular Measurement". In the following, only the coordinate transformation function is described.

(1) Fixed coordinate setting

① Origin: Indicates the position of the actual coordinate origin on the image. The default is the top left corner as the origin (0, 0).
② Rotation Angle: Indicates the counterclockwise rotation angle of the *X* axis. The *X* axis is horizontal to the right at 0°.
③ *Y* Axis Direction: Indicates the direction of the *Y* axis of the coordinate. Based on the *X* axis, when facing the *X* axis direction, the direction of the *Y* axis is indicated as "left" or "right". See the following explanation.

Initially indicated as "Initialize", after clicking, it sequentially indicates "Origin at Top Left", "Origin at Bottom Left", "Origin at Bottom Right", "Origin at Top Right", etc. The origin, rotation angle, and *Y* axis direction automatically change accordingly with the setting key content, as shown in Fig. 19.4.

(2) Method for Freely Setting Coordinate Orientation

① Select "Coordinate Transformation".
② Set *X* Axis Direction: Left-click two points on the image, and the direction of the line connecting the two points is the *X* axis direction. Right-click to cancel the setting.
③ Set Origin: After setting the *X* axis direction, move the mouse to the origin position and left-click. Right-click to cancel the setting. After setting, the relevant parameters are displayed in the coordinate transformation column.

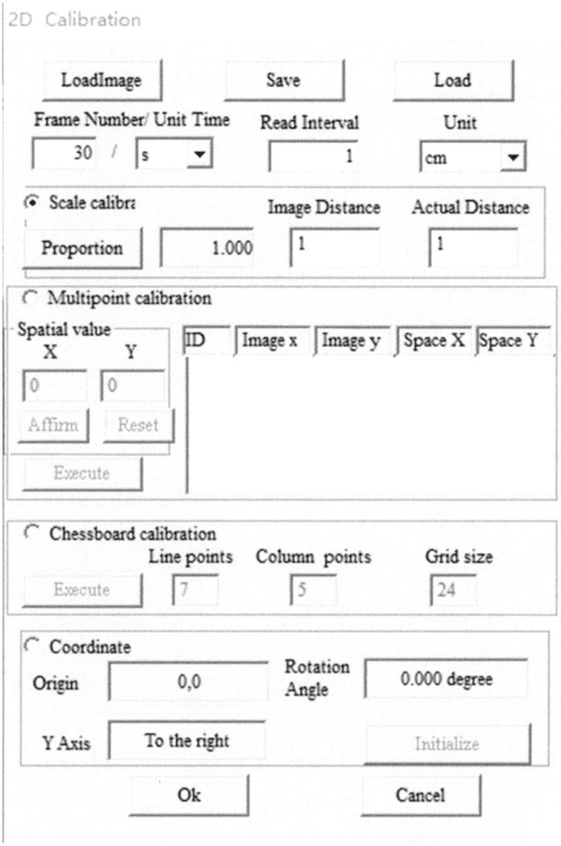

Fig. 19.3 2D calibration interface

19.4 Motion Measurement

Track the trajectory of the selected target in the motion image. The MIAS system provides three methods: automatic measurement, manual measurement, and identification measurement.

19.4.1 Automatic Measurement

Automatic measurement automatically tracks the running trajectory of the set target and measures motion parameters. Generally speaking, the automatic measurement method is suitable for motion targets with a good "recognition environment", such as when the target's RGB value or grayscale value has a good contrast with its surrounding background colour, making it relatively easy to distinguish, and when the environmental noise value is small.

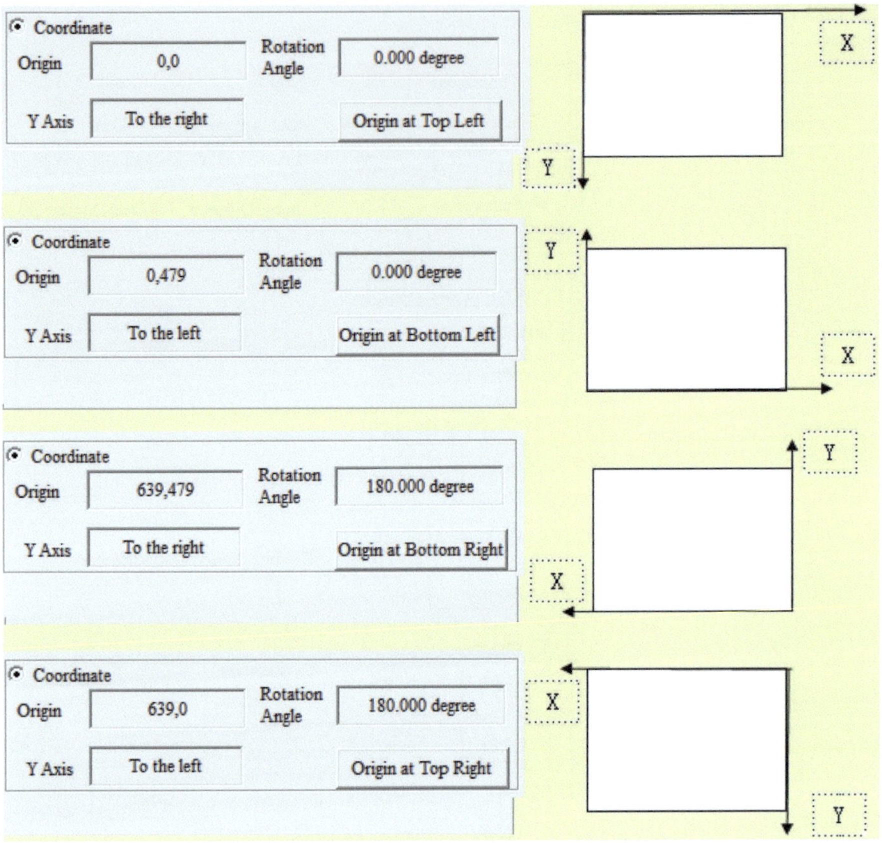

Fig. 19.4 Setting of origin position

After reading in the motion image, before performing measurement processing on the image, it is necessary to set the measurement target first. The system provides two methods for target setting: manual and automatic. The manual target setting method involves selecting the target range by dragging the mouse and then clicking the centre position of the target to be extracted. The automatic target setting method extracts the target according to the threshold set by the user and automatically measures the centre position of each target. Additionally, the system offers two tracking methods: semi-automatic and automatic. In the semi-automatic method, when automatic tracking is not possible, manual clicks are used to indicate target points on the frame, whereas the automatic method tracks throughout without any manual operation.

Figure 19.5 shows the automatic measurement interface, and its functions are introduced below.

(1) Motion Image File

The window shown in Fig. 19.5 indicates the path of the measurement file.

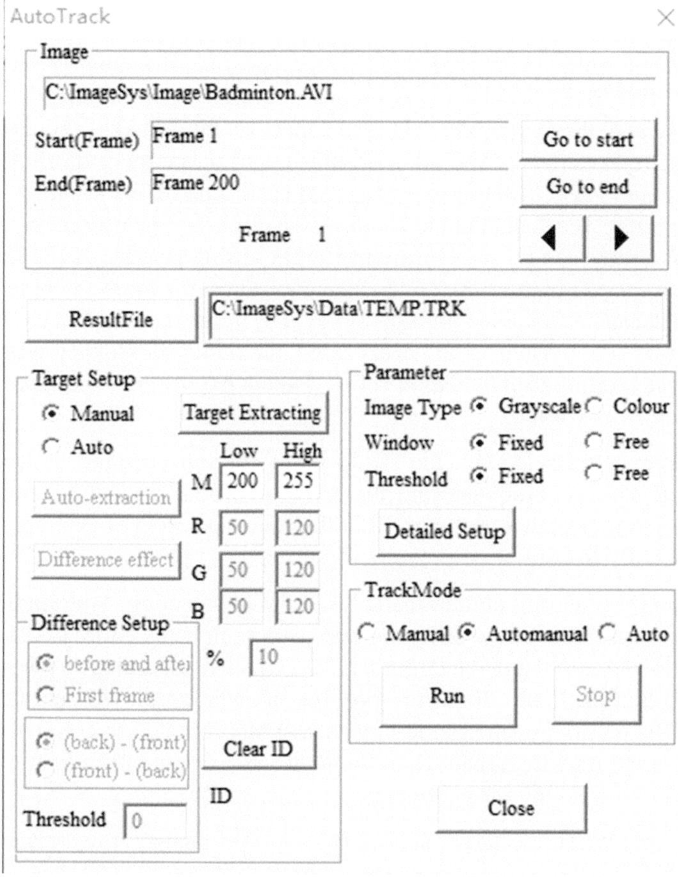

Fig. 19.5 Automatic measurement interface

① Start File (Frame). Represents the start file (continuous file) or start frame (video file) of the motion image to be measured.
② End File (Frame). Represents the end file (continuous file) or end frame (video file) of the motion image to be measured.
③ To Start File (Frame). Displays the start file (continuous file) or start frame (video file) of the motion image to be measured.
④ To End File (Frame). Displays the end file (continuous file) or end frame (video file) of the motion image to be measured.
⑤ Frame. Displays the frame represented by the current window. Clicking the flip button on the right can change the displayed image.

(2) Select Result File Set the save path and file name for the measurement result file.
(3) Target Setting

① Manual. Manually set the centre position of the measurement target point.

The method for setting a manual target: after selecting "Manual," hold down the "Shift" key, then hold down the left mouse button, drag the mouse to select the target range, and then click the centre position of the target to be extracted. If there are multiple targets, click multiple times, and the number of clicked targets is displayed after "ID." If the range size is set each time before clicking the target, and "Free Format" is selected in the window size, different measurement range sizes for different targets can be set. During the target setting process, if the target is set incorrectly, right-click anywhere on the image to cancel the most recent target range setting, which can be cancelled multiple times. The number of clicked targets will be displayed after "ID." When "Run" is executed, the target is extracted within the set target range according to the method set in "Detailed Settings," and target tracking is automatically performed.

② Automatic. After selecting, the "Auto Point" button becomes active. Execute the "Auto Point" command, and the target will be extracted according to the method set in "Detailed Settings," and the centre position of each target will be automatically measured.

The method for setting an automatic target: after selecting "Automatic," set the automatic measurement range using method ① (default is the entire image), execute the "Auto Point" button, and the target will be extracted according to the method set in "Detailed Settings," and the centre position of each target will be automatically measured. The number of measured targets will be prompted to confirm if correct. If correct, set the tracking area size of the target using method ①, then execute the "Run" button for tracking measurement.

③ Auto Point. When the target setting selects "Automatic," this button becomes active. Please refer to the explanation in ②.
④ Differential Effect. This button becomes active when "Differential" is selected in "Detailed Settings." Move the motion image forward by more than 1 frame, and after executing this button, the differential effect is displayed.
⑤ Differential Setting. Set the differential method and the binarization threshold after differentiation. After setting, "Differential Effect" can be executed. If the differential effect is poor, the parameter settings can be changed.
⑥ Target Extraction. Extract the set target. After execution, a window for image segmentation similar to Fig. 18.27 in Chap. 18 pops up. After executing "OK" in the target extraction window, the segmentation threshold is automatically displayed in each threshold window. For specific usage methods of the items in the image segmentation window, please refer to Chap. 18.

(4) Parameters

① Image Type. After reading the image, the system automatically determines whether the image is colour or grayscale and sets it automatically. If a colour image is read and a grayscale image is manually selected, the system will use the R component of the colour image for measurement.

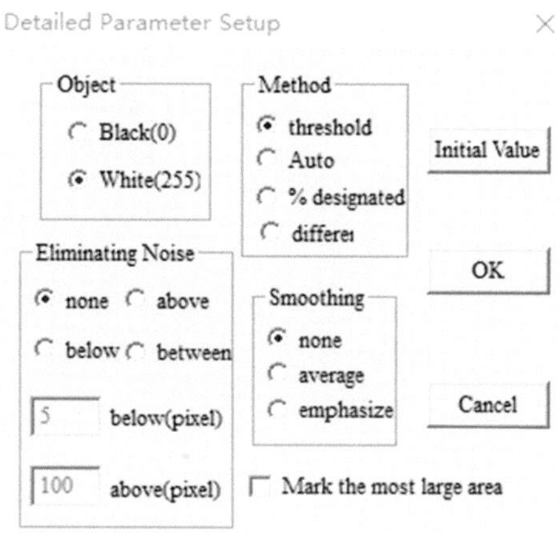

Fig. 19.6 Parameter details window

② Window Size. Select the format for the target extraction window size: fixed or free. Fixed: During tracking execution, the target window size is automatically unified to the size set for the last target; Free: During tracking execution, the target window size remains as the originally set size.
③ Threshold. Select the format for setting the threshold during target extraction: fixed or automatic. Generally, fixed is selected.
④ Detailed Settings. After execution, a parameter detail window as shown in Fig. 19.6 appears, where parameters such as "Binarization Method" are set. After setting, execute "OK" to close the window, and the settings become effective.

19.4.2 Manual Measurement

Manual measurement involves tracking the trajectory of a set target through manual operations. It is generally applicable when the moving target to be measured has a relatively complex "recognition environment" and is not easily distinguishable from the surrounding background. The target's trajectory is tracked frame by frame through manual operations.

During tracking, the target is manually clicked at the corresponding position in each frame. If an incorrect position is clicked during tracking, you can return to the previous tracking point, and multiple returns are possible. After returning, it is necessary to re-track the current point and all subsequent points. Figure 19.7 shows the manual measurement interface, with the items and functions within the window explained as follows.

① Motion file and select result file. Please refer to section "19.4.1 Automatic Measurement."
② Set the number of targets. Set the quantity of targets.

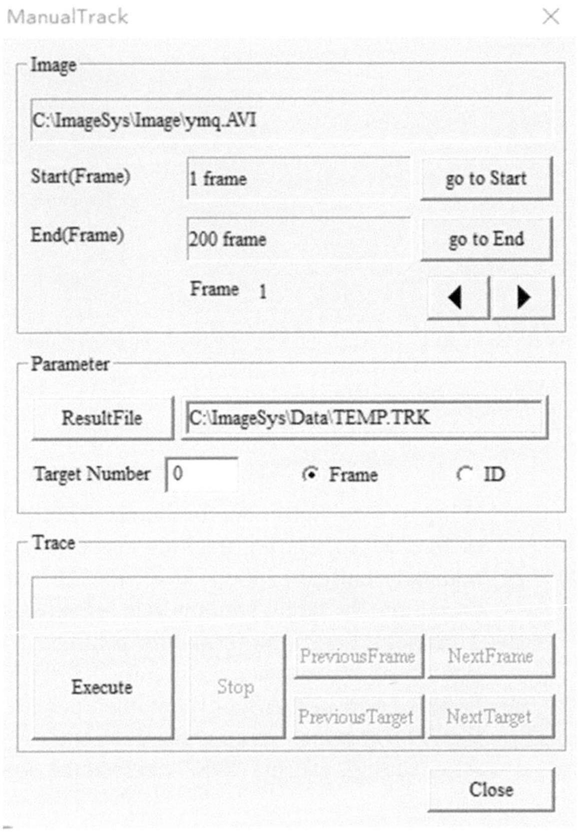

Fig. 19.7 Manual measurement interface

③ Frame unit tracking. In the process of manual measurement, each ID target in each frame is tracked one by one, and then the corresponding tracking of each target in the next frame is performed.
④ ID unit tracking. Track the entire trajectory of a single ID target across all frames, and then proceed to track the next target.
⑤ Execute. Run the above settings to execute tracking.
⑥ Status bar window. Displays the current frame and ID being represented.
⑦ Stop. Interrupt execution.
⑧ Previous frame. Return to the previous frame.
⑨ Next frame. Flip to the next frame.
⑩ Previous target. Flip to the previous.
⑪ Next target. Flip to the next target.
⑫ Close. Exit the manual measurement window.

19.4.3 Identification Measurement

Identification measurement involves tracking a set identification. Before tracking, coloured identification points need to be attached to the measurement object. There are two tracking methods: controllable tracking and fast tracking. Figure 19.8 shows the identification detection interface.

① Motion file and select result file. Most functions please refer to section "19.4.1 Automatic Measurement."

a. Play: Play continuous image files or video files.
b. Stop play: Stop playing continuous image files or video files.

② Tracking method: Divided into "controllable tracking" and "fast tracking."

a. Controllable tracking: Control the tracking speed through the player, and adjust the position of each point during the tracking process by clicking the mouse. When "controllable tracking" is selected, the options "distance correction," "selected correction," and "correct target number" are effective.

Fig. 19.8 Identifies measurement interface

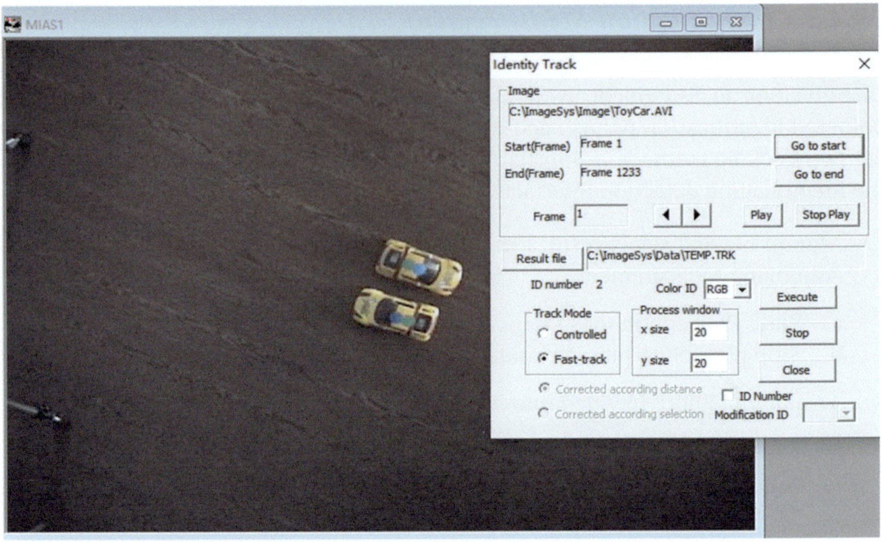

Fig. 19.9 RGB tracking measurement marked on the car

- Distance correction: After selecting the correction position, automatically move the target closest to the clicked position in this frame to the clicked position. Used for dispersed targets.
- Selected correction: In the correct target number column, select the target to be corrected, and after clicking the mouse, move the selected target to the clicked position. Used for concentrated targets.

b. Fast tracking: Fully automatic tracking in the fastest way.

③ Processing window size: Set the size of the tracking window.
④ Colour: Divided into four modes: RGB, R, G, B. Choose one reasonably based on the colour of the identification target and the background colour.

Figures 19.9, 19.10 and 19.11 are three examples of tracking measurement

19.5 Result Browsing

After completing the motion measurement of the target, the measurement results of more than ten items can be browsed in the form of charts, data, etc.

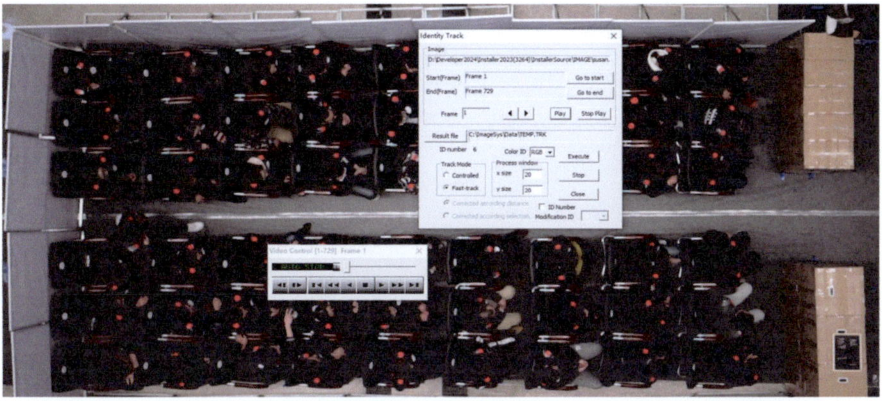

Fig. 19.10 Tracking measurement of red marks on human body

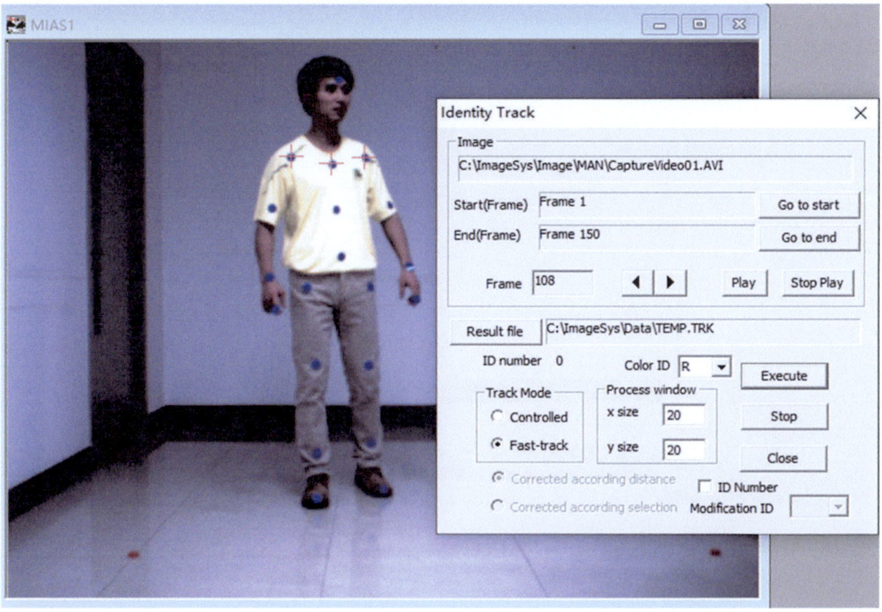

Fig. 19.11 R-tracking measurement of blue marks on human body

19.5.1 Result Video Representation

Result video representation mainly involves displaying the measurement results in charts, viewing data, copying, printing, etc., and the visual effects such as display colour and line type can be changed. Figure 19.12 is the result video representation interface, and its functions are introduced below.

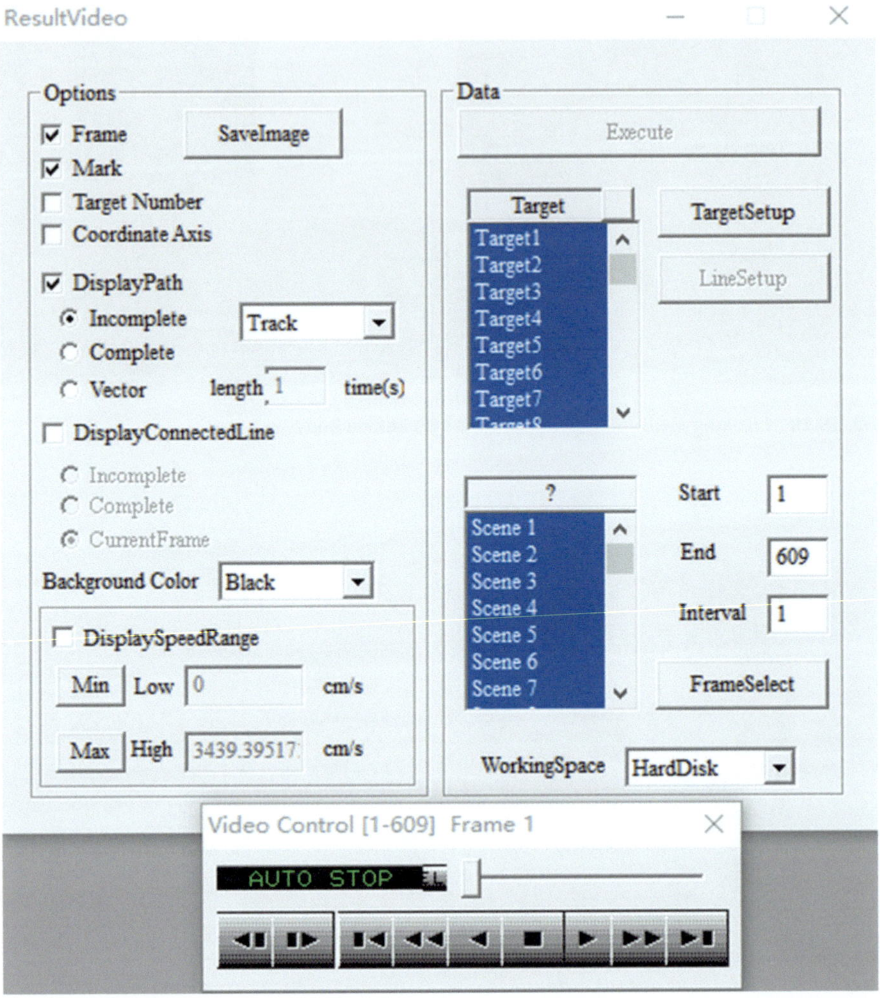

Fig. 19.12 Result video presentation interface

(1) Data setting

① Set target. Effective when "display trajectory" is selected, set the colour and line type of the target's motion trajectory. After execution, the window shown in Fig. 19.13 will pop up.

a. The top left of the window is the target list display box.
b. The first option box in the top right of the window indicates the current object target number.

Fig. 19.13 Setting goals

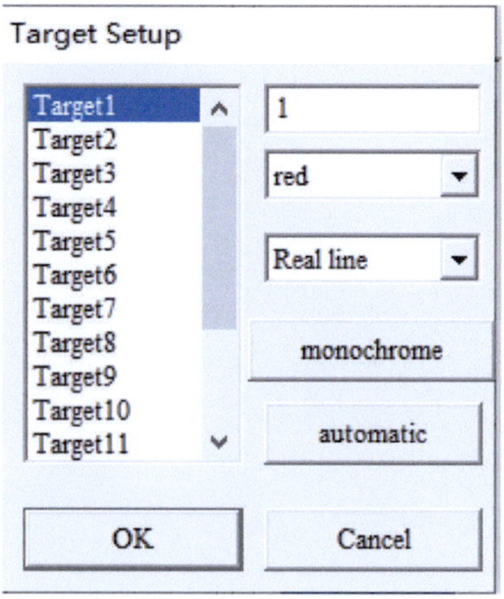

c. The second option box in the top right of the window indicates the currently selected colour. Colour options include: red, green, blue, purple, yellow, cyan, grey.
d. The third option box in the top right of the window indicates the current line type. Line type options include: solid line, dashed line, dotted line, dash-dot line, double dash line.
e. Monochrome Initialisation: Unify the colour and line type of all object target trajectories to the selected target's colour and line type.
f. Automatic Initialisation: Automatically set the colour of each target trajectory.
g. Confirm: Execute the set items.
h. Cancel: Do not execute the set items, exit the window.

② Set connections. Effective when "Connection Display" is selected, set, add, or delete connections between any two targets. After execution, the window in Fig. 19.14 pops up.

a. Measurement.

- Connection Line: The line connecting target to target, with the target connection list box below.
- Delete: Delete the target connection specified in the target connection list box.
- Delete All: Delete all target connections in the target connection list box.

b. Connection Line Setting.

The two option boxes at the top and middle are used to set the two object targets to be added. The lower option box sets the colour of the connection line. Connection line colour options include: red, green, blue, purple, yellow, cyan, grey.

- Add: Execute the settings of the above three option boxes, add the target connection.

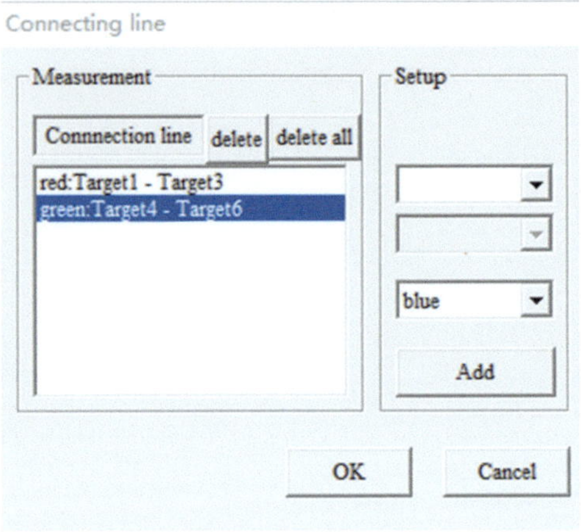

Fig. 19.14 Setting connection line

c. OK: Execute the settings of the connection line window.
d. Cancel: Exit the connection line window.

③ Target. Display the target list. The content in the box of Fig. 19.12 is the display list of 2 targets, the current operation objects are target 1 and target 2.
④ Start Frame. Set the starting frame to be displayed. The starting frame shown in Fig. 19.12 is frame 1.
⑤ End Frame. Set the ending frame to be displayed. The ending frame shown in Fig. 19.12 is frame 19.
⑥ Frame Interval. Set the interval frame number between frames to be displayed. The frame interval shown in Fig. 19.12 is 1.
⑦ Frame Selection. Execute the frame settings of items ④, ⑤, ⑥ above.
⑧ Frame. Display the frame list. The content in the box of Fig. 19.12 is the frame list after executing "Frame Selection".
⑨ Working Area. Options: Hard disk or memory.
⑩ Execute Settings. Run the project settings within the "Data Settings" range.

(2) Display Options

① Frame. Represents the continuous image frames read into the current window. Click the radio button to set whether to display "Frame".
② Mark. Represents the target's mark. Click the radio button to set whether to display "Mark".
③ Target Number. Represents the sequential number of the target. Click the radio button to set whether to display "Target Number".
④ Coordinate Axis. Click the radio button to set whether to display "Coordinate Axis".
⑤ Display Trajectory.

  a. Afterimage. Display the motion trajectory before the current frame. Options: Trajectory, Trajectory plus vector, Continuous vector.

b. All. Display all motion trajectories of the target.
c. Vector. Represents the direction of the target's motion trajectory. The small box on the right is used to set the length multiplier of the vector. The setting shown in Fig. 19.12 is for vector display at 1 × length.

⑥ Connection Display.

a. Afterimage. Display the connections on the frames that have been moved.
b. All. Display the connections from the specified start frame to the end frame.
c. Current. Display the connections on the current frame.

⑦ Background Colour. The background colour of the current display window, black or white. Note: When "Frame" is selected to be displayed, the background colour selection is invalid.
⑧ Speed Interval Highlight. Select the speed interval of interest, and the target's trajectory in this interval will be displayed with a thick solid line. After selecting "Speed Interval Highlight", the minimum and maximum settings are effective. "Minimum": Set the target's minimum speed, "Maximum": Set the target's maximum speed. The default low value for "Minimum" is the lowest speed of all targets, and the default high value for "Maximum" is the highest speed of all targets.
⑨ Save Image. Save the display image (continuous) within the current image window, which can be saved as continuous bmp type files and avi video type files.

When saving as bmp image type, set the file name to execute the save, and the system will automatically store the continuous motion images from the first frame to the last frame in sequence with increasing numbers.

When saving as an AVI video type, set the file name to execute the save. The system will prompt you to select a compression program, which can be chosen based on actual needs. If the quality of the saved result is highly important, it is best to choose the "(Full Frame) Uncompressed" option; conversely, if the image quality requirement is lower (resulting in relatively smaller storage space usage), other compression methods and their compression rates can be selected. After clicking "OK," the system will store the continuous motion images from the first frame to the last frame as a video file.

During the storage process, if you need to interrupt the storage task, you can click "Stop" on the processing progress interface. The saved BMP or AVI result file will have image content consistent with the current "Display Options" and "Data Settings."

Figure 19.15 lists several effects of the aforementioned display methods. Among them, Fig. 19.15a shows all motion trajectories in a continuous vector display mode; Fig. 19.15b displays all markers, target numbers, coordinate axes, all trajectories, and all connections, with the window background set to white; Fig. 19.15c highlights the speed range of interest, with the left image showing an arbitrarily selected speed range of interest, and the right image showing the motion trajectory of the target in that range with a thick solid line.

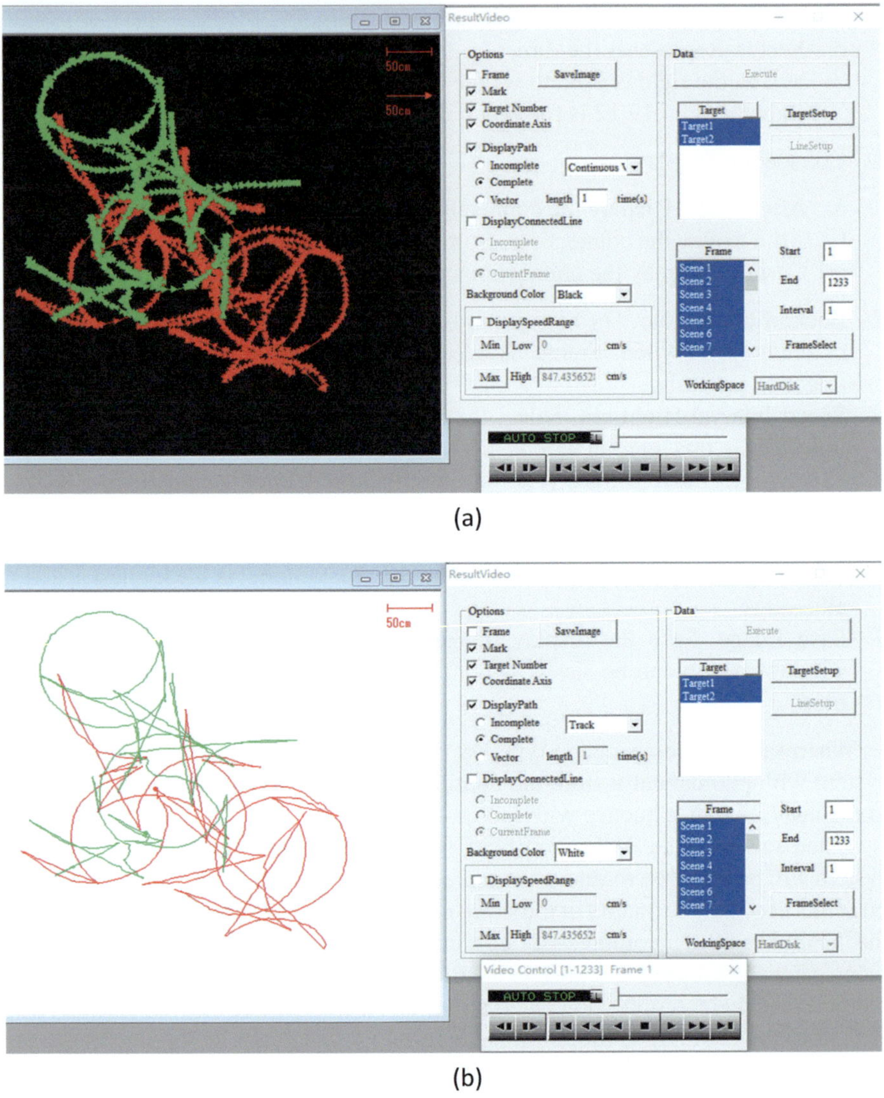

Fig. 19.15 Several display methods of track tracking results

19.5.2 Position Rate

Position rate refers to the position and rate of the target trajectory in different frames. In this section, you can view, copy, and print charts and data of various parameters, as well as change the display colour, line type, and other visual effects. The displayed parameters specifically include the target's coordinates X, coordinates Y,

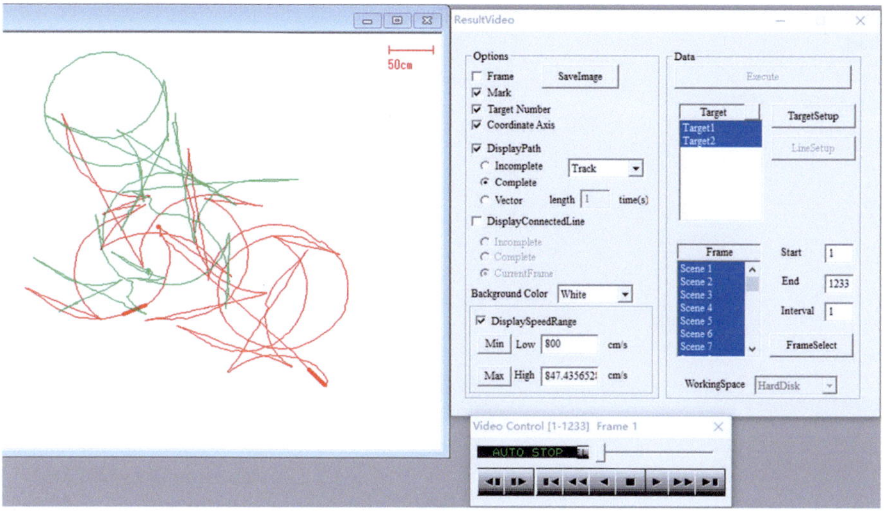

(c)

Fig. 19.15 (continued)

movement distance, speed, and acceleration, with two viewing methods: charts and data. Figure 19.16 is the operation interface for position speed.

① Set the target. Set the display colour and line type of the target marker and its motion trajectory line. Clicking will bring up the settings window shown in Fig. 19.13.

Fig. 19.16 Position speed interface

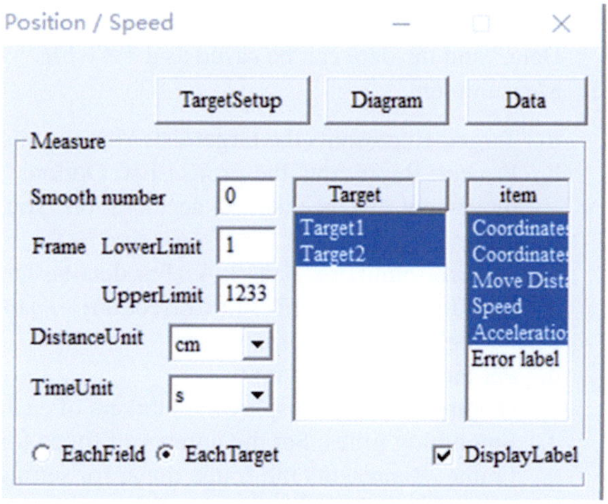

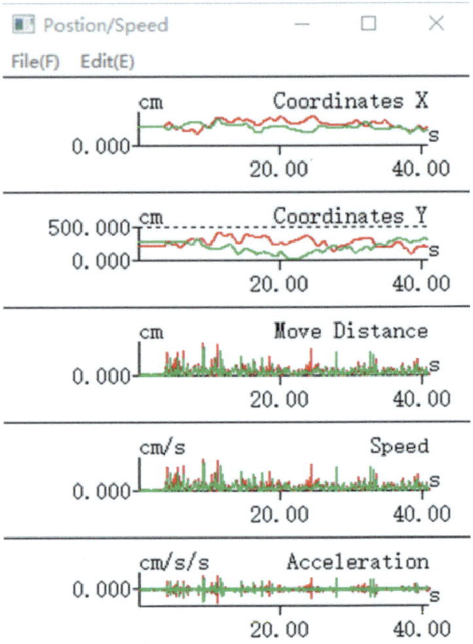

Fig. 19.17 Viewing the chart

② View charts. View the graphical representation of the target and project within the measurement parameter setting range. Figure 19.17 is the result window after executing "View Chart" for an already opened 2D result file. The red, green, and blue curves in the figure represent the corresponding values of three targets, and this chart can be saved and copied.

③ View data. View the numerical values of the target and project within the measurement parameter setting range. Figure 19.18 is the interface for executing "View Data," and the data can be saved as a TXT file.

④ Measurement

a. Target: Represents the target list. You can click to select the target object.
b. Project: Represents the project list. Options: coordinates *X*, coordinates *Y*, movement distance, speed, acceleration. You can click to select the project object.

   Error numbers: 1, 2, 3, 4. (For details, see the error prompt information introduction in the [Result Correction] → [Interpolation] interface).
c. Per field: By field unit.
d. Per target: By target unit.
e. Display markers: Display the markers of each target.
f. Smoothing times: Set the number of times for smoothing correction.
g. Frame: Represents the frame range for setting or viewing the object. Upper limit: Represents the starting frame. Lower limit: Represents the ending frame.

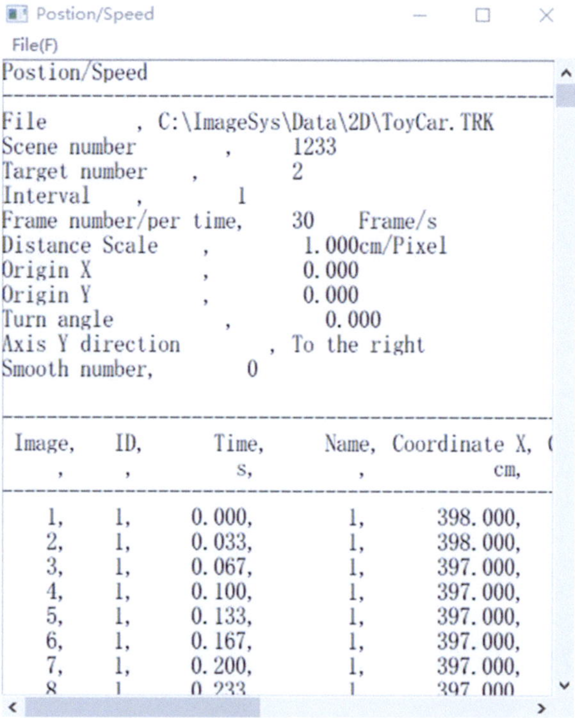

Fig. 19.18 Viewing data

h. Distance unit: Choose the unit of distance: pm, nm, um, cm, m, km.
i. Time unit: Choose the unit of time: ps, ns, us, ms, s, m, h.

19.5.3 Offset

Offset reflects the position change of the target trajectory in different frames. In this section, you can view the *X* direction offset, *Y* direction offset, and absolute value offset of a specified target relative to a set reference, with two viewing methods: charts and data. Figure 19.19 is its operation interface, where the functions of setting targets, viewing charts, viewing data, and measurement are the same as in Fig. 19.16. The reference position function is explained as follows.

① Smoothing times. Set the number of times for executing smoothing correction.
② Reference frame. After selection, calculate the offset of each target based on the set frame.
③ Reference position. After selection, calculate the offset of each target based on the set position.
④ Reference target. After selection, calculate the offset of each target based on the set target.

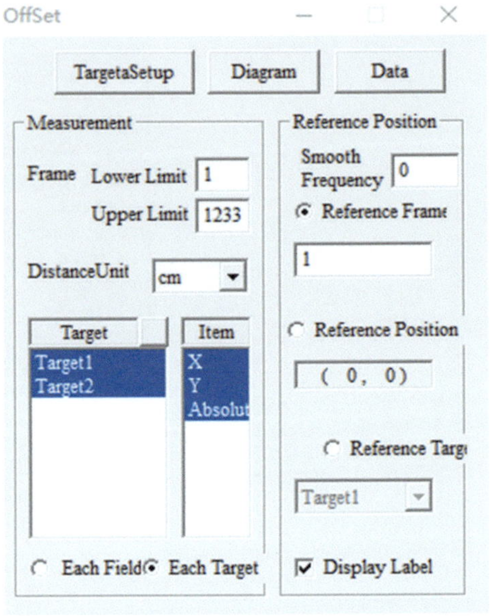

Fig. 19.19 Offset interface

19.5.4 Distance Between Two Points

The distance between two points refers to the straight-line interval between targets. Users can add multiple target lines on the operation interface, setting them to different colours and line types for distinction. Figure 19.20 is its operation interface, and the functions on the interface are basically the same as those in the previous operation interfaces, so they will not be explained in detail.

19.5.5 Angle Between Two Lines

This refers to the angle between lines formed by two or more targets, including angles between three points, angles between two lines, angles with the *X* axis, and angles with the *Y* axis, comprising four types. Figure 19.21 shows its operation interface, where most functions are similar to previous items, and only the sections that differ from previous ones are explained here.

① Angle between three points. Represents the clockwise or counterclockwise angle between three points.
② Angle between two lines. Represents the angle between two lines formed by three or four points.
③ *X* axis angle. Represents the angle between a line formed by two points and the *X* axis.

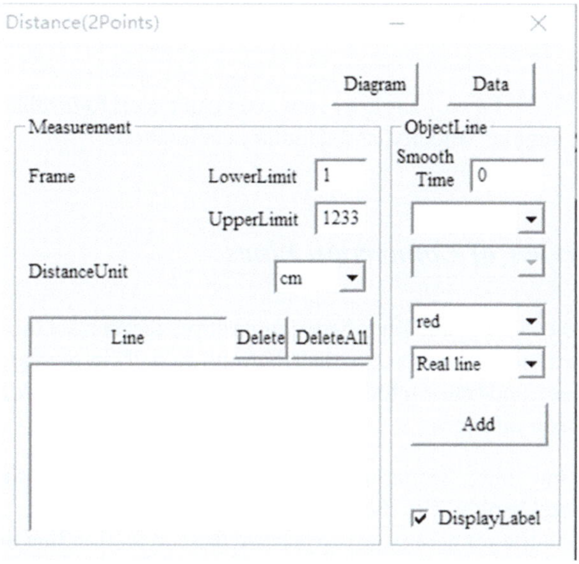

Fig. 19.20 Distance interface between 2 points

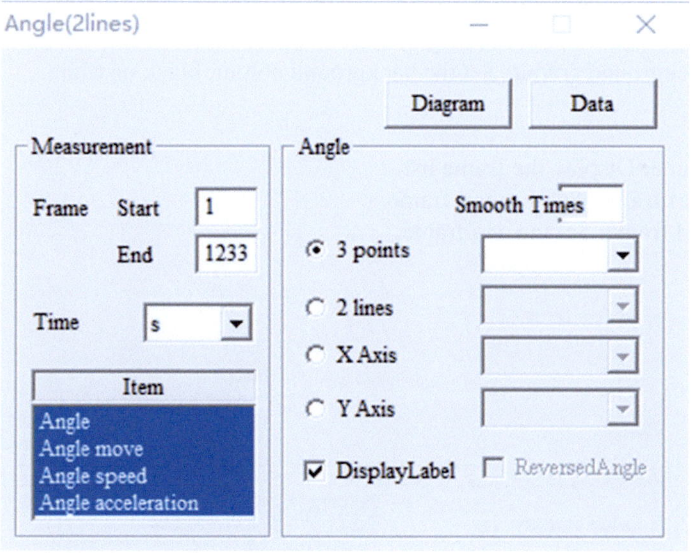

Fig. 19.21 Angle interface between 2 lines

④ *Y* axis angle. Represents the angle between a line formed by two points and the *Y* axis.

After selecting the type of angle to view, you can check four related items: angle, angle variation, angular velocity, and angular acceleration.

19.5.6 Overview of Connection Lines

This section allows you to add lines between multiple targets; set the colour of target lines; and set parameters such as the distribution interval (in pixels), magnification, background colour, and frame interval for lines in the *X* and *Y* directions. Figure 19.22 shows its operation interface.

① Set connection lines. Set target lines. You can refer to [Result Browsing] → [Result Video Display] for "Set Lines."
② View. Execute the set parameters to browse the connection line display diagram.
③ Options.

- *X* movement: Set the *X* movement.
- *Y* movement: Set the *Y* movement.
- Magnification: Set the magnification.
- Background colour: Set the background colour, black or white.

④ Frame.

- Frame: Display the frame list.
- Start frame: Set the start frame.
- End frame: Set the end frame.

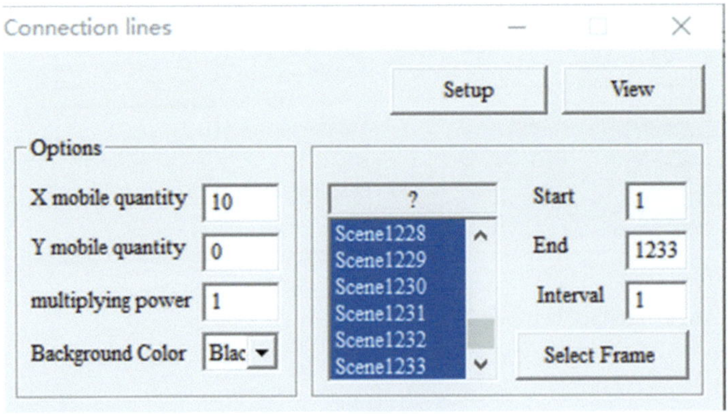

Fig. 19.22 Connecting line's general view

- Frame interval: Set the frame interval.
- Frame selection: Execute the above frame settings.

19.6 Result Correction

This system provides various ways to correct measurement results, specifically including: manual correction, modifying and correcting specified target trajectories; smoothing, removing angular noise from target motion trajectories to make them more curve-like; interpolation, spline curve interpolation to eliminate jagged appearances in images (trajectories); frame coordinate transformation, changing the reference coordinates of frames; human centre of gravity, measuring the location of the human centre of gravity; setting items, allowing the setting of reference frames, and adding or deleting target frames.

19.6.1 Manual Correction

Click the "Manual Correction" menu to pop up the manual correction interface shown in Fig. 19.23.

① Magnification. Select magnification: 2×, 4×, 8×, 16×.
② Move target. Move the target object to the centre position within the video window.
③ Target setting box. Select the target object.
④ Correct. Execute the above settings.
⑤ Cancel. Cancel the above settings and close the window.

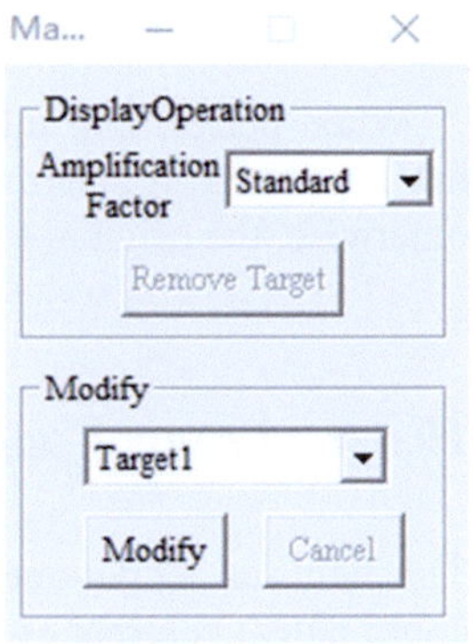

Fig. 19.23 Manual correction interface

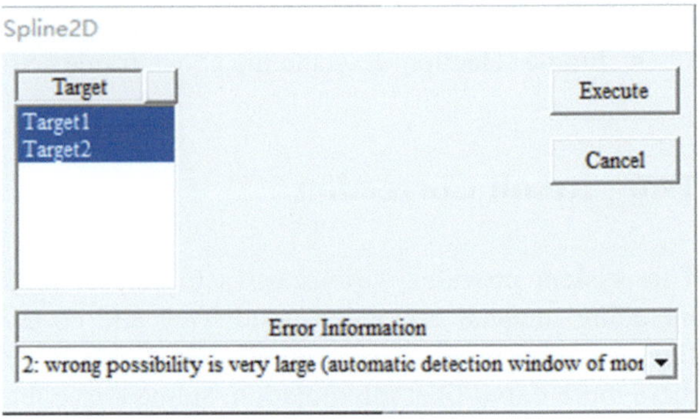

Fig. 19.24 Interpolation interface

19.6.2 Smoothing

Each time the "Smoothing" menu is clicked, a three-step trajectory data smoothing is performed for each target. Smoothing can be executed multiple times as needed.

19.6.3 Interpolation

After execution, the window shown in Fig. 19.24 will pop up. Interpolation can correct the following four errors:

① There may be an error (more than two objects appeared in the automatic detection window).
② The likelihood of error is high (noise in the automatic detection window exceeds 60).
③ The likelihood of error is high (no objects in the automatic detection window).
④ Error (not specified during manual or semi-automatic tracking).

When performing interpolation correction, if the current trace to be corrected has any of the four errors, it is effective when correcting the selected error type; otherwise, the original trace and related data remain unchanged.

19.6.4 Frame Coordinate Transformation

This item can set parameters such as the standard frame (frame number to transform coordinates), reference position, and reference axis to achieve frame coordinate transformation. Figure 19.25 is the coordinate transformation interface. In the figure, the project parameters are set as follows: standard frame: 5, reference position: target 1, reference axis: the line connecting target 2 and target 3.

Fig. 19.25 Coordinate transformation interface

19.6.5 Human Body Centre of Gravity Measurement

The human body centre of gravity measurement project can simultaneously measure the centre of gravity of multiple parts of the human body, such as the whole body, upper limbs, right upper arm, left lower leg, etc. Figure 19.26 is the human body centre of gravity measurement interface. The centre of gravity trajectory and motion trajectory of the selected part are displayed together.

Fig. 19.26 Human body center of gravity measurement interface

19.6.6 Setting Items

The setting items project can set the reference frame, add target frames, and delete target frames. Operate "Video Control" to change the current display frame, set the current display frame as the reference frame as needed, or add the current display frame as a target frame. The reference frame set here will serve as the reference frame for the entire measurement and calculation, displayed in the distribution of various data and charts; the set target frame will have a "+" identifier added in front of it on the data distribution screen.

19.7 View

Includes: pixel values, image zoom, and status bar.

① Pixel values. Displays the pixel values within a 7×7 range centred on the mouse position. In colour display mode, it shows RGB values, and in grayscale mode, it shows brightness values.
② Image zoom. Represents the enlargement and reduction of the screen. Six magnification ratios from 50 to 500% are represented, namely: 1/2 times, 1 time, 2 times, 3 times, 4 times, 5 times.
③ Status bar. Controls the switch of the status window.

19.8 Real-Time Measurement

Based on the MIAS system, a real-time tracking and measurement system RTTS for moving targets was developed. Compared with MIAS, this system mainly adds two functions: real-time target measurement and real-time identification measurement.

19.8.1 Real-Time Target Measurement

The operation interface displays the effective camera device connected to the computer for user selection, and users can set the functions of the camera device. After video image input, dynamic images can be previewed in the window, or the preview can be stopped, leaving the last frame image in the window for tracking settings. Before executing tracking, the background and tracking target type need to be set.

① Background setting. When the image is not dynamically displayed, background information is obtained by drawing a line on the background.

② Target type setting. When the image is not dynamically displayed, a "+" sign is drawn on the target to obtain a type of target information.

After setting the background information and target type information, start executing target tracking, and multiple targets can be selected for non-identification tracking simultaneously.

19.8.2 Real-Time Identification Measurement

The difference from real-time target measurement is that in real-time identification measurement, coloured identification points need to be attached to the tracked target before measurement, and then the identification points are tracked. Other functions and tracking processes are similar to real-time target measurement and will not be repeated here. Additionally, for real-time identification measurement, users can set whether to display the target number. If you want to increase or decrease the number of tracking targets, you can set the target and use the left and right keys to add or delete targets; if you temporarily do not want to increase or decrease the number of targets, you can lock the target, meaning any operation of the mouse in the view window will not affect the number of targets. Figure 19.27 shows the result of real-time tracking measurement of the colour identification points on the trolley.

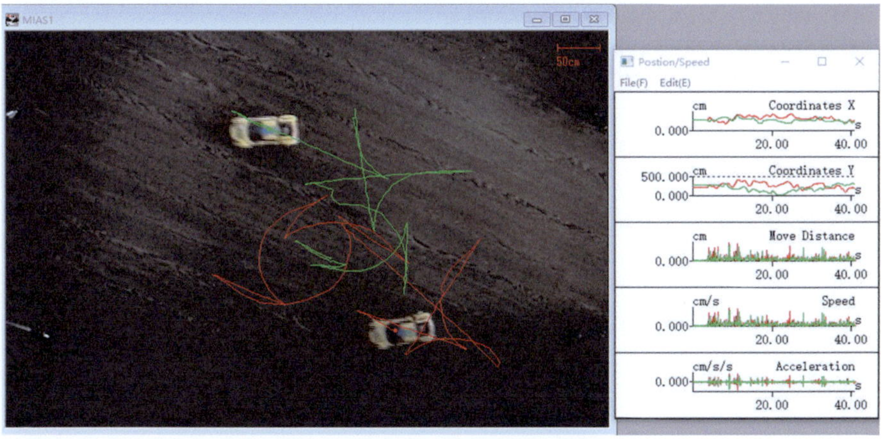

Fig. 19.27 Real-time tracking measurement results of the car

19.9 Development Platform MSSample

The MIAS system provides a development platform, MSSample, which serves as a framework source program. This framework platform includes various file operation functions, such as saving the current image, and offers a demonstration of differential processing of an AVI video file, allowing users to gain a more intuitive understanding of this development platform. Users can freely add their own image processing interfaces and functions on this platform to achieve more functionalities. Additionally, MSSample is by default connected to the large image processing function library equipped with the system, allowing users to directly call functions from the library during development. The function library provided by this platform is the same as the one provided by the general system development platform Sample in Chap. 18, which encapsulates over 400 practical functions for image processing, image display, and image access, offering users a wide range of options for development.

The initial settings of this system, system language settings, and image acquisition functions are basically the same as those of the general image processing system ImageSys, and will not be reiterated here.

Reference

1. Chen Bingqi. Detailed Explanation of Machine Vision Technology and Application Examples [M]. Beijing: Chemical Industry Press, 2014.

Chapter 20
Three-Dimensional Motion Measurement Analysis System MIAS 3D

20.1 Introduction to the MIAS 3D System

The MIAS 3D system is a comprehensive software system integrating functions such as multi-channel synchronous image acquisition, two-dimensional motion image measurement, three-dimensional data reconstruction, data management, and three-dimensional trajectory linkage representation [1].

Primary application areas include human motion analysis, measurement of human center of gravity, animal and insect behavior analysis, rigid body posture analysis, vibration and impact analysis of floating objects, robot vision feedback, scientific research, and education.

Main functional features: Simplified Chinese and English interfaces, easy to operate; featuring multi-channel synchronous image acquisition and single-channel image acquisition switching; a complete set of two-dimensional motion image measurement functions; three-dimensional scale setting capability; capability to synthesize three-dimensional data from two-dimensional measurements; multi-visual dynamic representation of three-dimensional motion trajectory and trajectory-image linkage representation function; free representation of 3D motion trajectory based on OpenGL; emphasis on representing specified speed interval trajectory function; various measurement results chart and document representation function; three-dimensional and two-dimensional measurement representation function of the centre of gravity trajectory of various parts of the human body; merging and connecting functions of multiple three-dimensional measurement result data. The measured two-dimensional and three-dimensional parameters encompass position, distance, speed, acceleration, angle, angular velocity, angular acceleration, angular displacement, relative coordinate positions, etc.

The MIAS 3D system's graphic window has an initial default setting of 640 × 480 pixels, which can be modified through the system's initial settings. When opening a 3D result file or a 2D tracking file, if the image size of the file to be read differs from the current system's image size, a system setting window will appear

© Chemical Industry Press 2026

B. Chen and S. Chen, *Machine Vision Technology*, Advanced and Intelligent Manufacturing in China, https://doi.org/10.1007/978-981-95-1184-6_20

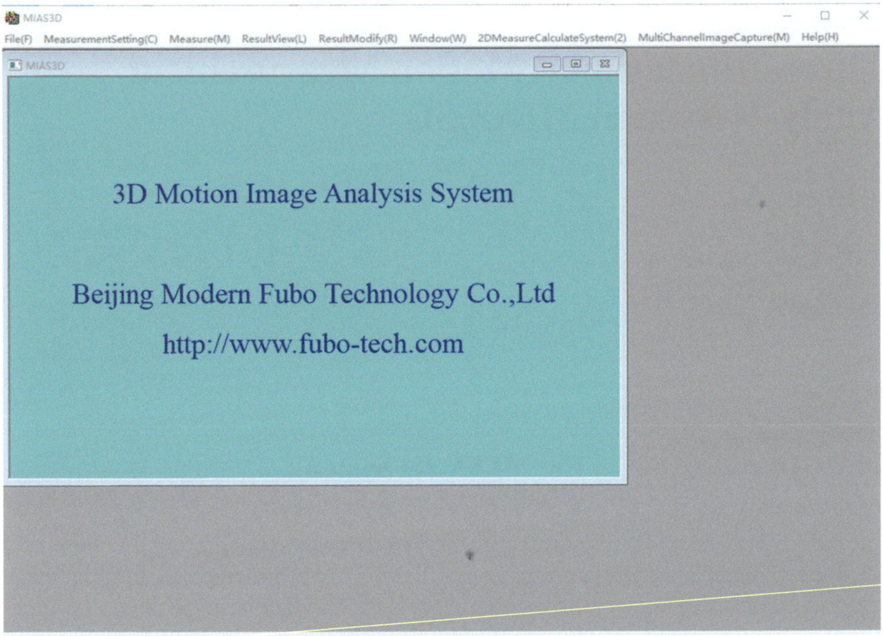

Fig. 20.1 Initial interface of MIAS3D system

prompting the user to input the correct image size. After selecting OK, the system will automatically close and restart according to the set image size and system frame number. The initial interface of the system is shown in Fig. 20.1. This system includes the two-dimensional motion measurement analysis system MIAS (refer to Chap. 19) and a set of independent multi-channel synchronous image acquisition. Here, only the interface functions of MIAS 3D are introduced.

20.2 Files

The MIAS 3D system has rich file processing functions, allowing further processing of saved result files and tracking files. Specific functions include: reading previously saved 3D measurement result files; changing the tracking file of a specified camera, changing the connection path of 3D measurement data and tracking files; merging multiple 3D measurement files; exporting 3DS motion data, so that the saved files can be read by software such as 3DMax, AutoCAD; saving the currently displayed image in bitmap file format; previewing the effect of the image before printing, setting the printer and printing the currently displayed image; displaying the most recent historical work files, etc.

20.3 2D Result Import, 3D Calibration and Measurement

The MIAS 3D system synthesises 3D measurement results from the measurement result (tracking) files of 2 or more 2D synchronous images and a 3D calibration file. Before 3D measurement, it is necessary to read in 2 or more 2D measurement result files for 3D calibration or read in a saved 3D calibration file. The specific operations are as follows.

(1) Open 2D tracking files

To synthesise 3D data, it is necessary to read in two or more 2D synchronous measurement result files.

(2) 3D Calibration

When synthesising 3D data, it is necessary to import a 3D calibration file. The 3D calibration function can generate a 3D calibration file. Read in the start and end files of the calibration images for each camera, set the storage path and file name of the calibration result file, select the scale unit, and then perform 3D calibration in a semi-automatic or manual manner. After calibration is completed, the system will prompt the calibration error, and for points with large calibration errors, recalibration can be performed.

Figure 20.2 is the 3D calibration interface, and the following explains its functions.

① Calibration images. Select the initial and final calibration image files.

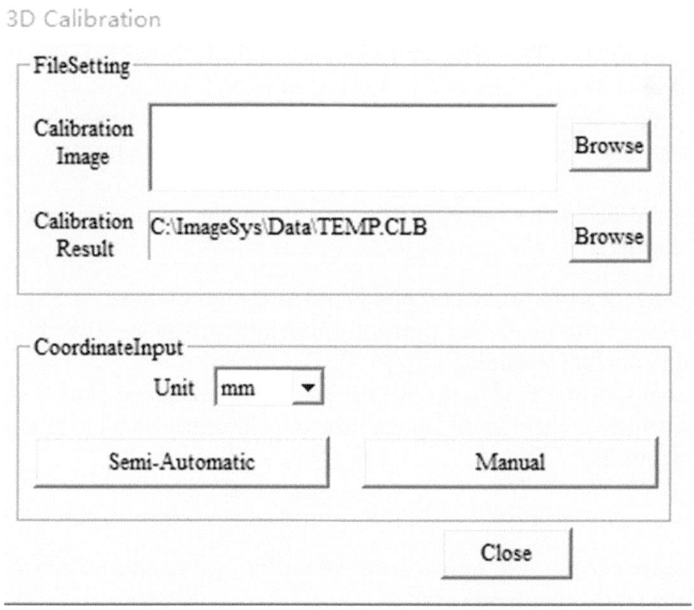

Fig. 20.2 3D calibration interface

② Result options. Set the path and file name for the calibration result file, file type .CLB.
③ Units. Select the scale unit: pm, nm, um, cm, m, km.
④ Manual. Manually determine the image coordinates of the calibration points and input the spatial coordinates of each point.
⑤ Semi-automatic. Supplemented by manual operation during execution, using image segmentation methods to determine the position of calibration points.
⑥ Close. Exit the 3D calibration window.

(3) 3D Chessboard Calibration

Chessboard calibration is generally used for calibrating small fields of view, such as desktops indoors. It is convenient to operate and has high calibration accuracy.

For photographing the chessboard, the following points should be noted.

① The two cameras should remain parallel;
② The placement position of the chessboard in the calibration space should be evenly distributed;
③ The angle between the chessboard plane and the camera lens plane should be kept within 45°, as too large an angle will affect accuracy;
④ If using a printed paper chessboard, it should be affixed to a hard object to ensure the flatness of the chessboard.

Figure 20.3 is the 3D chessboard calibration interface, and its functions are introduced below.

① Checkerboard file—Execute "Generate Save" to generate the checkerboard image and save it. The generated checkerboard pixel size is 100*100, and the actual size is 25 mm*25 mm (printed on A4 paper).
② Calibration images. Select the initial and final calibration image files.
③ Result options. Set the path and file name for the calibration result file, file type .CLB.
④ Chessboard parameter settings. After execution, the parameter setting interface shown in Fig. 20.4 will pop up.

   a. Number of corner points in chessboard rows and columns: Chessboard corner points refer to the corner points formed by four squares (two black squares and two white squares).
   b. Actual size of chessboard squares: The size of each chessboard square.
   c. Scale unit of chessboard square size: Available scales include pm, nm, um, mm, cm, m.

⑤ Origin:

   a. Left eye center—the optical center of the left eye camera is used as the origin of the world coordinate system.
   b. Chessboard corners—Take the first top-left corner of the first chessboard image as the origin of the world coordinate system.

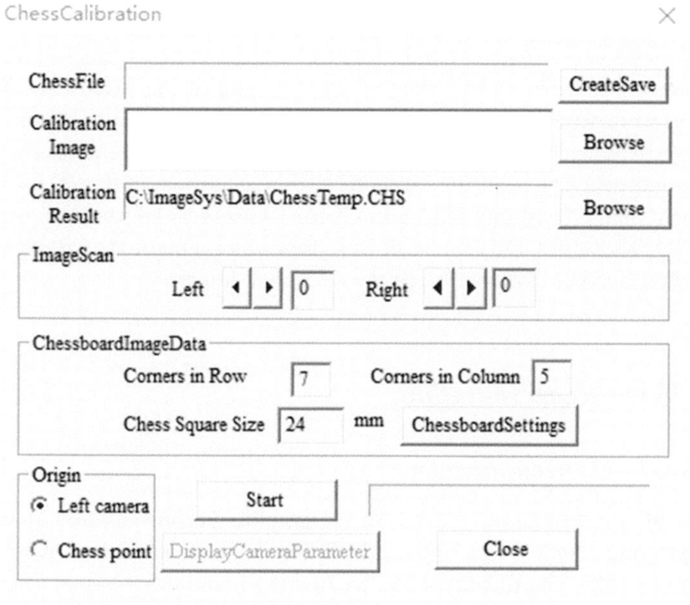

Fig. 20.3 3D checkerboard calibration interface

Fig. 20.4 Board parameter setting interface

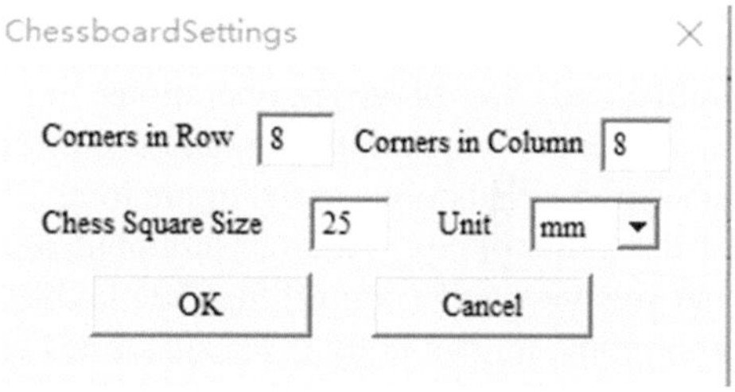

⑥ Start calibration. The system begins camera calibration.
⑦ Display parameters. After calibration, click "Display Parameters" to view the internal and external parameters of the camera.
⑧ Close. End calibration and close the dialog box.

Execute "Motion Measurement" to synthesize 3D data from the imported 2D trajectory file and 3D calibration file, generating a 3D trajectory result file.

20.4 Display Results

After motion measurement, the measurement results of the MIAS 3D system can be represented in various ways, such as video, point rate, offset, distance between points, angle between lines, and overview of connection line diagrams. The operation interfaces for displaying results are roughly the same as those of the 2D motion image measurement analysis system MIAS in Chap. 19, except that the data has changed from 2 to 3D. Therefore, only the various representation methods will be introduced below, without further explanation of the operation interface.

20.4.1 Video Representation

(1) Multi-view 3D Representation

Multi-view 3D representation allows for the graphical representation, data viewing, copying, printing, etc., of the imported 3D result file from above, front, rotating, side, and any angle. The display's colour, line type, and other visual effects can be changed. The connections of the trajectory and target points can be displayed in afterimage, vector, and other ways. For the trajectory, an area of interest in terms of speed can be selected, and after selection, the trajectory of the target in this area will be represented by a thick solid line. Additionally, during the representation process, the playback operation panel can be controlled to achieve fast forward, rewind, single frame, and other playback operations.

An example of multi-view 3D representation results is shown in Fig. 20.5. The measured target points in the figure total 20, distributed sequentially at various joints of the human body. The measurement results are displayed in top, front, rotating, and side views. By controlling the playback operation panel, the movement of each joint of the human body at various times can be observed.

(2) OpenGL 3D Representation

OpenGL 3D representation allows for the opening of the imported 3D result file in OpenGL, and 3DS files can be exported, which can then be read by software such as 3DMax, AutoCAD, and ProE. When in use, the display's colour, line type, target point sphere size, and other visual effects can be set. For the trajectory, an area of interest in terms of speed can be selected, and after selection, the trajectory of the target in this area will be represented by a thick solid line. Additionally, during the representation process, the playback operation panel can be controlled to achieve fast forward, rewind, single frame, and other playback operations.

An example of OpenGL 3D representation results is shown in Fig. 20.6, where the target point sphere size is 3, and the background is black. For the targets and trajectories displayed within the OpenGL window, the mouse can be used for zooming in, zooming out, arbitrary rotation, and other flexible operations, thereby achieving comprehensive observation of the target points and their motion trajectories.

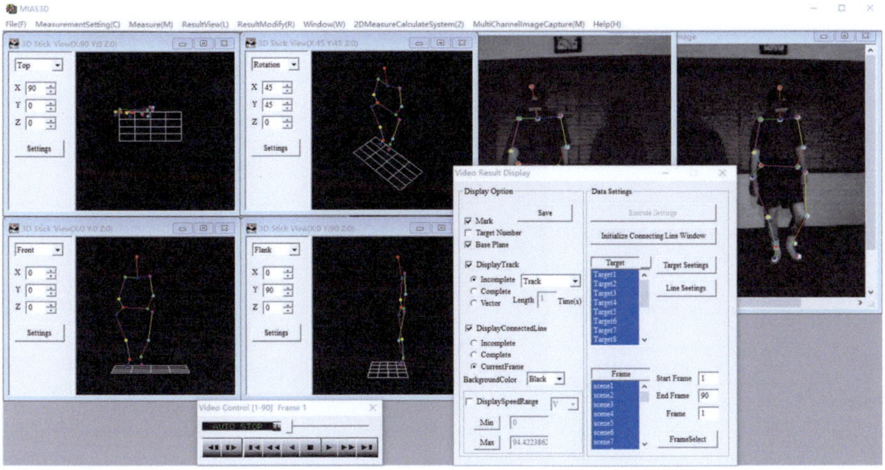

Fig. 20.5 Example of multi-directional 3D representation results

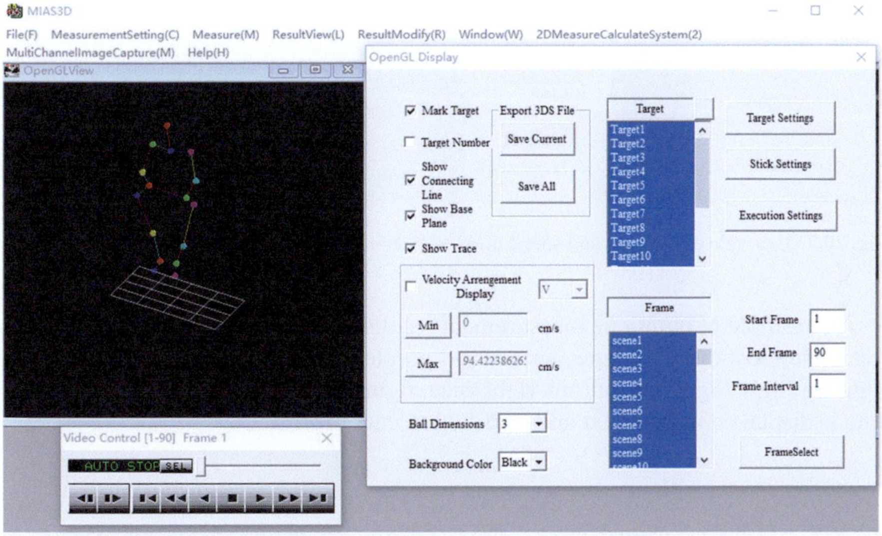

Fig. 20.6 OpenGL3D shows an example of the result

20.4.2 Point Rate

The point rate function can obtain parameters such as the position coordinates, movement distance, speed, and acceleration of a target at any given time. The resulting data can be displayed, saved, and printed in text form, and can also be visually displayed, copied, and printed in the form of distribution curves.

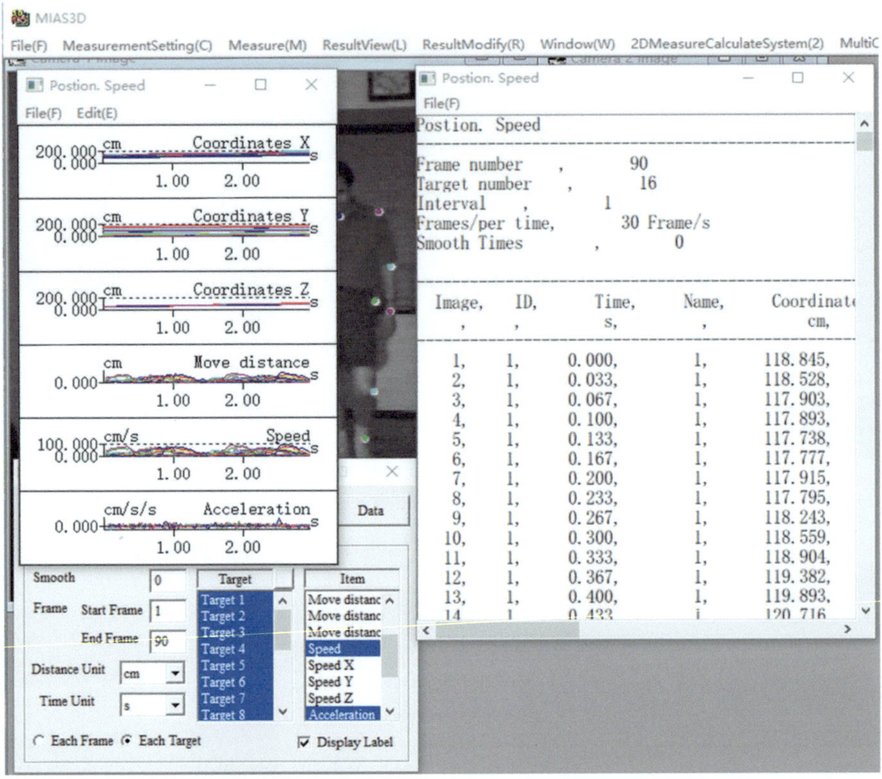

Fig. 20.7 Example of position and speed measurement results

An example of point rate measurement results is shown in Fig. 20.7, which illustrates the movement distance, speed, and acceleration of four target points on the right leg (right big toe, right foot, right knee, right hip joint). The measurement result data is displayed in both text and distribution curve forms.

20.4.3 Displacement

The displacement function can obtain the displacement of a target point relative to a reference frame, reference point, or reference target at any given time. The measurement result data can be displayed, saved, copied, and printed in text or distribution curve form.

An example of displacement measurement results is shown in Fig. 20.8. The figure measures the displacement in the *X*, *Y*, *Z* axes and the absolute value of four target points on the right leg (right big toe, right foot, right knee, right hip joint) relative to

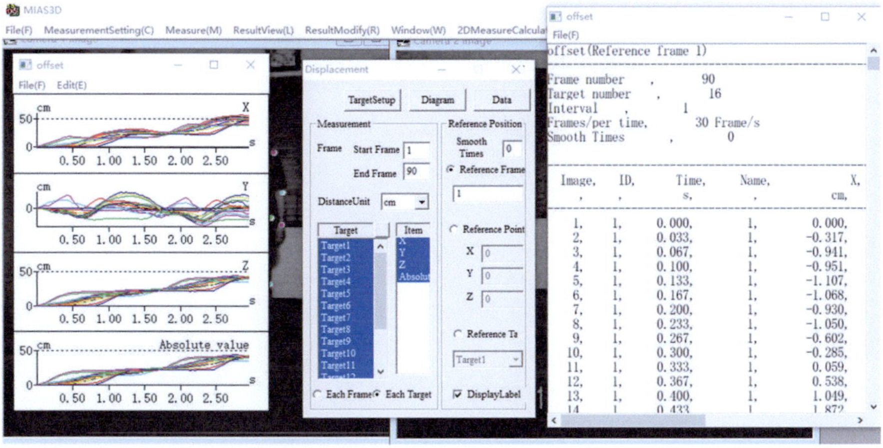

Fig. 20.8 Example of displacement measurement results

the first frame of the reference frame. The measurement result data is displayed in both text and distribution curve forms.

20.4.4 Distance Between Two Points

The distance between two points function can obtain the distance between specified targets. The measurement result data can be displayed, saved, copied, and printed in text or distribution curve form.

An example of the distance between two points measurement results is shown in Fig. 20.9, where the measurement target is the distance between the left and right big toes. The measurement result data is displayed in both text and distribution curve forms.

20.4.5 Angle Between Two Lines

The angle between two lines function can obtain the angle between targets, including types such as the angle between three points, the angle between two lines, and the angle with the *X*, *Y*, and *Z* axes. Additionally, when measuring angles, different angle calculation references can be selected, such as actual spatial angles, *XY* plane projection angles, *ZY* plane projection angles, and *XZ* plane projection angles.

An example of the angle between two lines measurement results is shown in Fig. 20.10, where the measurement target is the angle between the line connecting the right big toe and right foot and the line connecting the right foot and right knee

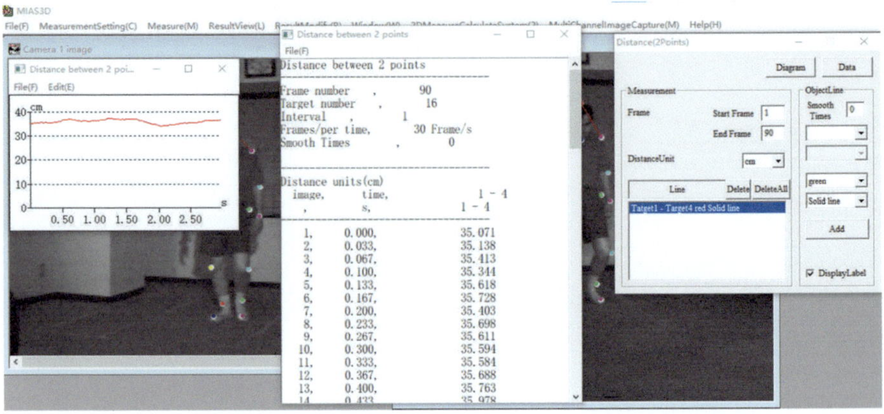

Fig. 20.9 Example of measurement results of distance between two points

on the right leg. The angle calculation reference is the actual spatial angle, and the measured parameters include angle, angle variation, angular velocity, and angular acceleration.

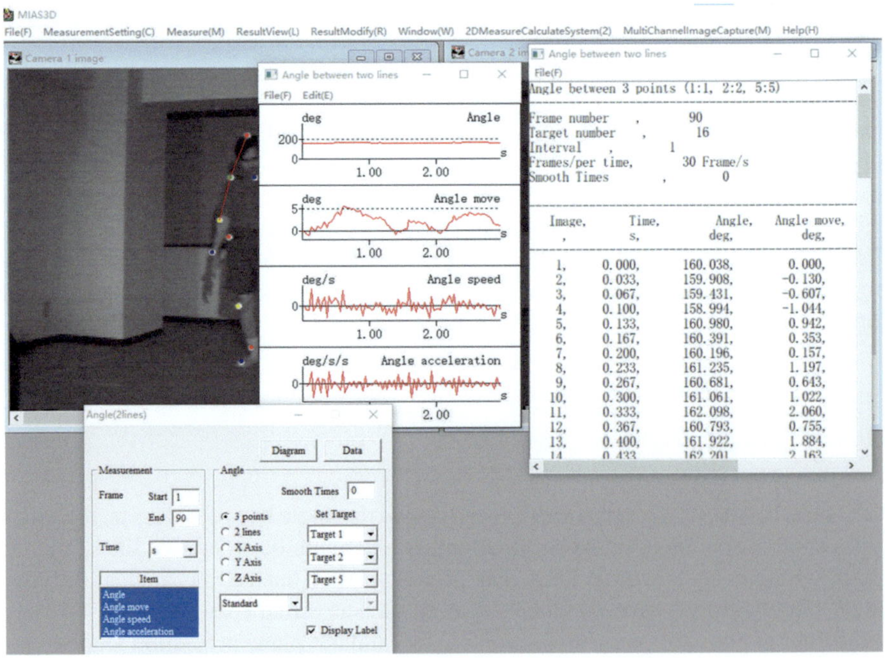

Fig. 20.10 Example of measurement results of the angle between 2 lines

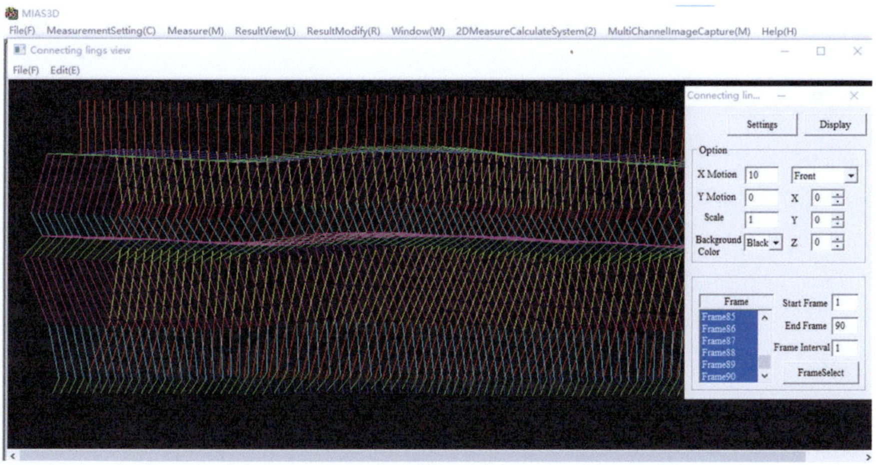

Fig. 20.11 An example connection line at a glance

20.4.6 Connection Line Overview Diagram

The connection line overview diagram function provides an overview of the connection lines between targets. When displaying, different frame intervals can be set, and different projection planes can be selected, such as top view, rotation view, front view, side view, etc.

An example of the connection line overview diagram results is shown in Fig. 20.11, where the parameter settings are a front view, black background, and a frame interval of 2.

20.5 Result Correction

The result correction function of the MIAS 3D system includes item setting and human body centre of gravity measurement, which are explained below.

(1) Item Setting

The item setting function can set the reference frame, add item frames, and delete item frames. An item frame refers to a frame that the user pays special attention to.

(2) Human Body Centre of Gravity

The measurement points for the human body centre of gravity function are the same as those of MIAS, except that the 2D data from MIAS has been converted into 3D data. Figure 20.12 shows a measurement example, where points 1 and 2 are the measured centre of gravity points. By replaying the results, parameters such as the movement

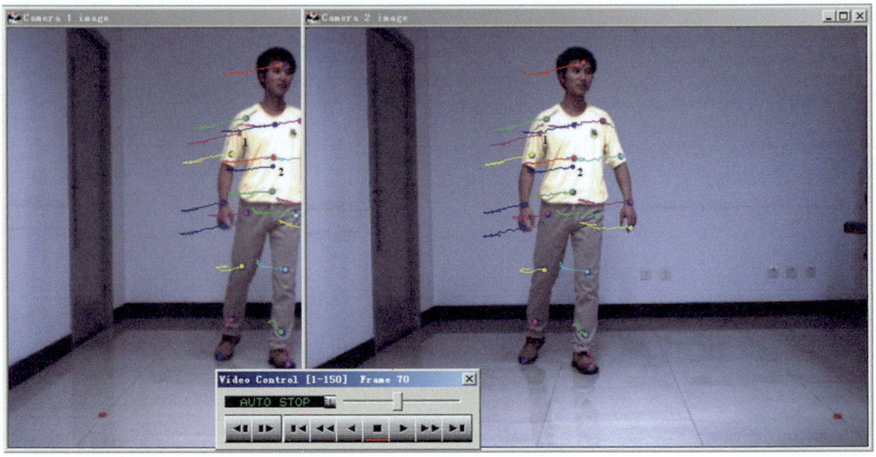

Fig. 20.12 An example of human center of gravity measurement

trajectory, point rate, displacement, and distance of centre of gravity points 1 and 2 can be obtained.

20.6 Other Functions

(1) Windows

The window menu of the MIAS 3D system can create a new 3D connection display window. If you wish to observe more than four stereoscopic sides simultaneously, you can execute this command. Additionally, it can set the size of the 3D connection display window and adjust the display scale, such as 1/4 times, 1/2 times, 1 time, 2 times, 8 times, 16 times, etc.

(2) 2D Measurement System

The 2D measurement system menu of the MIAS 3D system can open the MIAS 2D motion image measurement system.

(3) Multi-channel Image Acquisition

The multi-channel image acquisition menu of the MIAS 3D system can open the multi-channel image acquisition system.

Reference

1. Chen Bingqi. Detailed Explanation of Machine Vision Technology and Application Examples [M]. Beijing: Chemical Industry Press, 2014.

Chapter 21
Vehicle Vision Navigation System

21.1 The Development History and Trends of Autonomous Vehicles

In August 1925, the first documented autonomous vehicle in human history was unveiled. Francis P. Houdina, an electronic engineer from the U.S. Army, sat in a car and controlled the steering wheel, clutch, brakes, and other components of the car in front by emitting radio waves.

In 1956, General Motors of the United States officially unveiled the Firebird II concept car, the world's first concept car equipped with automotive safety and automatic navigation systems. It utilised new concepts such as titanium metal technology, power disc brakes, magnetic ignition keys, and independently controlled gas turbine power, resembling a "rocket car." In 1958, the third generation Firebird was launched, and the BBC broadcasted General Motors' test of the autonomous concept car on the highway. General Motors used embedded cables to send electronic pulse signals to cars equipped with receivers.

In 1966, the SRI Artificial Intelligence Research Center at Stanford University in the United States began developing a multifunctional machine named Shakey, which had a wheeled structure and could perform simple actions like turning lights on and off. Shakey operated indoors, and its built-in sensors and software system pioneered autonomous navigation capabilities.

In 1971, the UK's Road Research Laboratory (RRL) tested an autonomous vehicle similar in concept to General Motors' idea and released a video. The driver's seat was unoccupied, and the steering wheel continuously "shook" automatically to adjust direction. A specially designed receiving unit was located at the car's front bumper, through which computer-controlled electronic pulse signals were transmitted to the car to control steering.

© Chemical Industry Press 2026

B. Chen and S. Chen, *Machine Vision Technology*, Advanced and Intelligent Manufacturing in China, https://doi.org/10.1007/978-981-95-1184-6_21

In 1977, the Tsukuba Engineering Research Laboratory in Japan developed the first autonomous vehicle that utilized cameras to detect front markers or navigation information, abandoning the previously used pulse signal control method. This car was equipped with two cameras and used analogue computer technology for signal processing. It could reach a speed of 30 km/h but required the assistance of elevated tracks.

In 1983, the U.S. Defense Advanced Research Projects Agency (DARPA) initiated a new program named Autonomous Land Vehicle (ALV), aimed at granting full autonomy to vehicles through the use of cameras to detect terrain and computer systems to compute navigation and driving route solutions.

In the 1970s and 1980s, Professor Ernst Dickmanns' team at the Bundeswehr University Munich in Germany initiated a series of "dynamic vision computing" research projects, successfully developing several autonomous vehicle prototypes. In 1993 and 1994, the team modified a Mercedes-Benz S500 sedan, equipping it with cameras and various sensors to monitor the surrounding road environment and react in real-time. At that time, this Mercedes-Benz S500 successfully drove autonomously for over 1000 kilometres in normal traffic conditions.

In 1986, Carnegie Mellon University in the United States began exploring autonomous driving with a project named NavLab. Initially, the team modified a Chevrolet model, adding five portable computing devices to the vehicle, with a driving speed of only 20 km/h. In 1995, the team worked on a 1990 Pontiac Trans.

The Sport model was modified by attaching a portable computer, windshield camera, GPS receiver, and several other auxiliary devices to control the steering wheel and safety performance, successfully completing a "non-manual" driving journey from Pittsburgh to Los Angeles. Approximately 98.2% of the journey was fully autonomous, with only minimal human intervention for obstacle avoidance. The research achievements of Carnegie Mellon University provide significant reference value for current autonomous driving technology.

In 1998, the VisLab at the University of Parma in Italy completed the ARGO project with funding from EUREKA. This project used stereo vision cameras to detect the surrounding environment and computers to plan navigation routes, conducting a 2000-kilometre long-distance experiment, of which 94% was autonomously driven, with an average speed of 90 km/h and a maximum speed of 123 km/h. The success of this system demonstrated that autonomous driving could be achieved using low-cost hardware and imaging systems.

In 2004, DARPA initiated the most significant challenge in the history of autonomous vehicles. That year, the team successfully navigated an autonomous vehicle through the Mojave Desert. Three years later, DARPA moved the testing ground from the desert to the city, achieving second place in the Stanford Rally and first place in the Tartan Racing. Subsequently, Carnegie Mellon University, which gained fame in this field through the NavLab project, also achieved good results and received support from companies like General Motors, Continental, and Caterpillar.

In 2005, a modified Volkswagen Touareg from Stanford University more perfectly undertook the DARPA challenge of crossing the Mojave Desert. This vehicle was equipped not only with cameras but also with a laser rangefinder, radar long-range vision, GPS sensors, and an Intel Pentium M processor.

Since then, the functions of autonomous vehicles have become increasingly complex, requiring them to handle other vehicles, signals, obstacles, and learn how to coexist harmoniously with human drivers.

In 2009, Google, with support from DARPA, commenced its autonomous vehicle development project. That year, Google drove a modified Toyota Prius along the Pacific Coast for 14,000 miles over more than a year. Many engineers who researched at DARPA between 2005 and 2007 joined Google's team, using video systems, radar, and laser automatic navigation technology. Google's grand plan thus began.

In 2010, the VisLab team, the same team that conducted the ARGO project, embarked on an intercontinental journey with autonomous vehicles. Four autonomous vehicles set off from Parma, Italy, traversed nine countries, and successfully reached Shanghai, China. During this period, the VisLab team faced over 13,000 kilometres of everyday driving environment challenges. Notably, all onboard navigation systems were powered by solar energy, making it the first project of its kind to integrate sustainable energy into autonomous vehicles.

In 2013, Baidu's autonomous vehicle project began, led by Baidu Research Institute. Its core technology is the "Baidu Auto Brain," which includes four modules: high-precision maps, positioning, perception, and intelligent decision-making and control. Baidu's independently collected and produced high-precision maps record complete three-dimensional road information, enabling vehicle positioning with centimetre-level accuracy. Additionally, Baidu's autonomous vehicle relies on internationally leading traffic scene object recognition and environmental perception technology to achieve high-precision vehicle detection, recognition, tracking, distance and speed estimation, road surface segmentation, and lane line detection, providing a basis for intelligent decision-making in autonomous driving.

In 2014, Google publicly unveiled its fully self-designed autonomous vehicle. In 2015, the first prototype car was officially unveiled and ready for road testing. In this vehicle, Google completely abandoned the steering wheel design, allowing passengers to enjoy the convenience and fun of autonomous driving simply by sitting in the car.

In the latter half of 2016, Tencent's Autonomous Driving Lab was established. On December 15 of the same year, Tencent announced the signing of a strategic cooperation framework agreement with Shanghai International Automobile City, with both parties engaging in deep cooperation in areas such as autonomous driving, high-definition maps, and automotive intelligent network standards. Currently, Tencent's Autonomous Driving Lab focuses on the core technology development of autonomous vehicles and ground autonomous robots.

Vision serves as a crucial functional means for humans to observe and comprehend the world, with approximately 75% of the information humans obtain from the external environment originating from the visual system. Particularly, 90% of the information required by drivers while driving comes from vision. Among the environmental perception methods currently used in automotive assisted driving, visual sensors can obtain higher, more precise, and richer road structure and environmental information compared to ultrasound and LiDAR. Since the 1970s, developed countries like Japan have begun research on visual navigation, especially in the field of agricultural vehicle navigation, where visual navigation has received special attention from researchers. Compared to highway navigation, target identification in agricultural navigation is more complex and demands higher precision, yet there is no need to pay special attention to the surrounding environment. With the continuous development of machine vision technology, the software processing capability of visual navigation is becoming increasingly strong, while hardware costs are decreasing. Whether for highway navigation or agricultural navigation, visual navigation will become the mainstream technology for vehicle autonomous driving.

21.2 Hardware of the Visual Navigation System

The hardware of this visual navigation system includes an onboard industrial computer, a touchscreen, an angle sensor, a signal acquisition card, a camera, a motor driver, a motor, and a steering wheel rotation mechanism. Figure 21.1 is a schematic diagram of the hardware configuration of this visual navigation system. Figure 21.2 is a 3D diagram of the steering wheel rotation mechanism and a real object diagram installed on a tractor.

21.3 Software of the Visual Navigation System

The system software includes image signal acquisition and processing software and steering wheel rotation control software.

Image acquisition and processing mainly focus on the detection of navigation target lines for agricultural operations. Agricultural operations include ploughing, sowing, transplanting, fertilising, harvesting, and field management. Ploughing includes deep ploughing, harrowing, and land preparation; sowing includes wheat sowing, maize sowing, rice transplanting, cotton sowing, and soybean sowing; harvesting includes wheat harvesting, maize harvesting, soybean harvesting, and cotton machine picking; field management includes intertillage weeding, fertilising, and spraying. Each type of navigation target line has different characteristics, making the image detection of agricultural operation navigation target lines much more complex than that of highway lane lines. The common feature of the aforementioned agricultural operations is that the operation vehicles travel in a straight line along the

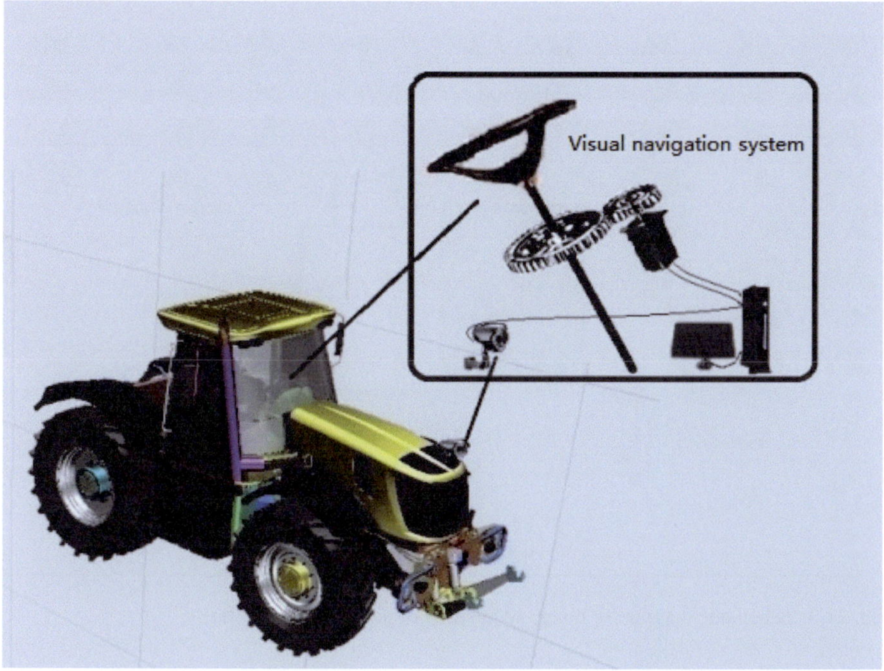

Fig. 21.1 Schematic diagram of the visual navigation system

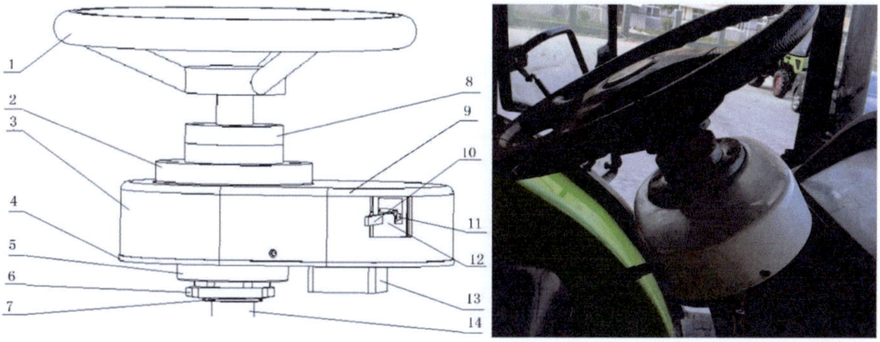

Fig. 21.2 3D drawing and physical drawing of steering wheel rotating mechanism. 1. Steering wheel, 2. Lower flange, 3. Upper housing, 4. Lower housing, 5. Open sleeve, 6. Mandrel locking nut, 7. Open threaded sleeve, 8. Upper flange, 9. Shell, 10. Dial rod, 11. Threaded rod, 12. Small dragon gate, 13. Stepper motor, 14. Steering column tube for steering wheel

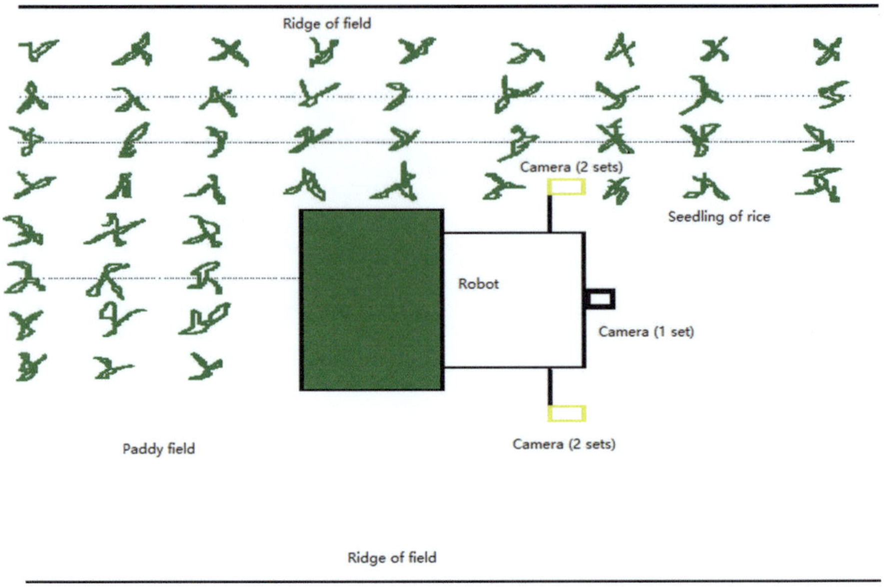

Fig. 21.3 Schematic diagram of image acquisition of navigation target line

boundary between the worked and unworked land. Figure 21.3 is a schematic diagram of navigation target line image acquisition, with the camera positioned directly above the navigation target or in the middle of the vehicle.

This system uses the known point Hough transform introduced in Chap. 7 for the image detection of navigation target lines, converting global line detection into the detection of candidate points for target lines and the determination of known points, achieving rapid line detection. The specific detection methods for each navigation target line can be referenced in another book by the author of this book, "Machine Vision Technology and Application Examples Detailed Explanation" (published by Chemical Industry Press in 2014). Figures 21.4, 21.5, 21.6, 21.7 and 21.8 show examples of the detection of different agricultural operation navigation target lines, with the image detection results of the navigation target lines represented by thin straight lines.

In addition to navigation target line detection, this system also collects the rotation angle of the guide wheel through an angle sensor. By combining the direction angle of the navigation target line, the centre offset, and the front wheel rotation angle, it determines the control motor's rotation direction and amount, then controls the steering and angle size of the steering wheel through the motor.

(a) Seedling line (b) Soil ridge line (c) Cement ridgeline

Fig. 21.4 Example of visual navigation target line detection for rice transplanter

(a) 2 days after transplanting (b) 1 week after transplanting (c) 2 weeks after transplanting

(d) 3 weeks after transplanting (e) 5 weeks after transplanting (f) 9 weeks after transplanting

Fig. 21.5 Examples of visual navigation target line detection of paddy field management machine in different growth stages

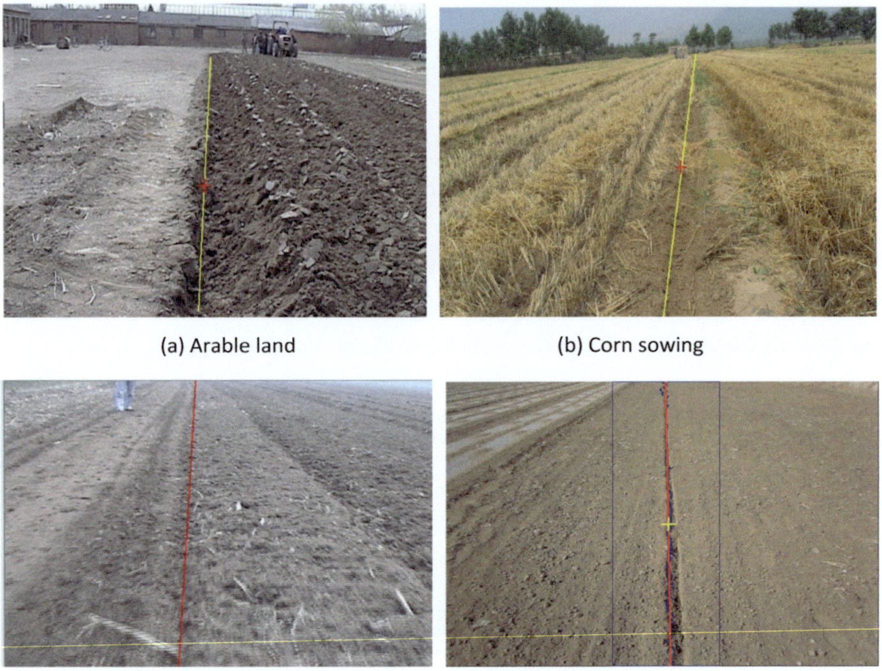

(a) Arable land (b) Corn sowing

(c) Wheat sowing (d) Cotton sowing

Fig. 21.6 Example of visual navigation target line detection in cultivated land seeding environment

21.4 Navigation Experiment and Performance Test Comparison

Figure 21.9 shows navigation test images in different environments, with identical hardware equipment, only different navigation line detection software selected for different navigation environments. Navigation test videos can be viewed on the website www.fubo-tech.com.

In July 2015, this visual navigation equipment conducted a cotton field spraying navigation test in Shihezi City, Xinjiang. According to the testing by the Agricultural Machinery Testing and Testing Centre of the Xinjiang Production and Construction Corps, the vehicle speed was 4.7 km/h, and the path tracking error was 20 mm. Figure 21.10 is the performance test report.

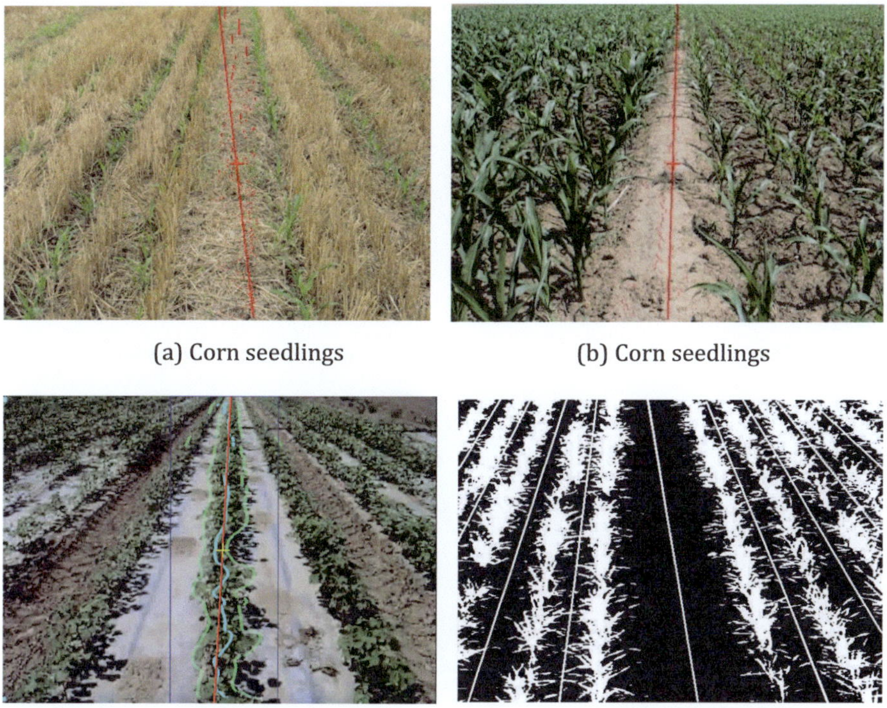

(a) Corn seedlings (b) Corn seedlings

(c) Cotton seedlings (d) Wheat seedlings

Fig. 21.7 Examples of navigation line detection in various management environments

(a) Wheat harvest (b) Cotton harvest

Fig. 21.8 Example of target line detection for harvesting environmental navigation

(a) Solid line of highway lane
(b) Dashed line of highway lane
(c) Highway lane bend line
(d) Highway lane offset line
(e) Arable land
(f) Wheat sowing

Fig. 21.9 Visual navigation trials in different environments

(g) Cotton sowing (h) Cotton field spraying

(i) Harvester seam navigation (j) Laser navigation

Fig. 21.9 (continued)

Table 21.1 is a performance comparison between this visual navigation and the currently used precision GNSS agricultural navigation devices.

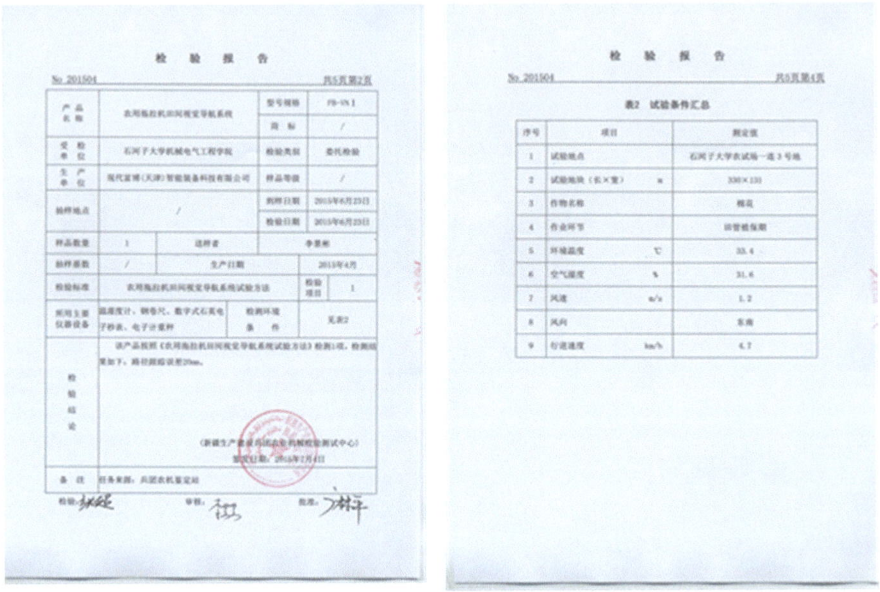

Fig. 21.10 Performance test report of this visual navigation system

Table 21.1 Comparison between this visual navigation and precision GNSS navigation

Performance comparison	This visual navigation	Precision GNSS
Navigation features	Bionic driver, can take curves, can be used indoors	Walk in a straight line according to the planned path outside
Range of adaptation	All farm operations and roads	Not suitable for seedling field management and highway
Navigation error	2 cm (measured accuracy)	5 cm (GNSS positioning accuracy)
Auxiliary facilities	No	Base stations need to be built
Weather effects	No	Sometimes the signal is bad
Geographic information	No	Need to fetch and import

Index

© Chemical Industry Press 2026
B. Chen and S. Chen, *Machine Vision Technology*, Advanced and Intelligent Manufacturing in China, https://doi.org/10.1007/978-981-95-1184-6

W

Z